Avian Evolution

Books in the **Topics in Paleobiology** series feature key fossil groups, key events, and analytical methods, with emphasis on paleobiology, large-scale macroevolutionary studies, and the latest phylogenetic debates.

The books provide a summary of the current state of knowledge and a trusted route into the primary literature, and act as pointers for future directions for research. As well as volumes on individual groups, the Series also deals with topics that have a cross-cutting relevance, such as the evolution of significant ecosystems, particular key times and events in the history of life, climate change, and the application of new techniques such as molecular paleontology.

The books are written by leading international experts and are pitched at a level suitable for advanced undergraduates, postgraduates, and researchers in both the paleontological and biological sciences.

The Series Editor is *Michael Benton*, Professor of Vertebrate Palaeontology in the School of Earth Sciences, University of Bristol.

The Series is a joint venture with the *Palaeontological Association*.

Previously Published

Dinosaur Paleobiology
Stephen L. Brusatte
ISBN: 978-0-470-65658-7 Paperback; April 2012

Amphibian Evolution
Rainer R. Schoch
ISBN: 978-0-470-67178-8 Paperback; April 2014

Cetacean Paleobiology
Felix G. Marx, Olivier Lambert and Mark D. Uhen
ISBN: 978-1-118-56153-9 Paperback; May 2016

Avian Evolution

The Fossil Record of Birds and Its Paleobiological Significance

Gerald Mayr

Senckenberg Research Institute Frankfurt
Senckenberganlage
Germany

WILEY Blackwell

Library of Congress Cataloging-in-Publication Data

Names: Mayr, Gerald.
Title: Avian evolution : the fossil record of birds and its paleobiological
    significance / Gerald Mayr, Senckenberg Research Institute Frankfurt,
    Ornithological Section, Frankfurt am Main, Germany.
Description: Chichester, West Sussex : John Wiley & Sons, Inc., 2017. |
    Series: Topics in paleobiology series | Includes bibliographical
    references and index.
Identifiers: LCCN 2016024809 (print) | LCCN 2016024993 (ebook) | ISBN
    9781119020769 (cloth) | ISBN 9781119020721 (pdf) | ISBN 9781119020738
    (epub)
Subjects: LCSH: Birds, Fossil. | Paleobiology.
Classification: LCC QE871 .M38 2017 (print) | LCC QE871 (ebook) | DDC
    568–dc23
LC record available at https://lccn.loc.gov/2016024809

10 9 8 7 6 5 4 3 2 1

Contents

Foreword

Paleobiology is a vibrant discipline that addresses current concerns about biodiversity and about global change. Further, paleobiology opens unimagined universes of past life, allowing us to explore times when the world was entirely different and when some organisms could do things that are not achieved by anything now living.

Much current work on **biodiversity** addresses questions of origins, distributions, and future conservation. Phylogenetic trees based on extant organisms can give hints about the origins of clades and help answer questions about why one clade might be more species-rich ("successful") than another. The addition of fossils to such phylogenies can enrich them immeasurably, thereby giving a fuller impression of early clade histories, and so expanding our understanding of the deep origins of biodiversity.

In the field of **global change**, paleobiologists have access to the fossil record and this gives accurate information on the coming and going of major groups of organisms through time. Such detailed paleobiological histories can be matched to evidence of changes in the physical environment, such as varying temperatures, sea levels, episodes of midocean ridge activity, mountain building, volcanism, continental positions, and impacts of extraterrestrial bodies. Studies of the influence of such events and processes on

the evolution of life address core questions about the nature of evolutionary processes on the large scale.

As examples of **unimagined universes**, one need only think of the life of the Burgess Shale or the times of the dinosaurs. The extraordinary arthropods and other animals of the Cambrian sites of exceptional preservation sometimes seem more bizarre than the wildest imaginings of a science fiction author. During the Mesozoic, the sauropod dinosaurs solved basic physiological problems that allowed them to reach body masses ten times those of the largest elephants today. Further, the giant pterosaur *Quetzalcoatlus* was larger than any flying bird, and so challenges fundamental assumptions in biomechanics.

Books in the **Topics in Paleobiology** series will feature key fossil groups, key events, and analytical methods, with emphasis on paleobiology, largescale macroevolutionary studies, and the latest phylogenetic debates.

The books will provide a summary of the current state of knowledge, a trusted route into the primary literature, and will act as pointers for future directions for research. As well as volumes on individual groups, the Series will also deal with topics that have a cross-cutting relevance, such as the evolution of significant ecosystems, particular key times and events in the history of life, climate change, and the application of new techniques such as molecular paleontology.

The books are written by leading international experts and will be pitched at a level suitable for advanced undergraduates, postgraduates, and researchers in both the paleontological and biological sciences.

Michael Benton,
Bristol

Preface

Curiously, birds generally belonged to those groups of fossil organisms that were always much neglected. Ever since we have been able to talk of a scientific paleozoology, only a few researchers have, strictly speaking, devoted their studies to this group of vertebrates. The reasons therefore are manifold; mainly they may perhaps be found in the fact that a formidable amount of morphological knowledge, that is, knowledge of the extant forms, is required to venture into the study of these remains. In other words, it is a cumbersome and long path that has to be walked until one can be considered a true expert in fossil birds and, hence, have the foundations on which successful research in the field of paleornithology is built.

(Abel 1936; translated from the German original)

Some 80 years after these laments in an obituary for the Hungarian paleontologist Kálmán Lambrecht, paleornithology is a booming area of research. Numerous fossils are reported each year, and it is fair to say that within the past 30 years more progress has been made in the study of the avian fossil record than in that of most other major vertebrate groups. Sparked by the discoveries of fascinating new fossils from Late Jurassic and Early Cretaceous localities in China and elsewhere, much of this research was devoted to archaic Mesozoic birds from the era of the dinosaurs. However, the past decades have also witnessed the description of many exceptional finds from later geological periods, which provided key insights into the evolutionary history of the extant bird groups. This ever-increasing avian fossil record is accompanied by novel hypotheses on the interrelationships of extant birds, in which analyses of molecular data played a central role, albeit not an exclusive one.

The last comprehensive survey of avian evolution, by contrast, was published nearly two decades ago (Feduccia 1999), and currently no textbook exists that covers both Mesozoic and post-Mesozoic fossils in some detail. The present work aims at filling this gap by providing a detailed picture of the evolutionary history of birds, even though the tremendously expanded avian fossil record necessitated a focus on the Mesozoic avian radiation and the evolutionary history of the major extant groups.

The first chapters give an overview of the early fossil record of basal avians. Here, one of the main research interests concerns the mode of character evolution in the lineage leading towards modern birds. The evolution of modern birds is then detailed in eight chapters, which give information on the phylogenetic interrelationships and evolutionary history of the extant avian groups. An integrative view is pursued, which takes into account the latest results of DNA-based phylogenetic analyses, and in some cases complementary data derived from current phylogenetic hypotheses and the fossil record shed new light on the evolutionary history of birds. Some aspects of the evolutionary significance of bird fossils from islands and quasi-insular regions are outlined in the last chapter. Not considered are geologically young fossils of species closely related to the living ones, which are only mentioned if they exhibit unusual morphologies or provide insights into general aspects of avian evolution. Fossils from the most recent geological periods are of great significance for an understanding of the extant species diversity. Adequate discussion of them would, however, be a book on its own, and is also hampered by the fact that many of these fossils are in need of critical revision before sound evolutionary conclusions can be drawn.

I am aware of the shortcomings of any attempt to squeeze the better part of avian evolution into a book with a limited page count, and some topics would have deserved a more detailed coverage. Not only were some accounts condensed to their essence, but to keep the literature section to a manageable size, an emphasis had to be placed on more recent publications, which can be consulted for earlier references. This book nonetheless brings together a great deal of detailed information, some of which may perhaps be considered to be of interest only to the specialist. At the moment, however, avian evolution is a very vivid research field, which attracts many research groups and individual scientists around the globe. In addition to the comparatively few paleornithologists, there are increasing numbers of molecular systematists, who study the interrelationships and evolutionary history of the extant bird groups. Not all have an in-depth knowledge of the avian fossil record, and the voluminous data are not readily gathered from the scattered literature. While I therefore expect the following chapters to be of interest to students of vertebrate paleontology and to avian systematists, I also hope to have succeeded in writing a coherent text that is intelligible for other readers with a moderate background in biology or geology. During the compilation of the present volume, I definitely learned much myself from the wealth of recently published studies, and these insights more than balance the efforts I put into writing this overview of the evolutionary history of birds.

Acknowledgments

Many colleagues and friends have contributed to the production of this book. Michael Benton is thanked for inviting me to contribute to the "Topics in Paleobiology" series, and Delia Sanford, Kelvin Matthews, Shummy Metilda, Sally Osborn, Rebecca Stubbs, and Ian Taylor provided editorial assistance. The comments of an anonymous reviewer improved the text. I am particularly indebted to Zhonghe Zhou for photos of Jehol Biota fossils. Further pictures were provided by James Goedert, Leonid Gorobets, Lance Grande, Daniel Ksepka, Nicholas Longrich, Hanneke Meijer, Jingmai O'Connor, Oliver Rauhut, Andreas Reuter, Marcelo Stucchi, Chris Torres, and Jakob Vinther. All other photographs were taken by Sven Tränkner or the author. Some skeletal reconstructions of Mesozoic paravians were kindly made available by Scott Hartman. For access to major fossil collections, I thank Elvira Brahm, Gilles Cuny, Michael Daniels, Amy Henrici, Carl Mehling, Norbert Micklich, and Stephan Schaal. Most of all, however, I am grateful to my wife, Eun-Joo, for her patience and understanding during the long period over which I compiled and wrote this book.

1

An Introduction to Birds, the Geological Settings of Their Evolution, and the Avian Skeleton

What is a bird? Just by looking at the extant world, this question is easily answered: a bird is a bipedal, feathered animal without teeth, which, with very few exceptions, is capable of flight. These and numerous other avian characteristics were, however, sequentially acquired in the more than 160 million years of avian evolution. As a result, the distinction between birds and their closest relatives becomes more blurred the further one goes back in time.

With about 10,000 living species, birds are the second most species-rich group of extant vertebrates, outnumbered only by teleost fishes. Owing to the constraints of their aerial way of life, most extant birds have quite a uniform appearance. Whereas the morphological diversity of mammals spans extremes like bats and whales, all present-day birds have two wings, two legs, and an edentulous beak, with most major external differences concerning plumage traits, neck and limb proportions, as well as beak shapes. This alikeness of bird shapes notwithstanding, their skeletons show a high diversity of morphological details. In this chapter, the reader is introduced to some of the main features of the avian skeleton. In addition, general terms and the geological setting of avian evolution are briefly outlined to aid understanding of the subsequent accounts.

Avian Evolution: The Fossil Record of Birds and its Paleobiological Significance,
First Edition. Gerald Mayr.
© 2017 John Wiley & Sons, Ltd. Published 2017 by John Wiley & Sons, Ltd.

Birds are Evolutionarily Nested within Theropod Dinosaurs

An understanding of avian evolution hinges on a robust phylogenetic framework, with a knowledge of the interrelationships of the studied groups being central to many evolutionary and paleobiological questions arising from the fossil record. The most rigorous method of reconstructing evolutionary trees is called **phylogenetic systematics**, or cladistics, and aims at identification of **monophyletic** groups or **clades** (readers who are not acquainted with phylogenetic terminology are referred to Figure 1.1 and the glossary at the end of this book, which explains words highlighted in the text). Organisms can be remarkably different from their closest relatives and the results of phylogenetic reconstructions are sometimes counterintuitive. Overall similarities may be misleading, because they are often based on the retention of primitive features (**plesiomorphies**) that were inherited from a common ancestor. Closely related organisms, on the other hand, can become profoundly different if they are on disparate evolutionary trajectories.

Birds are one of those animal groups that underwent particularly pronounced morphological transformations in their evolutionary history, and as a result their anatomy strongly departs from that of their closest living relatives. Even so, unanimous consensus exists that birds belong to the Archosauria. This clade also includes crocodilians and all non-avian dinosaurs and is characterized by a number of **derived characters** (**apomorphies**), such as teeth sitting in sockets of the jaw bones, a skull with an opening (antorbital

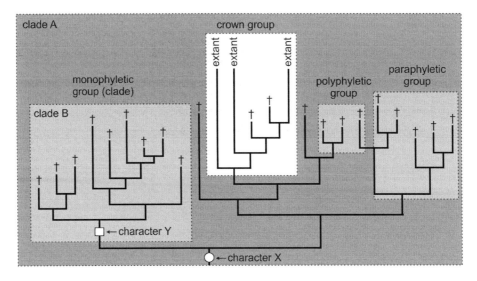

Figure 1.1 Illustration of some general phylogenetic terms used in this book. Phylogenetic systematics aims at identification of monophyletic groups (clades), which include an ancestral species and all of its descendants and are characterized by shared derived characters (apomorphies). Depicted is a hypothetical clade A with extant and extinct species, the latter being denoted by daggers. Character X is an apomorphy of this clade, whereas character Y represents an apomorphy of the subclade B. Groups are polyphyletic if they consist of only distantly related taxa, and paraphyletic if they do not include all of the taxa that descended from their last common ancestor. The white field marks the crown group of clade A, whereas all taxa in the dark and light gray areas are stem group representatives of this clade.

fenestra) between the orbits and the nostrils, and a four-chambered heart.

In the 19th century some scientists already assumed that the closest archosaurian relatives of birds are to be found among bipedal theropod dinosaurs. In its modern form, this hypothesis goes back to Ostrom (1976), who proposed an avian origin from one particular theropod clade, the Coelurosauria. At one time vigorously contested, a theropod ancestry for birds is now widely accepted. For space constraints and because an extensive literature already exists, these largely settled debates are not reviewed here (see, e.g., Prum 2002; Chiappe 2007; Makovicky and Zanno 2011; Xu et al. 2014).

Likewise, it is now generally appreciated that, within coelurosaurs, birds belong to the Maniraptora, which also include dromaeosaurs, troodontids, oviraptorosaurs, and a few other coelurosaurian theropods, such as ornithomimosaurs and therizinosaurs (Figure 1.2). Aside from features also present in some more distantly related dinosaurs (e.g., bipedal locomotion and a highly pneumatized skeleton), maniraptoran theropods are characterized by greatly elongated hands with only three fingers, a semilunate carpal bone, a bowed ulna, and thin radius, as well as an avian-like eggshell structure (Gauthier 1986; Makovicky and Zanno 2011). Most current phylogenetic analyses recognize oviraptorosaurs, dromaeosaurs, and troodontids as the closest avian relatives. Oviraptorosaurs are placed outside a clade formed by dromaeosaurs, troodontids, and birds for which the term Paraves was introduced (e.g., Makovicky and Zanno 2011; Turner et al. 2012).

A clade including oviraptorosaurs, dromaeosaurs, troodontids, and birds is robustly supported in most analyses, but, as will be detailed later, the jury may still be out on the exact interrelationships between these groups. Not only do various analyses show conflicting results, but some new

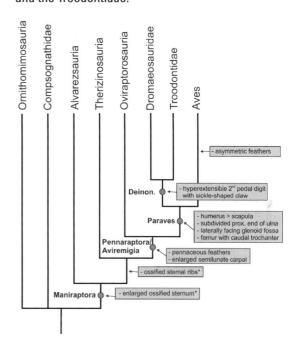

Figure 1.2 Phylogenetic interrelationships of birds and their closest theropod relatives, with some key apomorphies characterizing major groups (after Makovicky and Zanno 2011; Turner et al. 2012). The asterisked characters are absent in *Archaeopteryx* and the Troodontidae.

findings from the Early Cretaceous of China exhibit unexpected character mosaics, which challenge current phylogenetic hypotheses.

Aves, Avialae, or what constitutes a "bird"

Extant birds are classified in the **taxon** Aves, which is one of the traditional higher categories of vertebrates. If fossils are also considered, the content of Aves is a matter of considerable debate and depends on the underlying definition, which varies among current authors.

In phylogenetic discussions of groups, which include both fossil and extant species, it is import to distinguish between the **crown group** of a certain taxon and its **stem group** (Figure 1.1). At times when only a few Mesozoic birds were known, Aves was defined as the least inclusive clade comprising the earliest known bird, *Archaeopteryx*, as well as all extant species (i.e., the crown group), which were designated Neornithes. This terminology is still used by many authors and is also employed here. Following Gauthier (1986), who restricted the use of Aves to the crown group, the clade including *Archaeopteryx* and crown group birds is nowadays often termed "Avialae." This renders the well-established term Neornithes redundant and conflicts with common practice in paleontology, where crown group taxa are expanded to encompass fossil stem group representatives (e.g., in the case of Equidae, the clade including fossil and extant horses, or *Homo*, the taxonomic category for archaic and modern humans).

Restriction of the term Aves to the crown group would furthermore lead to the awkward classification of all Mesozoic birds outside the crown group as "non-avian avialans," no matter how similar to modern birds these may be, and would result in a discrepancy between the contents of the terms "avian" and "bird-like." As this is more counterintuitive than recognizing the avian affinities of aberrant fossil stem group taxa, Aves is used for the clade including *Archaeopteryx* and extant birds throughout this book.

The Geological Settings of Avian Evolution in a Nutshell

The known history of birds spans more than 160 million years, from the Late Jurassic until now. Avian evolution therefore extended over two geological eras, the Mesozoic and the Cenozoic, which showed profound differences in their paleogeographic, paleoenvironmental, and climatic regimes. Most readers of this book will probably have a basic acquaintance with these facts, so only some of the major geological settings are briefly summarized in the following (Figure 1.3).

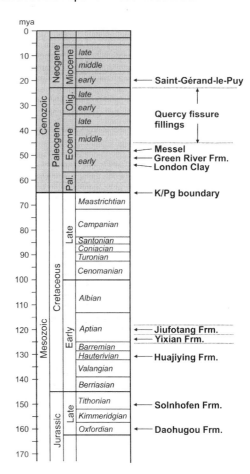

Figure 1.3 Time chart showing geological periods relevant for avian evolution and the stratigraphic position of some important fossil localities.

Geological eras are subdivided into periods, epochs, and stages. The **Mesozoic** era includes three periods, of which only the last two, the Jurassic and the Cretaceous, yielded avian fossils, with controversial reports of a Triassic *"Protoavis"* (Chatterjee 2015) being dismissed by most current researchers. All of the few Jurassic avian or avian-like fossils stem from the latest epoch of this period; that is, the Late Jurassic. Until recently, Jurassic birds were solely represented by the *Archaeopteryx* specimens from the Solnhofen limestone in southern Germany, which was deposited in the Tithonian stage, 150 million years ago (mya). In the past decade, however, a diverse array of somewhat earlier avian-like theropods was described from the Tiaojishan Formation of the Daohugou Biota in northeastern China, which stems from the Oxfordian stage of the Late Jurassic and has an estimated age of 160 million years (Liu et al. 2012; Sullivan et al. 2014).

Virtually all other early avians, however, are from Cretaceous deposits. The Cretaceous period lasted from 145–66 mya and includes two epochs, the Early and Late Cretaceous, each of which is further divided into several stages (Figure 1.3). It is beyond the scope of this brief account to review all localities that yielded Cretaceous avian fossils. The exceptionally fossil-rich Jehol Biota in northeastern China, however, is of particular relevance for avian evolution, and has a quite complex stratigraphy that is briefly outlined to aid understanding of the geological context of the finds discussed later.

The fossiliferous strata of the Early Cretaceous, lacustrine sediments of the Jehol Biota cover a large area in Inner Mongolia, Hebei, and Liaoning provinces. They can be divided into three stratigraphic units, which altogether span some 10 million years (Pan et al. 2013; Zhou 2014). The earliest of these is represented by the Huajiying ("Dabeigou") Formation (130.7 mya), which is mainly exposed in Hebei Province. Most fossils of the Jehol Biota, however, stem from the Barremian to early Aptian (125–122 mya) Yixian Formation and from the Aptian (120 mya) Jiufotang Formation. The Jehol Biota not only yielded great numbers of early avian taxa, but also numerous well-preserved fossils of small non-avian maniraptorans, which shed further light on the dinosaurian ancestry of birds.

The end of the Mesozoic is characterized by large-scale extinction events at the Cretaceous/Paleogene (K/Pg) boundary, which involved both marine and terrestrial organisms and ushered in the **Cenozoic**. This era falls into two major periods (Figure 1.3), the **Paleogene**, which includes the Paleocene, Eocene, and Oligocene epochs, and the **Neogene**, with the Miocene and Pliocene epochs. The latest Cenozoic period is the Quaternary, which encompasses the Pleistocene and Holocene epochs.

Numerous Cenozoic fossil sites yielded avian fossils, but some localities stand out with regard to the number of bird fossils found, the quality of their preservation, and the insights into avian evolution that can be gained from these specimens. Of particular significance are several Eocene sites, which yielded many of the fossils mentioned in this book. Among the most important of these are the marine deposits of the early Eocene (53 mya) London Clay in England, as well as the lacustrine sediments of the early Eocene (51 mya) Green River Formation in North America, and the slightly younger early Eocene (48 mya) Messel oil shale in Germany (until recently, this latter site was considered to be of middle Eocene age, but see Lenz et al. 2015). Many Paleogene bird fossils were also retrieved from karstic fissure fillings in the Quercy region in France, which cover a long temporal range, from the middle Eocene to the late Oligocene, and yielded thousands of avian bones. As early as the 19th century, numerous early Miocene avian fossils were furthermore described from lacustrine deposits of the Saint-Gérand-le-Puy area in France, one of the classical localities for Neogene birds.

Avian evolution was accompanied by the final break-up of the southern supercontinent Gondwana, which involved three major paleogeographic events. The earliest of these was the split of South America and Africa in the Early Cretaceous, which led to the opening of the South Atlantic. Australia and South America separated from Antarctica in the latest Cretaceous and late

Eocene, respectively. In the early Cenozoic, finally, the North Atlantic opened between Europe and North America. All of these geographic events seem to have had impacts on marine and atmospheric circulation systems, which in turn affected the prevailing climatic regimes (e.g., Smith et al. 1994; Haug and Tiedemann 1998; Scher and Martin 2006).

The global average temperatures in the Mesozoic were much higher than today. Although they decreased towards the Cenozoic, global temperatures in the earliest Cenozoic were still high, and subtropical vegetation flourished even in northern latitudes. After the Thermal Maximum at the Paleocene–Eocene boundary, global climatic cooling commenced in the Oligocene, but was interrupted by a warm period during the middle Miocene Climatic Optimum, some 15–17 mya (e.g., Zachos et al. 2001; Jenkyns 2003).

As will be detailed in later chapters, major biotic events that may have influenced avian evolution include the Cretaceous radiation of angiosperm (flowering) plants (e.g., Friis et al. 2011), as well as the rise of placental mammals, with the emergence and initial diversification of mammalian crown group taxa showing a broad temporal coincidence with avian evolution. For various avian groups, broad-scale Cenozoic habitat changes were also of great significance, which is especially true for the spread of open grasslands towards the mid-Cenozoic (e.g., Jacobs et al. 1999).

Characteristics of the Avian Skeleton

Avian evolution was characterized by the formation of compound bones through the fusion of individual skeletal elements, mainly in the skull and **distal** limb bones. Co-ossification of the limb elements probably went in parallel with another avian trait, the reduction of the distal limb muscles. Unlike in other tetrapods, the movements of the hand and foot of extant birds are largely due to the action of tendons of muscles that are situated near the body center, and in the course of avian evolution the muscle masses therefore came close to the center of gravity of these flying animals.

The aerodynamic demands of flight led to numerous other changes in the skeleton of birds, which set these animals apart from other vertebrates. In the following account of avian osteology, a focus was put on the characteristics of the skeleton of modern (neornithine) birds (Figure 1.4). The morphological diversity seen in their Mesozoic ancestors, as well as some of the key transformations that took place in the avian skeleton, are outlined in later sections, after the various fossil groups are introduced.

The avian skull

The skull of birds can be broadly divided into three units: the snout, the neurocranium, and the lower jaws (mandibles). The snout is formed by the paired nasal, maxillary, and praemaxillary bones. It is edentulous in neornithine birds and the bones are covered with a horny rhamphotheca. Upper and lower jaws therefore form a beak, which shows a great diversity of shapes within and across different neornithine taxa.

Neornithine birds have a kinetic skull, in which the beak is movable against the neurocranium. This movability is enabled by a flexible, sheet-like connection with the neurocranium, the nasofrontal hinge. A critical role in avian skull kinetics is fulfilled by the quadrate, which articulates with the jugal bar and the pterygoid and pushes these elements rostrally (i.e., towards the bill tip), so that the upper beak is uplifted. There are two different kinds of kinesis in Neornithes: In the prokinetic skull the beak does not change its shape, whereas

Figure 1.4 Skeleton of a domestic fowl (*Gallus*). Major bones and anatomical directions are labeled.

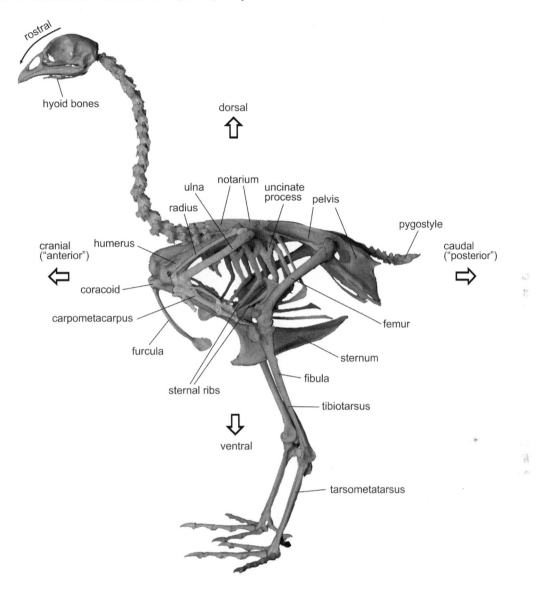

the rhynchokinetic skull exhibits additional bending zones within the upper beak.

Two basic types of nostril shapes can be distinguished in neornithine birds. The most widespread and presumably primitive condition is characteristic for the holorhinal beak, where the nostrils are ovate openings. In the schizorhinal beak, by contrast, the nostrils are elongated and have slit-like caudal ends, which reach beyond the nasofrontal hinge, the transition zone between beak and neurocranium (Figure 1.5). The schizorhinal condition increases the flexibility of the tip of the upper beak and is often found in particularly long-beaked species, which probe substrate for food.

Figure 1.5 Skulls of (a, d) a lapwing (*Vanellus*, Charadriidae) and (b, c) a moorhen (*Gallinula*, Rallidae) in lateral (left) and dorsal (right) views, with some major anatomical features. The arrows identify the caudal ends of the nostrils of the holorhinal moorhen and the schizorhinal lapwing. Not to scale.

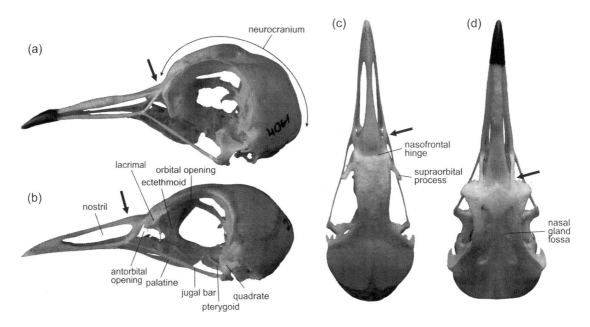

Besides the praemaxillary and maxillary bones, the osseous palate of birds is mainly formed by five bones: the vomer and the paired pterygoids and palatines (Figure 1.6). Anatomists of the 19th century noted that Neornithes can be divided into two major groups based on the structure of their palatal bones, which were termed Palaeognathae ("old jaws") and Neognathae ("new jaws"). Apart from differences in the proportions and relative positions of the involved bones, one of the major characteristics of the palaeognathous palate is a fusion of the pterygoids and palatines, which form a rigid unit and articulate with the braincase via well-developed basipterygoid processes. The neognathous palate, by contrast, exhibits a movable joint between pterygoid and palatine, which allows a greater mobility of the upper beak; basipterygoid processes are often reduced in neognathous birds (Figure 1.6).

The avian lower jaws, or mandibles, are composed of several bones, of which the dentaries bear teeth in many Mesozoic non-neornithine birds. In all Neornithes, the tips of the mandibles are fused and form an ossified mandibular symphysis.

Pectoral girdle and sternum

The avian pectoral girdle consists of six bones, the paired coracoids and scapulae, as well as the furcula and the sternum. These bones anchor the wing to the trunk, provide attachment sites of the flight muscles, and act as pulleys for the tendons of some of them. The coracoids articulate with the sternum, whereas the blade-like scapulae are situated laterodorsal of the ribcage. Where both bones meet, they form the glenoid fossa (see Figure 1.10), with which the **proximal** end of the humerus articulates. As will be detailed in Chapter 3, the position and orientation of this fossa changed in the course of

Figure 1.6 Palates of (a) a palaeognathous nandu (*Rhea*, Rheiformes), (b) a palaeognathous tinamou (*Rhynchotus*, Tinamiformes), and (c) a neognathous lapwing (*Vanellus*, Charadriiformes). In each image the left palate is highlighted by dotted lines. In palaeognathous birds palatine and pterygoid are fused, whereas both bones are separated by an intrapterygoid joint in neognathous birds. Not to scale.

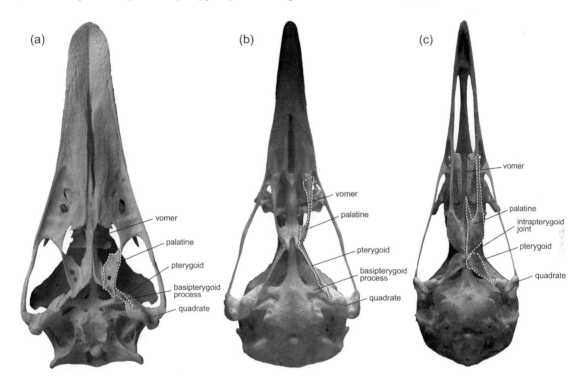

avian evolution, during the transition from gliding to flapping flight.

Initially, the coracoid of birds was a squarish bone, which closely resembled the coracoid of non-avian theropods. In the evolution towards the crown group, the bone became elongated and strut-like, although it regained a squarish shape in some palaeognathous birds (Figure 1.7). The coracoid has a broad sternal end, which articulates with the sternum. The opposite (upper) end of the bone is formed by the acrocoracoid process. Unlike in more basal avians, this upper end of the neornithine coracoid exhibits a well-developed procoracoid process, which projects medially from the shaft (Figure 1.7). Together with the cranial end of the scapula and the dorsal end of the

furcula, the procoracoid process contributes to the formation of the "triosseal canal," through which passes the tendon of the supracoracoideus muscle, the main elevator of the wing. In the neornithine **stem species**, the articulation facet of the coracoid for the scapula was cup-like, thereby forming a so-called scapular cotyla (Figure 1.7). In many extant taxa, however, it is only a shallow facet. This loss of a cup-like articulation facet occurred multiple times independently and is well documented in, for example, galliform, procellariiform, and psittaciform birds (Mourer-Chauviré 1992a; Mayr et al. 2010; Mayr and Smith 2012a). The functional significance of this character variability has not yet been studied. However, there appears to be a correlation with the shape

Figure 1.7 Coracoids of selected neornithine birds, to illustrate different morphologies of this bone and some major anatomical features. (a) Cassowary (*Casuarius*, Casuariiformes), (b) tinamou (*Tinamus*, Tinamiformes), (c) petrel (*Pterodroma*, Procellariiformes), (d) seriema (*Cariama*, Cariamiformes), (e) owl (*Strix*, Strigiformes), (f) woodpecker (*Dryocopus*, Piciformes). In (a) scapula and coracoid are co-ossified and form a scapulocoracoid. Not to scale.

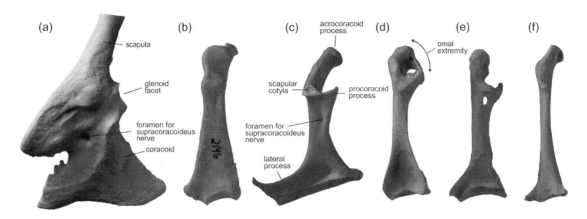

of the furcula, and taxa with a cup-like articulation facet usually feature a robust furcula. Bird groups with a shallow scapular articulation facet, by contrast, tend to have a furcula with narrow arms. The coracoid of many birds is pierced by a foramen for the supracoracoideus nerve. This foramen is primitively present in Neornithes, but it is lost in a large number of taxa, especially those with slender and strut-like coracoids. In flightless birds, especially species of the Palaeognathae, coracoid and scapula sometimes fuse to form a single bone, the scapulocoracoid.

The furcula is formed by the fused clavicles and primitively was a boomerang-shaped and robust bone, which may have contributed to the attachment of some pectoral muscles in early avians without an ossified sternum. In more advanced birds, the bone primarily acts as a spring and stores kinetic energy of the flight strokes. The sternal end often bears a ventral process, the furcular apophysis, which articulates with the tip of the sternal keel in some long-winged taxa. In a few birds,

the sternal end of the furcula is reduced, so that its arms are no longer connected.

The sternum of most neornithine birds bears a deep midline keel, which increases the attachment area of the greatly enlarged pectoral muscles, and which is one of the prerequisites of powered flapping flight. A keeled sternum distinguishes extant birds from all other living vertebrates. Paired ossified sternal plates can, however, already be traced back to non-avian theropods, in some of which they are even fused to form a single sternal plate. The sternal keel is lost or greatly reduced in a number of flightless taxa, in particular among palaeognathous birds. The proportions of the neornithine sternum vary greatly, and a long sternum is, for example, found in diving birds, such as loons or auks, whereas it tends to be short in soaring birds, such as frigatebirds. The caudal margin of the bone can be entire and straight, but usually it bears one or two pairs of incisions of varying depth, which may be closed to form fenestrae. These incisions are particularly marked in taxa that are capable of powerful burst take-offs, such as tinamous (Tinamiformes) and landfowl (Galliformes), whereas they are shallow in soaring birds.

The forelimb skeleton and the identity of the avian wing digits

The wing skeleton of birds is composed of three sections: a proximal one formed by the humerus, a midsection consisting of ulna and radius, and a distal hand section with the three digits, which are fused and – except for the central one – greatly reduced in neornithine birds.

The neornithine humerus shows a considerable variation in shape and proportions, being very long in soaring birds and greatly shortened in the stiff-winged swifts (Apodiformes; Figure 1.8). The tendons of various major wing muscles insert on the bone, and differences in the position and development of the tubercles, processes, and fossae at their attachment areas provide characters of phylogenetic relevance. The proximal humerus end bears the deltopectoral crest, on which the pectoral muscles insert. The

humerus of most Neornithes is pneumatized and exhibits openings for the entrance of air sacs into the hollow bone lumen. These foramina are situated on the bottom of a "pneumotricipital" fossa (Figure 1.8), which derives its name from the circumstance that it does not only receive air sac diverticula that are part of the complex avian lung system (e.g., Proctor and Lynch 1993), but also serves as an attachment site for the tendons of so-called tricipital pectoral muscles. On the distal end of the humerus, there are two condyles for the articulation of the bone with ulna (ventral condyle) and radius (dorsal condyle).

The ulna of birds primitively does not exceed the humerus in length, but the relative lengths of these bones are highly variable in Neornithes, with the ulna being much longer than the humerus in many long-winged birds and greatly shortened in, for example, some wing-propelled divers. The ulna serves for the attachment of the **secondary feathers**, and in some taxa the shaft of the bone forms distinctly raised

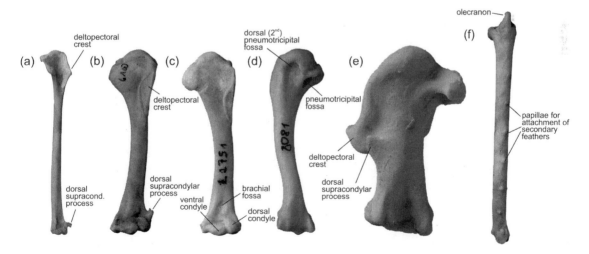

Figure 1.8 (a–e) Humeri of selected neornithine birds. (a) Albatross (*Diomedea*, Diomedeidae), (b) crow (*Corvus*, Passeriformes), (c) trogon (*Pharomachrus*, Trogoniformes), (d) partridge (*Arborophila*, Galliformes), (e) swift (*Apus*, Apodiformes). (f) Ulna of *Corvus* in cranial view. Not to scale (a–c: cranial view, d, e: caudal view).

papillae at the attachment sites of these feathers (Figure 1.8).

Birds are characterized by a skeletomuscular mechanism, which couples the elbow and wrist joints during wing movements. This "drawing parallel system," which is also known as "automatic extension" of the wing, results in a synchronization of the extension and flexion of the elbow and wrist joints, whereby the hand section is drawn in parallel to the humerus during wing movements.

The hand of the avian ancestor had three fingers, which consisted of a metacarpal bone and one, two, and three additional phalanges, respectively. Each of these fingers furthermore bore well-developed claws. There has been, and still is, much debate on the identity of the three fingers of birds (Feduccia 2012; Xu et al. 2014). Dinosaurs primitively had a hand with five fingers, and in phylogenetically basal theropods, such as the late Triassic *Herrerasaurus* and the Early Jurassic *Heterodontosaurus*, the fourth and fifth manual digits are reduced. This suggests that the digits of later three-fingered non-avian theropods, as well as those of birds, are the first, second, and third, with the fourth and fifth digits having been lost. Gene expression patterns also indicate that the three avian digits are the first to third ones (Z. Wang et al. 2011; Xu et al. 2014). From studies of the embryonic development of the hand of extant birds, however, it was concluded that the avian digits are the second, third, and fourth ones, and that the two outermost digits were lost (see Feduccia 2012). Developmental shifts in digital identity were proposed to explain this mismatch, and it was hypothesized that the embryonic tissue of digits 2–4 develops into digits 1–3 of the adults ("frameshift hypothesis"; e.g., Bever et al. 2011).

Resolving the issue of digital homology in birds and non-avian theropods is complicated by the fact that the adult morphology of fossil

taxa is compared with **ontogenetic** data from neornithine birds (Xu et al. 2014). Moreover, the two outermost manual digits – that is, the first and the fifth – are reduced in a least one theropod, the Late Jurassic ceratosaur *Limusaurus* (Xu et al. 2009a). Although this taxon is probably too distantly related to birds to bear directly on the identity of digital homology in birds and non-avian theropods, it indicates that there was some **homoplasy** in the reduction of the forelimb digits of theropods. In any case, the hand of *Archaeopteryx* is so similar to that of closely related non-avian theropods that there can be little doubt that the fingers of *Archaeopteryx* are **homologous** to those of, for instance, dromaeosaurs, irrespective of the exact digital identity (Zhou 2004). To account for the uncertainties in digital homology, the fingers of birds and closely related Mesozoic taxa are here termed alular (1st/2nd), major (2nd/3rd), and minor (3rd/4th) digits (Figure 1.9).

Neornithine birds have two free carpal (wrist) bones, which are traditionally considered to be the ulnar and radial carpals (see, however, Botelho et al. 2014a for ontogenetic data that challenge these identifications). Both carpals exhibit complex shapes and guide the movements of the hand. Two further carpal bones are fused with the metacarpals and contribute to the carpal **trochlea** of the carpometacarpus (Figure 1.9). One of these distal carpal bones, the semilunate carpal, is of particular phylogenetic interest, because it represents a key apomorphy (i.e., one of the defining characters) of maniraptoran theropods (e.g., Makovicky and Zanno 2011).

The carpometacarpus is one of the most characteristic avian compound bones (Figure 1.9). Its main portion is formed by the proximally and distally conjoined metacarpals of the major and minor digits, which delimit the intermetacarpal space. The carpometacarpus is the main attachment site for the proximal **primary feathers**. In

Figure 1.9 Major features of the hand skeleton of neornithine birds. (a) Palaeognathous nandu (*Rhea*, Rheiformes), (b) goose (*Anser*, Anatidae), (c, d) juvenile galliform currasow (*Crax*, Cracidae), (e) adult phasianid galliform (*Rollulus*, Phasianidae). Note the presence of wing claws in (a) and (b). All bones are from the right side and not to scale (a–c: ventral view; d, e: dorsal view).

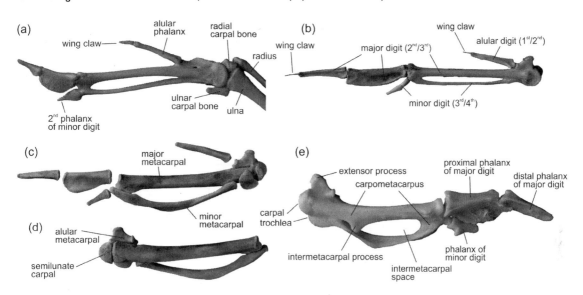

some birds the major metacarpal forms a distinct intermetacarpal process (Figure 1.9), which increases the leverage of a muscle flexing the hand, and which occurs in birds exposed to high aerodynamic forces on the hand section of the wing (Stegmann 1965).

The metacarpal of the alular digit is co-ossified with the carpal trochlea and bears the cranially directed extensor process, on which inserts the tendon of a muscle that extends the wing during flight stroke. The alular digit itself serves for the attachment of the feathers that form the alula. This separate "winglet" constitutes an important device to prevent stalling of the airflow, if the wings are held under a high angle of attack, especially during start and landing. Various groups retain vestigial wing claws on the alular and major digits, which sometimes reach a fair size (Figure 1.9). Whereas the alular and major digits of neornithine birds therefore still show the original phalangeal counts in some taxa, the minor digit is always greatly

reduced and is usually composed of only a single free phalanx, rather than three as in the avian ancestor.

The axial skeleton, pelvic girdle, and tail

The ribs of most neornithine birds are connected by transverse processes, which provide muscle attachment sites and play an important role in respiration. The length of these "uncinate processes" (Figure 1.10) also appears to be correlated to locomotory habits and they are short in terrestrial birds, but very long in diving taxa (Tickle et al. 2007). The distribution of uncinate processes shows some homoplasy. Such processes are found in several theropods currently placed outside Aves (e.g., some dromaeosaurs), but they are absent in some large flightless Neornithes, such as the extinct Dromornithidae, Gastornithidae, and Phorusrhacidae, as well as in the volant Anhimidae (screamers).

Birds primitively possessed rib-like gastral elements (gastralia) in the caudal portion

Figure 1.10 (a) The trunk skeleton of a roller (*Coracias*, Coraciidae) illustrates characteristics of the body plan of neognathous birds. (b) The pelvis of a palaeognathous tinamou (*Rhynchotus*, Tinamiformes); note the open ilioischiadic foramen, and the boundaries between the pelvic bones are indicated by dotted lines. Not to scale.

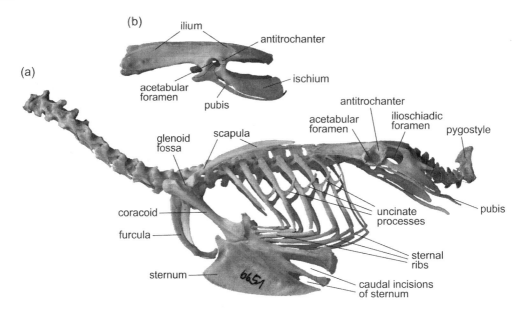

of the trunk (see Figure 2.1), for which a respiratory function has likewise been assumed (Claessens 2004; see, however, O'Connor et al. 2015a, who hypothesized an involvement in the attachment of pectoral muscles). Gastralia are widespread among archosaurs, but were lost in the evolution towards the crown group. In non-neornithine birds their numbers range between 16 and 4 pairs (O'Connor et al. 2015a).

In *Archaeopteryx* and other basal Mesozoic avians, the pelvis consists of three paired bones: the ilium, ischium, and pubis. The contact zones between these bones delimit the acetabular foramen, which forms the socket for the femur head. In Neornithes, all pelvic bones are fused into a single solid unit; usually they also co-ossify with the synsacrum, a bone formed by fusion of the sacral vertebrae. The pubis is caudally directed in crown group birds, but was more vertically oriented in the earliest avians, in which the

two pubic bones were furthermore conjoined at their tips and formed a pubic symphysis. Neornithes show some diversity of pelvis morphologies, and whereas foot-propelled diving birds have a very long and narrow pelvis, the pelvis is short and wide in highly aerial birds that make little use of their legs.

The earliest avians had a long bony tail, which is greatly foreshortened in neornithine birds, where it consists of only a few vertebrae. The caudalmost ones of these form a plate-like bone, the pygostyle, to which the central tail feathers are anchored.

The hindlimb
Like the wing, the avian leg is divided into three major units; that is, the femur, the tibiotarsus and fibula, and the foot. The major visible leg joint of birds, sometimes falsely assumed to be a reversed "knee," is an intertarsal joint, which develops between the tarsal (ankle) bones. In Neornithes, these

tarsals are completely fused with the distal end of the tibia and the proximal ends of the metatarsals of the foot (Figure 1.11). The resultant compound bones are termed tibiotarsus (tibia and proximal tarsals) and tarsometatarsus (metatarsals and distal tarsals). Unlike in most other tetrapods, where it swings back and forth, the femur of birds remains largely fixed in the same position during locomotion, and hindlimb movement is mainly driven by the knee (e.g., Allen et al. 2013).

The proximal end of the tibiotarsus bears two so-called cnemial crests, which serve as attachment sites of the shank musculature and are absent in early Mesozoic avians, in which the leg musculature may not yet have been so centralized. The cnemial crests are proximally elongated in swimming and diving birds, which use their legs for aquatic locomotion and therefore require larger attachment sites of the hypertrophied foot muscles. In long-legged, terrestrial birds the cnemial crests are also well developed, albeit not proximally elongated, whereas they are reduced in short-legged, arboreal taxa. The distal end of the tibiotarsus bears two condyles that articulate with the tarsometatarsus. Just above these condyles, the cranial surface of the bone exhibits a marked sulcus, through which pass the tendons for extensor muscles of the toes (Figure 1.11). This "extensor sulcus" occurred late in avian evolution. In the majority of neognathous birds it is bridged by an osseous arch, the supratendinal bridge, which is secondarily lost in some taxa and primitively absent in most palaeognathous birds.

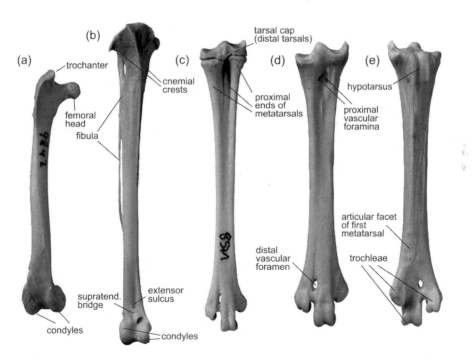

Figure 1.11 Major features of the leg bones of neornithine birds. (a) Femur and (b) tibiotarsus of a rock partridge (*Alectoris*, Galliformes). (c) Tarsometatarsus of a juvenile pheasant (*Lophura*, Galliformes), which shows the incomplete proximal fusion of the metatarsals and the cap formed by the distal tarsals. Tarsometatarsus of an adult rock partridge (*Alectoris*) in (d) dorsal and (e) plantar view. All bones are from the right side and not to scale.

Birds derive from an ancestor with five metatarsals. Even in the earliest avians, however, the fifth metatarsal is only a rudimentary splint, and the bone was completely lost in the evolution towards the crown group. The first metatarsal, which in neornithine birds carries the hind toe, is distally positioned and likewise reduced (Figure 1.12). In the course of avian evolution, the three other metatarsals – that is, those of the second, third, and fourth toes – completely fused with each other and the distal tarsals, thereby forming the tarsometatarsus.

The tarsometatarsus is another characteristic compound bone of extant birds, which greatly varies in its proportions across different taxa. It is particularly short and stout in highly aerial birds and elongated and slender in wading or **cursorial** ones (Figure 1.12). The proximal end of the neornithine tarsometatarsus bears the hypotarsus, a bony structure that guides the tendons of the flexor muscles of the toes. Early Mesozoic avians lack a hypotarsus and in more advanced non-neornithine birds its structure is very simple, consisting of a single low crest. This plesiomorphic hypotarsus morphology is still found in palaeognathous birds, whereas the hypotarsus of most neognathous birds exhibits well-developed furrows and canals, the configuration of which is often of taxonomic significance (Figure 1.12).

The distal end of the tarsometatarsus also shows distinctive morphologies that characterize various neornithine taxa. The bone usually exhibits a foramen between

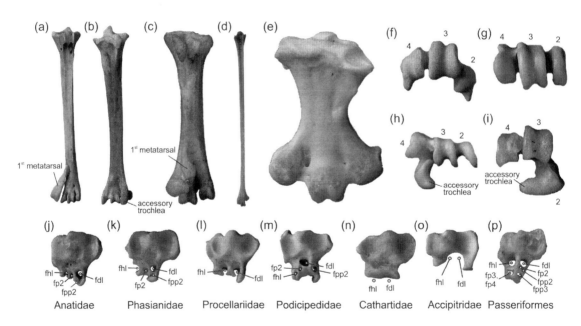

Figure 1.12 Different morphologies of the neornithine tarsometatarsi. Depicted are (a, g) an anisodactyl songbird (*Corvus*, Passeriformes), (b, h) a zygodactyl woodpecker (*Dryocopus*, Piciformes), (c, i) a heterodactyl trogon (*Pharomachrus*, Trogoniformes), (d) a flamingo (*Phoeniconaias*, Phoenicopteriformes), (e) a potoo (*Nyctibius*, Nyctibiiformes), and (f) a phasianid francolin (*Pternistis*, Galliformes). All bones are from the right side and not to scale (a–c: plantar view, d, e: dorsal view, f–i: distal view). (j–p) Different patterns of the sulci and canals of the hypotarsus on the proximal tarsometatarsus end; indicated are the passages for tendons of the muscles flexing the hind toe (fhl), all three fore toes (fdl), and the second (fp2, fpp2), third (fp3, fpp3), and fourth toes (fp4).

the fused metatarsals of the third and fourth toes, the distal vascular foramen, and forms trochleae for the articulation with the three fore toes. Very rarely is one of these trochleae reduced, and, if it is, it is always that of the second toe, as in the cursorial ostriches (Struthioniformes). The configuration of the tarsometatarsal trochleae is variable and reflects different locomotory adaptations. In birds with reversed fore toes (see next paragraph), the corresponding tarsometatarsal trochleae are deflected and often bear accessory trochleae (Figure 1.12). Embryological studies suggest that the arrangement and morphology of the tarsometatarsal trochleae are shaped by developmental constraints, and the occurrence of short and parallel trochleae with narrow incisions, for example, seems to be correlated with the reduction of certain toe muscles (Botelho et al. 2014b).

Besides its locomotory function, the avian foot often serves as a grasping tool and many different kinds of foot specializations exist. Most Neornithes have four toes, three of which direct forwards and one, the hind toe or hallux, is turned backwards. This toe arrangement is termed anisodactyl and represents the primitive condition for neornithine birds. Very rarely are all four toes directed forwards (pamprodactyl foot). To increase the grasping capabilities of the foot, the fourth toe became reversed in some groups, resulting in a zygodactyl foot. This condition is developed to various degrees in different taxa, and the fourth toe can be moved back and forth in facultatively zygodactyl birds, is laterally spread in semizygodactyl ones, or is permanently directed backwards in fully zygodactyl taxa. Trogons are the only neornithine group in which the second toe is permanently reversed, forming the heterodactyl foot (a putative heterodactyl foot in one specimen of a Cretaceous bird reported by Zhang et al. 2006 is likely to be an artifact of the preservation of the feet in this fossil). The ontogenetic mechanism involved in the formation of zygodactyl or heterodactyl feet is an asymmetric development of some of the muscles of the fourth or second toe, which causes the reversion of these toes (Botelho et al. 2014b).

In summary, therefore, the avian skeleton exhibits numerous specializations and shows a high degree of variation across neornithine birds. Still, it was a long evolutionary way until this diversity developed, and the fossil record allows us to trace many of the steps in between.

2 | The Origin of Birds

The origin of birds involves one of the major evolutionary transitions in the history of vertebrates, from an earthbound biped to a winged airborne animal, and fervid controversies surrounded the beginnings of flight and the identity of the closest avian relatives. In this chapter the very beginnings of avian evolution are outlined and hypotheses on the origins of two key avian attributes, feathers and flight capability, are reviewed.

Only a few years ago, organization of an account of bird evolution would have been straightforward, starting with *Archaeopteryx* as the earliest known avian taxon. Since then, however, not only were earlier *Archaeopteryx*-like fossils discovered, but the rapidly increasing menagerie of Cretaceous birds and bird-like theropods has also made identification of "the" avian ancestor virtually impossible. As is to be expected in an evolutionary continuum, no clear-cut boundary exists between birds and their "non-avian" ancestors. The homoplastic distribution of avian-like characteristics in non-avian theropods furthermore aggravates a straightforward identification of the closest avian relatives and a reconstruction of the earliest avian diversifications.

This overview nevertheless begins with *Archaeopteryx*, because it is comparatively well known and its morphology may therefore serve as a template for comparisons with other taxa near the avian base. Moreover, the interrelationships of other candidate taxa are not yet sufficiently well established for these fossils to challenge the central position of *Archaeopteryx* for an understanding of the origin of birds.

Avian Evolution: The Fossil Record of Birds and its Paleobiological Significance, First Edition. Gerald Mayr.
© 2017 John Wiley & Sons, Ltd. Published 2017 by John Wiley & Sons, Ltd.

Archaeopteryx: The German "Urvogel" and Its Bearing on Avian Evolution

Termed the "icon of evolution" (Wellnhofer 2009), the "Urvogel", or "primordial bird," *Archaeopteryx* from the Late Jurassic (150 mya) Solnhofen limestone in southern Germany constitutes one of the most impressive transitional forms in the fossil record. Its evolutionary relevance, which makes it the literal textbook example of a "missing link," rests in the demonstrative mosaic distribution of plesiomorphic theropod features and derived avian ones. To some degree, the significance of this taxon is, however, also due to the fact that the first fossils had already been described in 1861, just two years after the publication of Charles Darwin's monumental work on the origin of species.

Archaeopteryx is currently known from 11 published skeletal specimens, all of which stem from quarries in the Eichstätt and Solnhofen areas (Plate 1). Most of these fossils are referred to by the cities of the institutions that house them. The exact number of species and supraspecific taxa represented by these very differently sized skeletons is controversial. Some authors assumed that all represent growth series of a single species, but others emphasized morphological differences that justify the recognition of more than one species, or even of different "genera" (Elzanowski 2002; Mayr et al. 2007; Wellnhofer 2009). The thrush-sized Eichstätt specimen is only half the size of the largest *Archaeopteryx* fossil, the Solnhofen specimen. Equivalent intraspecific size variations are known from other Mesozoic birds and non-avian theropods (see Chapter 4), but several further differences between the *Archaeopteryx* specimens are unlikely to be due to allometric changes in a single species. The Munich and Eichstätt specimens, for example, differ in their limb proportions

from the other skeletons (Wellnhofer 2009). The ischium of the London specimen is much narrower than that of, for example, the Thermopolis specimen, which indicates a distinctness of these two specimens on the species level (Mayr et al. 2007; contra Wellnhofer 2009). The Solnhofen specimen is furthermore characterized by an anomalous phalangeal formula of the foot, with the fourth toe consisting of only four, instead of five, phalanges (Elzanowski 2002). Owing to the preservation of the fossils, this foot morphology is not visible in the equally large and possibly conspecific London and Maxberg specimens (Mayr et al. 2007).

Overall, the skeleton of *Archaeopteryx* (Figure 2.1) is very similar to that of basal paravians, especially taxa currently assigned to the Troodontidae. The skull has a long and narrow snout, with 12–13 teeth in each upper jaw (4 in the praemaxilla and 8–9 in the maxilla) and 11–12 in the lower jaws. All teeth are small, with recurved tips, and a constricted root (Plate 1f). As in many non-avian theropods, but unlike other toothed Mesozoic birds, there are interdental plates between the teeth of the lower jaws. The palate resembles that of non-avian theropods (see Chapter 3). The cervical vertebrae exhibit pneumatic openings, which indicate the presence of an air-sac system.

The furcula is robust and boomerang shaped and the coracoid is squarish. The scapula and coracoid of *Archaeopteryx* are often described as being "fused," but although the two bones appear to have been tightly adjoined, they are clearly two separate bones in most specimens (Wellnhofer 2009). An ossified sternum is absent and the ribs lack uncinate processes.

Archaeopteryx has long forelimbs, which reach the length of the hindlimbs. Carpal and metacarpal bones show no signs of fusion and the minor digit is very long. The phalangeal formula of the manual digits corresponds with that of most non-avian maniraptorans

Figure 2.1 Skeleton of *Archaeopteryx* from the Late Jurassic of Germany.

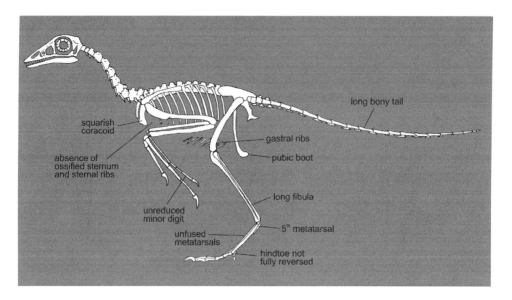

in that the alular digit has two phalanges, the major digit three, and the minor digit four, including large claws on all three digits.

The synsacrum consists of five fused vertebrae. In the hindlimbs, neither the proximal nor the distal tarsal bones are fused. The fibula is still very long and reaches the distal end of the tibia. The three longest metatarsals are merely fused at their very proximal ends, and this only in the largest individuals. A rudimentary fifth metatarsal is present. Contrary to many earlier reconstructions and unlike in extant birds, the first toe is not reversed, but inserts medially on the foot (Figure 2.2; Mayr et al. 2005). As indicated by the shape of its phalanges, the second toe was dorsally hyperextensible (Mayr et al. 2005, 2007). Unlike in the paravian dromaeosaurs and troodontids (Chapter 1), however, the penultimate phalanx is not foreshortened and the claw is not greatly enlarged. The long tail consists of 21–23 vertebrae, which probably formed a rather stiff unit.

The forelimb feathering of *Archaeopteryx* has a very modern appearance, both concerning the shape of the individual feathers and that of the entire wing. The number of primary feathers is nevertheless debated, and estimates vary between 8 and 12, with the controversial low count resulting from an interpretation of faint shaft impressions in the Berlin specimen as coverts (Wellnhofer 2009; Longrich et al. 2012; Foth et al. 2014; Nudds 2014). The 16–17 pairs of long, serially arranged tail feathers exhibit slightly asymmetric vanes and are likely to have fulfilled an aerodynamic function. Contrary to most of the earlier reconstructions, the tip of the tail may have been forked rather than rounded (Foth et al. 2014). The contour feathers of the body have likewise become better known only recently, and the tibia bears well-developed **pennaceous feathers** (see Figure 2.10; Foth et al. 2014). Alula feathers were not reported for *Archaeopteryx*, but the alular digit itself may have fulfilled a function in preventing stalling of the airflow across the wing surface under high angles of attack (Meseguer et al. 2012).

Because of its presumed implications for the origin of avian flight, the lifestyle

Figure 2.2 (a) Left foot of the Thermopolis specimen of *Archaeopteryx*. (b) Schematic drawing of the foot of *Archaeopteryx* in comparison to that of (c) a pigeon to show the different orientation of the first toe. In (b) and (c) the toes are numbered.

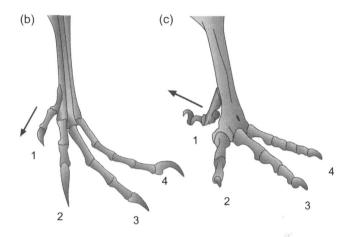

of *Archaeopteryx* was a matter of much controversy. It is now, however, generally acknowledged that the "Urvogel" was already too advanced to reveal immediate insights into the earliest stages of the transition between flightless and flying theropods.

None of the *Archaeopteryx* fossils shows stomach contents, but the low teeth are suggestive of a diet consisting of insects or other small invertebrates. Continuing debates surrounding *Archaeopteryx* concern its flight capabilities, which are discussed further later in this chapter, and the question of whether it was a terrestrial or an arboreal, possibly even a trunk-climbing, animal (Elzanowski 2002; Wellnhofer 2009). The trunk-climbing hypothesis was supported with the shape of the large forelimb claws, which are preserved with their horny sheaths in most specimens. The curvature of both the manual and pedal claws of *Archaeopteryx* does indeed correspond with that of the foot claws of extant arboreal birds and mammals (Feduccia 1993), but similar claw geometries are found in various non-arboreal birds (Wellnhofer 2009). The morphology of the hindlimb skeleton

of *Archaeopteryx* does not indicate climbing habits (Wellnhofer 2009), and most likely it was neither a cursorial nor a strictly arboreal animal. However, although the "Urvogel" may have been predominantly terrestrial, it probably also had some perching capabilities (Elzanowski 2002; Chiappe 2007; Wellnhofer 2009).

In recent years, the significance of *Archaeopteryx* for an understanding of avian evolution has repeatedly been challenged, be it through discovery of slightly older *Archaeopteryx*-like fossils (see later in this chapter) or through phylogenetic hypotheses, which considered the "Urvogel" to be more closely related to theropods traditionally considered "non-avian" (Xu et al. 2011, 2015). New fossil finds of the past few years have furthermore shown that some allegedly avian features are either not present in *Archaeopteryx* (e.g., a triradiate palatine bone and a reversed first toe; Mayr et al. 2005) or also occur in other maniraptorans (e.g., feathers). However, if Aves is defined as the least inclusive clade comprising *Archaeopteryx* and extant birds (see Chapter 1), the "Urvogel" remains the critical benchmark for avian affinities of extinct taxa. After all, the importance of *Archaeopteryx* does not so much depend on its exact placement in the phylogenetic tree, but on whether it represents a morphology that can be considered close to that of the avian stem species.

The Closest Maniraptoran Relatives of Birds

Birds are maniraptoran theropods, and there is a broad consensus that oviraptorosaurs, dromaeosaurs, and troodontids are their closest relatives (see Chapter 1). Dromaeosaurs and troodontids are usually united as Deinonychosauria, a key apomorphy of which is a dorsally hyperextensible second toe with a hypertrophied, sickle-shaped claw. This derived foot morphology is often interpreted as a slashing device, but it has also been proposed that it served to pin down prey items (Manning et al. 2006).

For a long time, oviraptorosaurs and deinonychosaurs were mainly known from Late Cretaceous stages of the Northern Hemisphere, mainly Asia and North America. The earliest of these fossils were therefore some 40–50 million years younger than *Archaeopteryx*, and this "temporal paradox" was cited as a major inconsistency of the presumed close relationships between deinonychosaurs and birds (e.g., Feduccia 1999). In the past few decades, however, deinonychosaur fossils were also reported from Early Cretaceous and Late Jurassic localities, and the oldest of these even precede *Archaeopteryx* in age. These new discoveries blurred the distinction between basal paravians, and close affinities between birds, oviraptorosaurs, and deinonychosaurs can no longer be seriously questioned. However, although a close relationship between birds and maniraptorans is now even acknowledged by former opponents

of the theropod ancestry of birds (Feduccia 2012), the exact interrelationships of the taxa involved are far less certain.

Oviraptorosaurs, caudipterygids, and scansoriopterygids

Oviraptorosaurs (Figure 2.3a) are currently considered the **sister group** of Paraves (e.g., Turner et al. 2012), although some earlier authors discussed a position within Aves (Elzanowski 1999; Maryańska et al. 2002; Paul 2002). Most fossils of these unusual animals stem from the Cretaceous of Asia, but one group, the Caenagnathidae, occurs in the Late Cretaceous of North America. Identification of typical oviraptorosaurs is straightforward, because these presumably herbivorous theropods are characterized by an unmistakable morphology of the skull, which is very high, with dorsally positioned nostrils and a short snout that in more advanced species forms an edentulous beak (Osmólska et al. 2004). Oviraptorosaurs have ossified sternal plates and relatively long forelimbs with three fingers. In a few taxa, such as the Late Cretaceous *Nomingia*, the distal tail vertebrae are fused into an incipient pygostyle.

Putatively basal oviraptorosaurs were also described from the Early Cretaceous Jehol Biota in China. The best known of these belong to the Caudipterygidae, which include *Caudipteryx* from the Yixian Formation (Figure 2.3b, Plate 3b) and *Similicaudipteryx* from the Yixian and Jiufotang formations (Ji et al. 1998; Zhou et al. 2000; Xu et al. 2010). The turkey-sized caudipterygids have short snouts and only the praemaxillaries bear peg-like teeth (see Figure 4.3). Like in other oviraptorosaurs, the edentulous lower jaws of caudipterygids exhibit large fenestrae and their tips are fused and form a symphysis. There is a pair of sternal plates, but these are smaller than those of Late Cretaceous oviraptorosaurs, in which the sternal plates furthermore are often fused (see

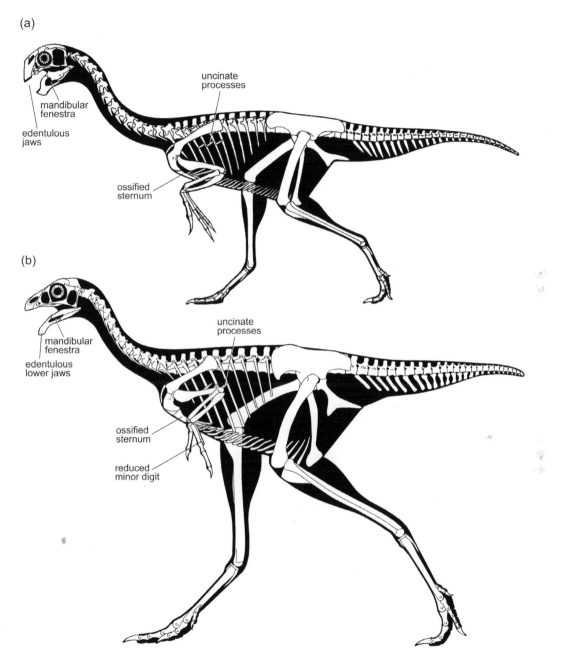

(a)

uncinate
processes

mandibular
fenestra

edentulous
jaws

ossified
sternum

(b)

uncinate
processes

mandibular
fenestra

edentulous
lower jaws

ossified
sternum

reduced
minor digit

Figure 2.3 Skeletons of (a) the oviraptorosaur *Khaan* from the Late Cretaceous of Mongolia and (b) the caudipterygid *Caudipteryx* from the Early Cretaceous Jehol Biota. Not to scale. Reconstructions © Scott Hartman.

Figure 4.6). The morphology of the forelimbs of *Caudipteryx*, however, is more derived than in typical oviraptorosaurs, in that the minor digit consists of the metacarpal and only two small phalanges without a claw. The minor digit of *Similicaudipteryx*, by contrast, has the full phalangeal count and is not reduced.

Caudipterygids exhibit modern-type pennaceous forelimb and tail feathers. However, they were clearly flightless, because the forelimb feathers are much too short to have provided enough lift for these fairly large animals to have become airborne. Unlike in volant birds, the vanes of the forelimb feathers of caudipterygids are furthermore not asymmetric. In *Caudipteryx* they only insert on the proximal section of the hand, and the fingers seem to have been free of long feathers. The tail of *Caudipteryx* bears a feather frond on its tip, whereas that of *Similicaudipteryx* has serially arranged tail feathers. Two of the three published *Similicaudipteryx* specimens are juveniles or subadults, which differ in integumentary features (Xu et al. 2010). The ontogenetically younger specimen exhibits unusual, "proximally ribbon-like" forelimb and tail feathers, which may actually represent growing feathers (Prum 2010). The preservation of **gastroliths** (gizzard stones) in several *Caudipteryx* fossils indicates a predominantly herbivorous diet.

Another taxon placed at the base of Oviraptorosauria is *Protarchaeopteryx* from the Yixian Formation, which is only known from the holotype skeleton and was initially considered to be an archaeopterygid (Ji et al. 1998; but see Balanoff et al. 2009 concerning presumed affinities to the basal oviraptorosaur *Incisivosaurus*). *Protarchaeopteryx* has proportionally shorter forelimbs than *Archaeopteryx*, from which it also differs in that the skull is short and tall, with long and peg-like teeth that are restricted to the tip of the snout. Unlike in *Caudipteryx* but as in *Similicaudipteryx*, the minor manual digit of *Protarchaeopteryx* is not reduced. *Protarchaeopteryx* exhibits pennaceous forelimb and tail feathers, but judging from the short forelimbs it was also flightless.

Recognition of oviraptorosaurian affinities of caudipterygids and *Protarchaeopteryx* suggests the presence of pennaceous forelimb and tail feathers in later and more derived oviraptorosaurs of which the integument is unknown. As will be detailed later in this chapter, it is here considered likely that pennaceous feathers evolved in an aerodynamic context, although other opinions exist, which assume, for example, a primary function for signaling. The key question surrounding the occurrence of pennaceous feathers in early oviraptorosaurs is therefore whether these animals were primarily flightless, or whether they secondarily lost flight capabilities, as was assumed by a few authors, some of which even proposed a position of caudipterygids within Aves (Maryańska et al. 2002; Paul 2002).

The majority of current researchers do not consider likely a secondary flightlessness of caudipterygids, and from an evolutionary point of view one would also have to address how these animals could have lost flight capabilities in a paleoenvironment with terrestrial predators. However, the hypothesis of avian affinities of caudipterygids and, hence, oviraptorosaurs is not entirely far-fetched. The unusually short and high skulls of *Caudipteryx* and *Protarchaeopteryx*, whose dentition is restricted to the tip of the snout, strikingly resemble those of some basal avians, such as *Jeholornis* and *Sapeornis* (see Chapter 3 and Figure 4.3). The brain of oviraptorosaurs also shows avian-like attributes, the significance of which is disputed, however (Balanoff et al. 2014). Unlike in non-avian theropods and *Archaeopteryx* but as in more derived avians, the palate of oviraptorosaurs furthermore features a triradiate palatine, whereas this bone is tetraradiate in most non-avian theropods.

Most current analyses found oviraptorosaurs to be outside Paraves, and presumptive paravian apomorphies, which are absent in oviraptorosaurs, include forelimb features functionally correlated with the origin of flight (Turner et al. 2012). Other presumably primitive attributes of oviraptorosaurs include the lack of a retroverted pubis (Makovicky and Zanno 2011), although the pubis of at least *Archaeopteryx* was probably not significantly more reversed than that of oviraptorosaurs (Wellnhofer 2009). If oviraptorosaurs were secondarily flightless, some of the characters listed in support of their position outside Paraves, such as the shorter forelimbs and the absence of vane asymmetry of the forelimb feathers (Makovicky and Zanno 2011), may be due to the flightlessness of these animals. Because flightlessness often involves **paedomorphosis** – that is, a delayed ontogenetic development, which results in the retention of juvenile characters in the adults (Feduccia 1999, 2012) – other putatively primitive features of oviraptorosaurs may represent secondary reversals into the primitive condition. On the other hand, however, many of the characters that were proposed to establish a placement of oviraptorosaurs within

Aves show a high degree of homoplasy (e.g., the loss of maxillary teeth and the presence of a mandibular symphysis), and previous analyses suggesting avian affinities of oviraptorosaurs (Maryańska et al. 2002) suffered from a very limited taxon sampling.

Further light on the affinities of oviraptorosaurs was perhaps shed by the recognition of derived similarities to the Scansoriopterygidae, a peculiar group of theropods from the Late Jurassic Chinese Daohugou Biota, which includes the taxa *Scansoriopteryx* ("*Epidendrosaurus*"), *Epidexipteryx*, and *Yi* (Zhang et al. 2002, 2008a; Sullivan et al. 2014; Xu et al. 2015). *Scansoriopteryx* (Figure 2.4) is only known from two sparrow-sized juvenile specimens, and the larger, pigeon-sized *Epidexipteryx* is likewise based on a juvenile or subadult individual (the possibility that the former represents an earlier ontogenetic stage of the latter was dismissed because of differences in the morphology of the tail, which consists of more vertebrae in *Scansoriopteryx* than in *Epidexipteryx*). The only scansoriopterygid taxon represented by an adult individual is *Yi*, which features an unusual rod-like skeletal element in the wing (Xu et al. 2015). This bone was considered evidence for bat-like wing membranes in scansoriopterygids, but its identification and functional interpretation certainly

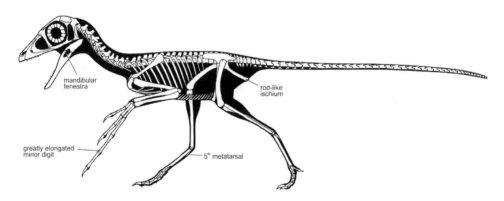

Figure 2.4 Skeleton of *Scansoriopteryx* (Scansoriopterygidae) from the Late Jurassic Daohugou Biota. Note the very long minor digit of this peculiar animal. Reconstruction © Scott Hartman.

deserve further study. The forelimbs of *Scansoriopteryx* and *Yi* are very long and their length exceeds that of the hindlimbs (unfortunately, the forelimbs of *Epidexipteryx* remain poorly known). A peculiar characteristic of scansoriopterygids – that is, of *Scansoriopteryx* and *Yi* at least – is a greatly elongated minor digit, which exceeds the major digit in length, therefore belying digital terminology in the case of these animals. The skull is short and high, with peg-like procumbent teeth, and shows a resemblance to the skull of caudipterygids (see Figure 4.3). The *Epidexipteryx* holotype exhibits two pairs of elongated, "ribbon-like" tail appendages, which were assumed to have

had a display function (Zhang et al. 2008a). Unusual integumentary structures on the body consist of parallel, barb-like filaments. The very long forelimbs, as well as the long minor digit and the shape of the claws, were taken as indicative of arboreal habits of scansoriopterygids (Zhang et al. 2002).

The affinities of scansoriopterygids are contentious, not least because most specimens represent juveniles or subadults (Figure 2.5). Scansoriopterygids were obtained as the sister taxon of Aves in some analyses (Zhang et al. 2008a; Turner et al. 2012; Xu et al. 2015), but they were also considered to be the sister taxon of oviraptorosaurs, which they resemble in skull

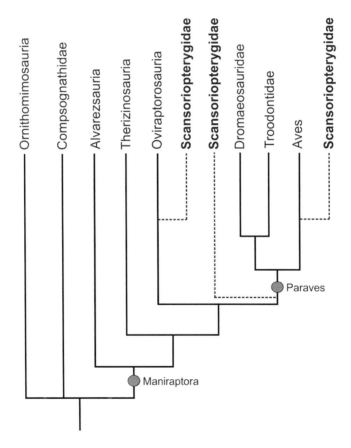

Figure 2.5 Current consensus phylogeny of the Maniraptora and the different phylogenetic positions proposed for the Scansoriopterygidae (based on Xu et al. 2011, 2015; Godefroit et al. 2013a; O'Connor and Sullivan 2014).

shape, the dorsal position of the nostrils, and the presence of a large mandibular foramen (see Figure 4.3; Agnolín and Novas 2013; O'Connor and Sullivan 2014). A sister group relationship between scansoriopterygids and oviraptorosaurs, if confirmed in future studies, would have potential implications for the status of oviraptorosaurs. Not only would the very long forearms of scansoriopterygids and the occurrence of modern-type pennaceous feathers in at least the earliest oviraptorosaurs lend further support to the hypothesis that oviraptorosaurs were secondarily flightless animals with degenerated wings, but because scansoriopterygids are placed within Paraves in most current analyses, close affinities between these animals and oviraptorosaurs would also entail paravian affinities of the latter, and would therefore potentially open a new perspective on the close similarities between the skulls of scansoriopterygids, oviraptorosaurs, and some early avians.

Another poorly known taxon from the Early Cretaceous Yixian Formation, *Zhongornis*, is based on a single specimen of a juvenile individual. This animal appears to be edentulous and has a short tail, which does not form a pygostyle. The minor digit consists of only two phalanges and a claw. An initial phylogenetic analysis suggested avian affinities of *Zhongornis* (Gao et al. 2008), but more recently it was identified as a "scansoriopterygid-like non-avian" (O'Connor and Sullivan 2014). A definitive assessment of its relationships, however, will probably only be possible once adult or better-preserved specimens are discovered.

There are further theropod taxa of controversial affinities, which may be close to the avian ancestry. One of these is *Avimimus*, an avian-like theropod that is known from several skeletons from the Late Cretaceous of Mongolia (Vickers-Rich et al. 2002). *Avimimus* was a large animal with an edentulous praemaxilla, proximally fused metacarpals,

an ulna that seems to exhibit quill knobs, and long and slender legs with fused metatarsals. Owing to the incomplete knowledge of the skeleton, however, its phylogenetic position is poorly resolved and closer relationships to ornithomimosaurs, oviraptorosaurs, or birds were proposed (Maryańska et al. 2002; Vickers-Rich et al. 2002).

Dromaeosaurs – the "raptors"

Dromaeosaurs include some of the most iconic theropods, and it was the study of the dromaeosaur *Deinonychus* (Figure 2.6a) that led to a renaissance of the hypothesis of a theropod ancestry for birds (Ostrom 1976). Definitive dromaeosaurs were only found in Cretaceous rocks, and most of the best-known taxa, such as *Velociraptor*, *Deinonychus*, and *Bambiraptor*, stem from North American or Asian fossil sites. However, there are also fossils from the Southern Hemisphere, especially South America, for which dromaeosaurid affinities are assumed (Turner et al. 2012). These South American records are now assigned to the taxon Unenlagiinae, which is set apart from other dromaeosaurs by several skeletal characteristics, and which some authors consider to be more closely related to avians than to dromaeosaurs (Agnolín and Novas 2011; see, however, Makovicky et al. 2005; Turner et al. 2012). Typical dromaeosaurs, such as *Deinonychus*, *Dromaeosaurus*, and *Velociraptor*, were fairly large animals, but a small size is likely to be plesiomorphic for the group and is found in phylogenetically basal taxa, such as *Mahakala* from the Late Cretaceous of Mongolia and *Microraptor* from the Early Cretaceous of China (e.g., Turner et al. 2007a, 2012).

Overall, the skeletal morphology of dromaeosaurs agrees well with that of *Archaeopteryx*, and the difficulties in distinguishing basal deinonychosaurs from basal avians are exemplified by *Rahonavis* from the Late Cretaceous of Madagascar.

Figure 2.6 Skeletons of (a) the dromaeosaur *Deinonychus* and (b) the troodontid *Troodon*. Not to scale. Reconstructions © Scott Hartman.

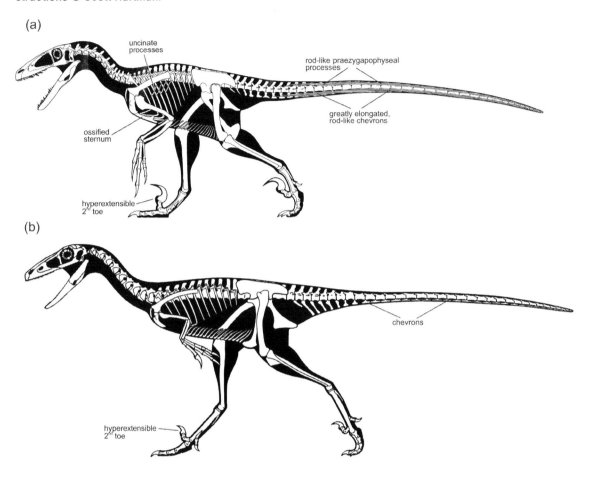

(a)

uncinate processes

rod-like praezygapophyseal processes

ossified sternum

greatly elongated, rod-like chevrons

hyperextensible 2nd toe

(b)

chevrons

hyperextensible 2nd toe

Known from a partial skeleton, this taxon was first described as a Mesozoic bird with a deinonychosaur-like sickle-shaped claw of the second toe (Forster et al. 1998). In more recent analyses, however, the avian affinities of *Rahonavis* were challenged and it was placed within dromaeosaurs, as the sister taxon of the South American Unenlagiinae (Makovicky et al. 2005; Turner et al. 2012).

Unlike *Archaeopteryx* and toothed Mesozoic birds, most dromaeosaurs have serrated teeth, which resemble those found in many other carnivorous coelurosaurs. The coracoid and the scapula are tightly sutured in most of the larger species, but in *Rahonavis* both elements are separate (Forster et al. 1998). Dromaeosaurs have at least two ossified sternal plates, which are fused into a sternum in some taxa (e.g., *Microraptor*, *Bambiraptor*; Burnham 2004). The relative length of the forelimbs varies within dromaeosaurs. In some taxa, such as *Microraptor*, they are elongated and approach the relative forelimb length of *Archaeopteryx*. In others, such as *Mahakala* or the Early Cretaceous Chinese *Tianyuraptor* and *Zhenyuanlong*, the forelimbs are short, although the fingers are still comparatively long relative to the lengths of humerus and ulna (e.g., Lü and Brusatte 2015). With the exception of the basally branching *Mahakala* and the Southern Hemispheric unenlagiines (as well as *Rahonavis*), dromaeosaurs are characterized by a derived tail morphology with greatly elongated chevrons (or haemapophyses, ventral ossicles between the caudal vertebrae) and very long, rod-like, and intertwined praezygapophyseal processes, which span several vertebrae and serve to stiffen the tail against dorsoventral flexion (Figure 2.6a; Norell and Makovicky 2004; Turner et al. 2012; Persons and Currie 2012).

Some spectacular findings from Early Cretaceous fossil localities in China provided key insights into the anatomy of dromaeosaurs. Together with the compsognathid *Sinosauropteryx*, the dromaeosaur *Sinornithosaurus* from the Yixian Formation was among the first non-avian theropods for which integumentary appendages were described (Xu et al. 1999, 2001). Whereas only "fur"-like filaments of varying complexity could be identified in *Sinornithosaurus*, the slightly younger *Microraptor* from the Jiufotang Formation even exhibits asymmetric, pennaceous feathers. Remarkably, the presence of these feathers is not restricted to the forelimbs and the tail, but equally well-developed pennaceous feathers are found on the hindlimbs, where they extend onto the metatarsals (see Figure 2.9; Xu et al. 2003). Similar hindlimb feathers are also known from the closely related *Changyuraptor* from the Yixian Formation (Han et al. 2014), as well as from various taxa considered to be basal troodontids (see the next section).

Whether the absence of pennaceous feathers in *Sinornithosaurus* reflects a possible juvenile condition of the known fossils, is an artifact of preservation, or is due to true integumental differences still needs to be assessed. The integument of large dromaeosaurs is unknown, but for the ulna of *Velociraptor* quill knobs were reported (Turner et al. 2007b), which suggests that pennaceous feathers are an ancestral trait of dromaeosaurs.

In *Balaur* from the Late Cretaceous of Romania not only the second, but also the first toe appear to have been hyperextensible. *Balaur* was initially described as a dromaeosaur (see Turner et al. 2012), but shares several derived traits with avians, including a proximal fusion of the metacarpals, a greatly reduced minor manual digit, and co-ossified metatarsals. Some recent analyses indeed supported avian affinities of *Balaur* and placed the taxon in an early divergence at the base of Aves (Godefroit et al. 2013a; Foth et al. 2014; Cau et al. 2015). Many of the bird-like characteristics of *Balaur* do, however, exhibit a high degree of homoplasy within Aves and further studies may be needed to elucidate the relationships of this taxon.

Troodontids, "troodontid-like" Jurassic paravians, and the controversial paravian interrelationships

Even more similar to *Archaeopteryx* are some representatives of the second deinonychosaurian group, the Troodontidae, which mainly occurred in the Cretaceous of Asia and North America (Figure 2.6b; Makovicky and Norell 2004). Bird-like attributes of troodontids include large orbits and a similar brain morphology (Balanoff et al. 2013). Some taxa (e.g., *Byronosaurus*) also have avian-like teeth, which lack the serrated margins found in most deinonychosaurs and many other carnivorous non-avian theropods. Unlike dromaeosaurs but in agreement with *Archaeopteryx*, troodontids lack ossified sternal plates.

The skeletal similarities between troodontids and *Archaeopteryx* are exemplified by recently described *Archaeopteryx*-like paravians from the Late Jurassic Chinese Daohugou Biota, which, with an age of about 160 mya (Liu et al. 2012), predate the "Urvogel" fossils by 10 million years. Of these fossils, *Anchiornis* in particular, which was initially identified as a troontid, closely corresponds to *Archaeopteryx* in skeletal morphology (Hu et al. 2009; Xu et al. 2009b). Further *Archaeopteryx*-like taxa from the Daohugou Biota, such as *Xiaotingia*, *Aurornis*, and *Eosinopteryx*, have proportionally shorter forelimbs, but otherwise also show an "Urvogel-like" general body plan (Xu et al. 2011; Godefroit et al. 2013a, b; the provenance of the *Aurornis* holotype is controversial and it may actually come from the Early Cretaceous Yixian Formation: Sullivan et al. 2014).

Anchiornis, *Xiaotingia*, *Eosinopteryx*, and *Aurornis* not only differ in their relative forelimb lengths, but also in various other characteristics, such as skull proportions and details of the dentition, with the teeth being more bulbous in *Xiaotingia*. All four taxa have somewhat shorter and higher snouts than *Archaeopteryx*, with unserrated, low teeth that are more closely packed at the tips of the jaws (Xu et al. 2009b, 2011; Godefroit et al. 2013b). *Anchiornis* is known from more than 200 specimens, in none of which ossified sternal elements or sternal ribs are preserved (Zheng et al. 2014a), and these are also absent in the fossils of *Xiaotingia*, *Eosinopteryx*, and *Aurornis*.

Anchiornis features well-developed pennaceous forelimb feathers, but unlike in *Microraptor* and *Archaeopteryx* these have symmetric vanes and the longest forelimb feathers are close to the wrist rather than near the tip of the hand (Hu et al. 2009). The forelimb airfoil is notable in that it consists of multiple layers of very narrow feathers (Longrich et al. 2012). *Anchiornis* at least also sports long, pennaceous hindlimb feathers, which unlike in *Microraptor* have symmetric vanes (Plate 2c–e; Hu et al. 2009); the tail feathers are similar to those of *Archaeopteryx*. The fore- and hindlimb feathers of some *Anchiornis* fossils exhibit color patterns, and through study of fossilized **melanosomes** there have been attempts to

reconstruct the original coloration (Li et al. 2010; see also later in this chapter).

The phylogenetic affinities of all of the above taxa are far from being well established. *Anchiornis* was hypothesized to be most closely related to either Aves (Godefroit et al. 2013b; Xu et al. 2009b) or Troodontidae (Hu et al. 2009), or it was placed in a clade together with *Xiaotingia* and *Archaeopteryx*, with this clade in turn being the sister taxon of deinonychosaurs (Figure 2.7; Xu et al. 2011). Other analyses supported a sister group relationship between Troodontidae and a clade including *Xiaotingia* and *Anchiornis* (Turner et al. 2012); showed *Anchiornis*, *Archaeopteryx*, and *Xiaotingia* to be successive sister taxa

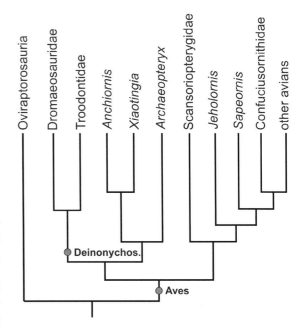

Figure 2.7 An alternative phylogenetic hypothesis of paravian interrelationships, in which *Archaeopteryx* is more closely related to deinonychosaurs than to the clade formed by *Jeholornis*, *Sapeornis*, and other avians (after Xu et al. 2011). Note that in this phylogeny Aves – if defined as the least inclusive clade comprising *Archaeopteryx* and neornithine birds – has the same content as Paraves.

of a clade including dromaeosaurs and birds (Godefroit et al. 2013a); or found a sister group relationship between deinonychosaurs and a clade including *Xiaotingia* and *Anchiornis* (Xu et al. 2015). *Eosinopteryx* was placed at the base of Paraves (Godefroit et al. 2013a) or within Troodontidae, as a sister taxon of *Anchiornis* (Godefroit et al. 2013b), whereas *Aurornis* resulted as a sister taxon of a clade including *Archaeopteryx* and other avians (Godefroit et al. 2013a).

The Daohugou Biota is also exposed in Inner Mongolia, where another Late Jurassic theropod with long pennaceous hindlimb feathers was found. *Pedopenna* is only known from hindlimb bones of a single individual (Xu and Zhang 2005), and its affinities to the coeval taxa from the Daohugou Biota of Liaoning Province still need to be scrutinized. A further poorly known *Archaeopteryx*-like taxon, *Jinfengopteryx*, was reported from the Early Cretaceous Huajiying Formation of Hebei Province (Ji and Ji 2007). Originally likened to *Archaeopteryx*, *Jinfengopteryx* is now considered to be a troodontid (Chiappe 2007; Turner et al. 2012).

With the inclusion of *Anchiornis* and *Xiaotingia* in phylogenetic analyses, the position of *Archaeopteryx* relative to troodontids and dromaeosaurs has become unstable, and the proliferation of names that were proposed in past years for various paravian subclades belies the fact that the precise relationships between dromaeosaurs, troodontids, *Archaeopteryx*, and other avians remain poorly resolved. As yet, no derived characters have been identified that convincingly support a clade including *Archaeopteryx* and crown group birds to the exclusion of deinonychosaurs. Often-cited avian characteristics, such as long forelimbs that exceed the hindlimbs in length and asymmetric forelimb feathers with well-developed vanes (Makovicky and Zanno 2011; Turner et al. 2012), also occur in the dromaeosaur *Microraptor*. Of further "avian" attributes (Turner

et al. 2012: 115), a long praeacetabular portion of the ilium is likewise present in some dromaeosaurs (*Bambiraptor*, *Rahonavis*, and the South American Unenlagiinae), whereas other traits are unknown from many basal avians.

Archaeopteryx possesses one of the deinonychosaurian key attributes, a hyperextensible second toe (Mayr et al. 2005, 2007), although it lacks the hypertrophied claw of this toe and pedal modifications are less advanced than in typical deinonychosaurs (Figure 2.8). The presence of a hyperextensible second toe in *Archaeopteryx* was contested (Turner et al. 2012: 139), but the less pronounced development of this feature in *Archaeopteryx* does not preclude it from being homologous to the more advanced morphology of typical deinonychosaurs. Turner et al. (2012: 139) raised the objection that a dorsally bulging distal trochlear surface of the penultimate phalanx of the second toe – the characteristic on which the assumption of an hyperextensible second toe in *Archaeopteryx* was based – is absent in deinonychosaurs. This is, however, erroneous (e.g., Turner et al. 2012: Figure 40A) and, apart from the smaller claw, the phalanges of the second toe of *Archaeopteryx* closely resemble those of *Anchiornis*, for which a hyperextensible second toe was considered present by Turner et al. (2012: character 204).

With regard to the presence of well-developed interdental plates, *Archaeopteryx* exhibits a more plesiomorphic morphology than most deinonychosaurs, in which these plates are either fused (all dromaeosaurs except *Austroraptor*; Agnolín and Novas 2011) or absent (troodontids). Once a robust paravian phylogeny is available, it remains to be assessed whether the occurrence of interdental plates in *Archaeopteryx* does indeed represent a secondary reversal into the primitive condition (Turner et al. 2012: 98), or whether it is a genuinely plesiomorphic character.

Figure 2.8 Detail of the right foot of the Thermopolis specimen of *Archaeopteryx* (left) and the dromaeosaur *Velociraptor* (mounted cast). Note the dorsally bulging distal end of the first phalanx of the second toe (arrows; the phalanx is highlighted by a dotted line in *Velociraptor*), which is a characteristic of hyperextensible toes.

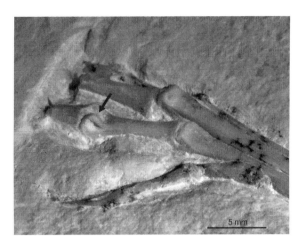
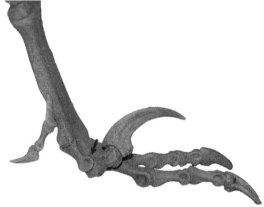

It is to be hoped that future studies will also identify unequivocal apomorphies that better characterize the clades near the base of Aves. In any case, it should be noted that some of the analyses discussed support avian ("avialan") affinities of deinonychosaurs or a part thereof, in which case "Paraves" would become a synonym for Aves (Figure 2.7).

Feather Evolution

Feathers are among the most characteristic attributes of birds and display a wide variety of morphologies, from the archetypical pennaceous ones to the more fluffy down feathers, and from striking ornamental appendages to hair-like eyelashes. Even the latter, however, are clearly distinguished in their chemical composition and ontogenetic development from mammalian hair, the only other filamentous integumentary structures of extant vertebrates.

As "contour feathers," pennaceous feathers cover most of the avian body and contribute to the airfoils, with the long wing and tail feathers being termed remiges and rectrices, respectively. Pennaceous feathers have a shaft (rachis), from both sides of which serially arranged barbs diverge that form the vanes. Each of these barbs bears smaller barbules with tiny hooks, which conjoin neighboring barbs and form an interlocking system similar to that of hook-and-loop fasteners. These feather structures enable a preening bird to smooth disarranged feather vanes in order to maintain a coherent airfoil. The interlocking system formed by the barbules is absent in down feathers, which therefore have a more "fluffy" appearance and a primarily insulating function.

Because Late Jurassic avians and avian-like theropods already exhibit perfectly modern feathers, these integumentary structures must have a long evolutionary history, and feather precursors are likely to have been present in much earlier Mesozoic archosaurs. In the past few years, various kinds of fossilized feather homologues were indeed described from a multitude of non-avian theropods. Their diversity ranged from simple filamentous appendages to true pennaceous feathers, and some morphologies were also described that have no counterpart in extant birds. These finds contributed to a better understanding of feather evolution and led to a revision of traditional hypotheses on the evolution of feathers.

The origin and evolution of feathers

Earlier authors assumed that feathers evolved from elongated scales. These first developed fringed margins, and progressive deepening of the incisions would have resulted in a pinnate proto-feather, which eventually formed barbs and barbules (e.g., Regal 1975). Under this scenario, pennaceous feathers would represent the primitive feather type, whereas down feathers would be more specialized. Some unusual feather types of Mesozoic birds were initially considered evidence for an origin of feathers from scales (Zhang and Zhou 2000), but these occur in taxa too far away from the avian origin and are now regarded as specialized morphologies.

Based on the discovery of filamentous integumentary coverings in close theropod relatives of birds, an alternative scenario for feather evolution was proposed (Prum and Brush 2002). In a simplified form, this model postulates that after the formation of a feather follicle, a single tubular structure emerged first. This hollow, cylindrical filament then became subdivided by barb ridges, which resulted in a tuft of barbs. In the next step, a rachis formed by fusion of barb ridges, and barbule development commenced. The barbules finally differentiated into distal and proximal ones, which led to the formation of the interlocking system and a closed feather vane. This hypothesis therefore postulates that pennaceous feathers represent the most derived feather type. The new model agrees well with the developmental origin of feathers, and fossil examples of possible transitional stages were identified in Cretaceous amber (Perrichot et al. 2008; McKellar et al. 2011).

The hypothesis that unbranched filaments rather than pennaceous feathers evolved first is also in concordance with the assumption that the integumentary coating of theropods originally had an insulating rather than an aerodynamic function. The objection was raised that such a filamentous integument would not be very effective in thermal insulation, since it would have lost its insulating function once wet (e.g., Feduccia 2012). The physical properties of such filaments are, however, unknown, and they may well have been coated with water-repellent lipids. Furthermore, the relatively long filaments of theropod dinosaurs are actually less similar to the downy coating of extant bird hatchlings – which are sensitive to wetness – than to mammalian fur, which fulfills its insulating functions very well.

Filamentous integumentary appendages were also reported from pterosaurs and from dinosaurs that are only distantly related to birds (Plate 2a, b; Mayr et al. 2002a; Godefroit et al. 2014). Within non-avian theropods, they are known from a wide range of taxa, including ornithomimosaurs, therizinosauroideans, compsognathids, and several paravians, such as the dromaeosaur *Sinosauropteryx* (Xu et al. 2009c; O'Connor et al. 2012a; Zelenitsky et al. 2012). Some of these integumentary appendages of theropod dinosaurs exhibit highly unusual morphologies. This is, for example, true for the so-termed proximally ribbon-like feathers of the caudipterygid *Similicaudipteryx*, which have a very wide shaft and may represent molting feathers (Prum 2010; Xu et al. 2010). Particularly unusual are the integumentary structures associated with the scansoriopterygid *Epidexipteryx*, which consist of parallel filaments (Zhang et al. 2008a). Whether these latter structures represent one of the earlier stages of the feather evolution model outlined earlier remains elusive, but it is notable that similar parallel filaments were reported from the ornithischian dinosaur *Kulindadromeus*, in which, however, they seem to arise from scales (Godefroit et al. 2014).

Other than in *Archaeopteryx*, true pennaceous feathers, with a shaft and a vane, occur in, for example, the oviraptorosaurs *Caudipteryx* and *Protarchaeopteryx*, the troodontid-like *Xiaotingia* and *Anchiornis*, and the dromaeosaur *Microraptor* (O'Connor et al. 2012a; Foth et al. 2014). The wing and tail feathers of these taxa have a modern-type morphology and the presence of barbules can be deduced from the fact that the vanes form a coherent surface. Because some of these animals are considered primarily flightless, it was assumed that pennaceous feathers initially evolved in a non-aerodynamic context, and an initial function for signaling or display was proposed (Xu and Guo 2009; Li et al. 2010; Dyke et al. 2013; Foth et al. 2014; Koschowitz et al. 2014).

Birds are diurnal animals with high visual capabilities, and their integument is more likely to evolve display structures than the fur of mammals, in which the olfactory sense dominates and many species are nocturnal. Sexual selection or signaling may therefore have played a role in the evolution of some unusual integumentary structures of Mesozoic theropods, such as the elongated "ribbon-like" feathers of scansoriopterygids (Zhang et al. 2008a; Xu et al. 2009c). However, as yet no selective advantage has been identified that would lead to a pennaceous feather with its complex interlocking system of barbs and barbules just for display or signaling reasons. A coherent vane increases the feather surface available for signaling (Koschowitz et al. 2014), but this is more easily achieved by a simple widening of the shaft.

An initial origin of pennaceous feathers for display or signaling purposes also conflicts with the circumstance that usually only limited parts of the plumage are involved in the display of birds and that pennaceous display feathers are often greatly modified (e.g., the ribbon-like feathers of some birds of paradise). Except for rare cases, where they are used for warning, feathers with signal functions are furthermore mainly of relevance in courtship behavior, and in most cases they are therefore restricted to one sex, usually males. If pennaceous feathers primarily evolved for signaling, it would be difficult to understand why female birds exhibit the same feather morphologies as males. Moreover, it stretches any evolutionary and physical plausibility that such a vaned feather, if having primarily evolved in a non-aerodynamic context, would have fulfilled an aerodynamic function without any further major modifications. A primarily aerodynamic function of pennaceous feathers is finally supported by the fact that feather vanes often lose their coherence in flightless birds, owing to a reduction of the barbules (Feduccia 1999).

For all paravians with pennaceous feathers, such as *Microraptor* and *Anchiornis*, at least the capabilities of gliding flight are assumed. As already detailed above, oviraptorosaurs and, hence, caudipterygids may stem from an arboreal, scansoriopterygid-like ancestor, in which case an aerodynamic origin of their feathers would be likely. Hypotheses for the evolution of pennaceous feathers in a non-aerodynamic context are therefore not only evolutionarily implausible, they are probably also unnecessary to explain the origin of these most fascinating integumentary structures of living vertebrates.

Although a multiple independent origin of pennaceous feathers cannot be excluded a priori, it is not the most obvious hypothesis. Here it is considered more likely that pennaceous feathers were already present in the stem species of Paraves, and that they evolved only once and fulfilled aerodynamic functions from the beginning (contra, e.g., Foth et al. 2014).

"Hindlimb wings" and the feather–scale transition

Certainly one of the most intriguing surprises of the avian fossil record was the discovery of various Late Jurassic and Early Cretaceous paravians with long pennaceous hindlimb feathers. These were first reported for the dromaeosaur *Microraptor* from the Chinese Jehol Biota, where these feathers extend onto the metatarsals and seem to have formed a second pair of "wings" (Figure 2.9), for which an aerodynamic function was inferred from the asymmetry of the feather vanes (Xu et al. 2003). Very similar hindlimb feathers occur in *Changyuraptor*, a close relative of *Microraptor* from the Jehol Biota (Han et al. 2014). Well-developed pennaceous hindlimb feathers, albeit with symmetric vanes, were furthermore described for *Pedopenna* and *Anchiornis* from the Daohugou Biota (Xu and Zhang 2005; Xu et al. 2009b, 2011); at least in *Anchiornis*, these feathers are less "wing-like" and show a closer resemblance to the contour feathers on the legs of extant emus (Plate 2f).

Asymmetry of the hindlimb feathers of *Microraptor* was considered to be indicative of an aerodynamic function (Xu et al. 2003). Controversies exist, however, about the orientation of these feathers in the living animals, and both a "four-winged" ("*Tetrapteryx*") and a staggered "biplane" configuration were proposed (Xu et al. 2003; Chatterjee and Templin 2007). From an aerodynamic point of view, a "four-winged" condition with laterally spread hindlimbs was considered more plausible (Alexander et al. 2010). Whether *Microraptor* and other paravians were, however, anatomically able to spread their legs in such a position remains controversial (Padian 2003), and some recent authors regarded a display function of long hindlimb feathers as more likely (O'Connor and Chang 2015).

It has been hypothesized that pennaceous leg feathers first evolved distally on the hindlimbs (Hu et al. 2009) and that "hindlimb wings" are an ancestral trait of birds, with avian foot scales being secondarily derived from feathers (Zheng et al.

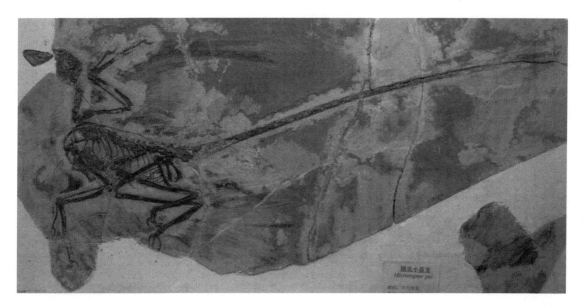

Figure 2.9 The dromaeosaur *Microraptor* from the Early Cretaceous Jehol Biota with pennaceous fore- and hindlimb feathers. Photograph by Jingmai O'Connor.

2013a). Long pennaceous hindlimb feathers are indeed present in some individuals of the basal Early Cretaceous avian *Sapeornis*, but here they form a tuft of long feathers at the intertarsal joint rather than a functional airfoil (Zheng et al. 2013a; O'Connor and Chang 2015). In *Archaeopteryx* and the more advanced Enantiornithes vaned feathers occur on the tibia alone (Figure 2.10; Zhang and Zhou 2004; Longrich 2009; Zheng et al. 2013a). In other early avians with preserved plumage remains, such as jeholornithids and confuciusornithids, "hindlimb wings" have not been reported, and the hypothesis that such feathers are primitive for birds is therefore only weakly based.

In extant birds with feathered tarsi and toes, such as owls and grouse, the foot scales often grow feathers on their tips (Blaszyk

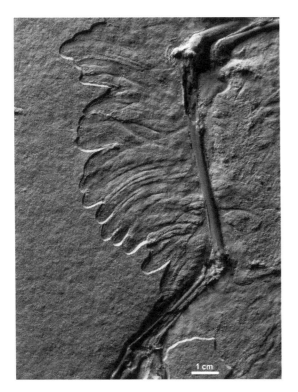

Figure 2.10 Hindlimb feathers of *Archaeopteryx* from the Late Jurassic of Germany (11th specimen). Photograph by Oliver Rauhut and Helmut Tischlinger.

1935), and on a molecular level feathers and scales can be transformed into each other by differential gene expression patterns (Zheng et al. 2013a). Both morphological and molecular data therefore document the possibility of a transformation of scales into feathers or vice versa, but the question of what came first, foot feathers or foot scales, cannot yet be definitely answered.

Melanosomes and the color of extinct animals

A number of fossil deposits are renowned for excellent soft tissue preservation, such as the Late Jurassic/Early Cretaceous limestones of the Chinese Daohugou and Jehol Biota, and the early Eocene oil shale of Messel in Germany. In these and other localities, avian feathers are often preserved as dark organic matter. Close examinations have shown this to consist of microscopic, rod-shaped or ellipsoid structures (Figure 2.11), which were first considered fossilized feather-degrading bacteria.

These structures are now known to represent fossilized melanosomes, the cell organelles involved in the synthesis of melanin (Vinther et al. 2008; Vinther 2015). By comparison with data from extant birds, their morphology was used to reconstruct the coloration patterns of extinct birds and non-avian theropods (e.g., F. Zhang et al. 2010; Li et al. 2010, 2012; Clarke et al. 2010; Carney et al. 2012). Different melanosome shapes allow a distinction between black feathers and those exhibiting various shades of brown (Clarke et al. 2010), and the shape and spatial arrangement of these organelles even suggests the former iridescence of some fossilized feathers (Vinther et al. 2010).

Arguments were raised in favor of the original bacteria hypothesis (Moyer et al. 2014). However, although there are still unresolved issues to be addressed in future studies, such as the nature of the matrix in which the melanosomes are preserved and the effects of the fossilization processes

Figure 2.11 Scanning electron microscope images of melanosome layers preserved in birds from the early Eocene German fossil site Messel. Photographs by Jakob Vinther.

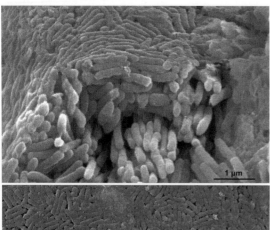
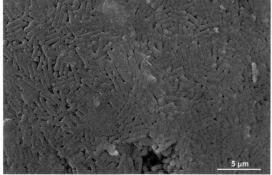

on their spatial arrangement (McNamara et al. 2013), the melanosome hypothesis is consistent with the size of the structures and their differential preservation in fossil feathers with color patterns (Vinther et al. 2008; Vinther 2015).

Nevertheless, more detailed studies of experimentally matured feathers (e.g., by charring under high temperatures and pressure; McNamara et al. 2013) have to be carried out to verify the accuracy of melanosome-based color reconstructions of extinct animals. In many cases, the coloration of birds is due to the enclosure of pigments in the feather barbs, and blue and green tones are usually the result of light scattering and interference by keratinous nanostructures of the feather barbs (e.g.,

Prum et al. 1998). These effects can involve melanosome layers, but because usually neither color pigments nor keratin are preserved in fossil feathers, color reconstructions based on the melanosomes alone may not always accurately reflect the original color spectrum. Surprisingly often, these analyses result in the assumption of black feathers, as in the cases of *Archaeopteryx* (Carney et al. 2012) and *Microraptor* (Li et al. 2012). At least among extant birds, a black plumage is, however, not very common and current color reconstructions may be all too simplistic (McNamara et al. 2013; Moyer et al. 2014). Whether the feathers of *Archaeopteryx* and *Microraptor* had, for example, a bluish or greenish hue we will never know, although it can be safely said that they were neither white (no melanosomes) nor did they exhibit a purely pigment-based coloration, which is often the case in red and yellow feathers.

Without doubt, however, the recognition of melanosome preservation is a fascinating discovery, which shows that far more soft tissue structures have a fossilization potential than is commonly thought. Other instances of unusual soft tissue preservation concern the waxes of the oil gland, which can be traced in some bird fossils from the Messel oil shale in Germany (Mayr 2006a), as well as the preservation of foot scales in fossil birds from various localities (Plate 16a–c).

The Origin of Avian Flight

More than in any other group of flying animals, there is an extensive and ongoing debate on the origin of flight in birds. Controversies mainly center around three questions: how flight evolved; the flight modes of *Archaeopteryx* and other winged theropods; and how often flight capabilities emerged; that is, which of the basal avians and avian-like theropods were able to fly (e.g., Feduccia 1999, 2012; Chiappe 2007; Wellnhofer 2009).

Before these questions are addressed, it is necessary to outline the two basic forms of flight employed by animals. The simplest and aerodynamically most straightforward way to become airborne is a passive one, through parachuting or gliding, which is performed by a fair number of species in all groups of land vertebrates, from frogs to lizards and rodents. To achieve parachuting or gliding capabilities, it is often sufficient to develop drag-increasing flight membranes, which reduce the fall velocity. This passive type of flight can only evolve from the top down, through the use of gravity.

Much higher are the requirements of the second flight type, powered flapping flight, in which the animal actively gains height against gravity and is able to start from the ground. This kind of locomotion necessitates the development of true wings and a powerful musculature, and among vertebrates it is only found in birds, pterosaurs, and bats.

How did avian flight evolve?

It has long been known that a small body size is a prerequisite for the evolution of flight, and various recent studies have shown that a miniaturization took place in the theropod lineage leading to birds (Turner et al. 2007a; Puttick et al. 2014; Lee et al. 2014). Then, however, the consensus about the origin of avian flight almost ends. The most controversial debates center around the fact that the aerodynamically most plausible model, an origin of flight through gliding arboreal forms ("tree-down hypothesis"), is difficult to reconcile with the presumed cursorial way of life of the closest theropod relatives of birds, which suggests a flight origin against gravity ("ground-up hypothesis").

An arboreal animal can use the energy provided by gravity for gliding, in which case an incremental formation of structures increasing drag or providing lift is positively selected for. Earlier authors therefore assumed that the avian ancestors were tree dwelling and depicted *Archaeopteryx* as a trunk-climbing animal. Analyses of morphological traits associated with a strictly arboreal way of life do not support such habits of *Archaeopteryx* and the paravian ancestor of birds, but indicate that arboreal adaptations evolved within the avian lineage (Dececchi and Larsson 2011). The explanatory power of such analyses is, however, challenged by the circumstance that they can make inferences on the habits of extinct taxa only by comparison with extant animals that live in very different ecosystems. Some maniraptorans near the base of Aves, such as the Scansoriopterygidae, furthermore exhibit morphological adaptations unknown among living taxa, which defy a straightforward functional interpretation.

Birds are deeply nested within bipedal theropods and, apart from the possibly arboreal Scansoriopterygidae, their closest theropod relatives are cursorial animals. Moreover, birds are the only group of extant flying vertebrates in which the hindlimbs do not contribute to the airfoils, and, unlike in most other tetrapods, their fore- and hindlimbs form discrete locomotor modules that act independently of each other (Gatesy and Dial 1996). This functional decoupling of the hindlimbs likewise indicates that bipedal locomotion played a role in the origin of flight; that is, that flight evolved in a cursorial animal that used its hindlimbs for terrestrial locomotion (Peters 1985).

Although the occurrence of "hindlimb wings" in some early paravians may have added a twist to these considerations, it is controversial whether these feathers did indeed have an aerodynamic function (see earlier discussion). Irrespective of that, it was noted that the long hindlimb feathers of some early paravians would have constituted a hindrance in a bipedal running animal.

If "hindlimb wings" were primitive for paravians, an origin of flight through a gliding stage would therefore be a more plausible hypothesis (Zheng et al. 2013a). However, all paravians for which hindlimb wings are currently known already had well-developed forelimb wings, so that their integumentary traits may not necessarily reflect those of the immediate (par)avian ancestor. Moreover, there are certain breeds of domesticated fowl and pigeons that exhibit very similar, long, and asymmetric feathers on their legs, especially the Silkie chicken and the Saxon Fairy Swallow pigeon (Figure 2.12). At least under husbandry conditions, the hindlimb feathers of these terrestrial, ground-dwelling birds do not constitute a severe hindrance in bipedal locomotion (the similarities of these extant "hindlimb wings" to those of Mesozoic paravians are notable, and detailed future comparisons would be expedient with respect to their developmental origin and the attachment and orientation of the feathers involved).

A cursorial origin of flight was long equated with a ground-up origin against

Figure 2.12 The Saxon Fairy Swallow, a domestic breed of the Rock pigeon (*Columba livia*) with a well-developed "hindlimb wing." Photograph by Andreas Reuter.

gravity. To explain the evolution of flight in a running animal, various hypotheses were suggested, including the idea that the avian ancestors captured prey with their forelimbs and wings evolved as insect traps (Ostrom 1974), or that "stability flapping" developed in deinonychosaurs in order to assist in pinning down larger prey items (Fowler et al. 2011). Usually, however, the hypothesis of a cursorial origin of avian flight is coupled with the assumption that a running animal used wing flaps to produce lift, and it was reckoned that the wings of *Archaeopteryx* could have acted as thrust generators to achieve the speeds required for take-off (Burgers and Chiappe 1999).

The physical and biological implausibility of a strict "ground-up" origin of flight against gravity has been extensively discussed in the literature, and the arguments against it are only briefly summarized here (see Feduccia 2012 for a more detailed discussion). One of the major points of criticism was the notion that rudimentary wings – that is, initial stages in wing evolution – would show little aerodynamic effects and that no selective advantages of such "half wings" can be identified. Furthermore, an earthbound animal mainly accelerates by running. Regardless of whether a small cursorial theropod was physiologically able to reach speeds high enough for sufficient lift production, the objection was raised that with increasing lift production there would have been a loss of traction between the feet and the ground, which would in turn have resulted in a loss of speed. The winged forelimbs could have been used for additional acceleration, but in this case they would have also produced drag, which counteracted forward thrust.

Meanwhile, however, it was shown that these considerations are not necessarily true, and an aerodynamically more convincing scenario of a cursorial origin of flight was termed the "wing-assisted incline running" hypothesis (Dial 2003a; Heers et al. 2014). This model is based on the observation that the wing flapping of hatchlings of extant galliform birds creates aerodynamic and inertial forces directed towards the substrate. These forces actually enhance hindlimb traction and allow these birds to run up almost vertical inclines. In analogy with this behavior, it was hypothesized that flapping with even short wings may have produced aerodynamic forces that enabled avian ancestors to move rapidly on uneven terrain.

Plausible as this hypothesis may seem, it also faces several problems. Not only would avian flight have begun with the energetically costly (Rayner 2001) flapping flight, rather than with an aerodynamically more plausible gliding stage, but, more importantly, "wing-assisted incline running" requires the anatomical capabilities of wing flapping. An early origin of sustained and powerful wing flapping with high wingbeat amplitudes is, however, not supported by the fossil record, which indicates that flapping capabilities evolved later in avian evolution, at the base of Ornithothoraces (Senter 2006). The genetic programs to develop wings are furthermore already manifested in galliform hatchlings, and selective advantages still need to be identified that in a cursorial animal favor the origin of wings against mere hindlimb specializations.

Other models were proposed that combine aspects of the arboreal and cursorial models for the origin of avian flight. According to one of these, avian ancestors were moving on uneven terrain and sought elevated places, such as tumbled logs, for escape glides (Peters 1985). It is possible that an understanding of the settings of the beginnings of avian flight might therefore be improved by a better knowledge of the vegetation structures of the Jurassic habitats of avian ancestors, which are likely to have been very different from extant angiosperm-dominated forests.

What was the flight mode of *Archaeopteryx* and other early winged theropods?

In volant birds, the distal wing feathers are asymmetric and exhibit narrow trailing vanes, which increase the rigidity of the feather against the air flow. Such asymmetric forelimb feathers are already present in the earliest Mesozoic avians and even in winged non-avian paravians, such as *Microraptor*, and substantiate some flight capabilities of these animals (Feo et al. 2015). For *Archaeopteryx* at least, flight capabilities are furthermore indicated by brain morphology, with areas related to an increased sense of vision and complex spatial locomotion being better developed than in flightless theropods (Domínguez Alonso et al. 2004). Simulation of flight aerodynamics under different parameters, as well as wind tunnel experiments with reconstructed life models, also suggests flight capabilities for both *Archaeopteryx* and *Microraptor* (e.g., Chatterjee and Templin 2003, 2007; Dyke et al. 2013).

That *Archaeopteryx* and other paravians with long forelimbs and asymmetric pennaceous wing feathers, such as *Microraptor*, were able to perform some sort of flight is therefore highly likely, but the exact degree of their flight capabilities remains contentious. This is particularly true for the question of whether these animals could already perform powerful and sustained flapping flight, one of the requirements for a take-off from the ground, or whether they were mere gliders, which had to seek elevated places to become airborne.

The most critical anatomical prerequisites of flapping flight are powerful muscles and anatomical correlates, which allow an elevation of the wing beyond the back of the animal. In extant birds, it is mainly the supracoracoideus muscle that lifts the wing during the upstroke. This muscle originates on the sternum and the furcula – that is, ventral of the wing – but its tendon runs through a canal formed by the pectoral girdle bones, the triosseal canal, and is deflected in such a way that contraction of the muscle elevates the humerus.

The architecture of the wing and pectoral girdle of *Archaeopteryx*, *Microraptor*, and other early birds or bird-like theropods profoundly differs from that of extant birds. In particular, the articulation between the pectoral girdle and the humerus – that is, the glenoid fossa – is less dorsally elevated in *Archaeopteryx* than in modern birds (Senter 2006). Because of these skeletal constraints, the "Urvogel" could probably not lift its wing as much as more advanced birds. Moreover, *Archaeopteryx* not only lacks an ossified sternum and, hence, a sternal keel, but the coracoid also does not exhibit acrocoracoid and procoracoid processes, which in extant birds contribute to the triosseal canal and act as a pulley system for the tendon of the supracoracoideus muscle. In *Archaeopteryx*, this latter muscle may have inserted on a cartilaginous sternum and on the very robust furcula and coracoids (Olson and Feduccia 1979). The anatomical constraints discussed nevertheless indicate that *Archaeopteryx* and similar paravians such as *Xiaotingia* and *Anchiornis* – which also lacked an ossified sternum (Zheng et al. 2014a) – were not capable of sustained and powerful flapping flight with high wing-beat amplitudes. Likewise, these taxa are unlikely to have been able to perform sophisticated aerial maneuvers, such as starting from and landing on the ground, as well as slow or long-distance flights (Wellnhofer 2009).

The assumption that *Archaeopteryx* and similar taxa mainly performed rather simple aerial locomotion over short distances, which did not involve powerful wing strokes, is consistent with the occurrence of profound changes in skeletal morphology in later avian evolution. These can be correlated with the origin of flapping flight capabilities in Ornithothoraces and include a strut-like coracoid with an acrocoracoid process,

an elevated humerus articulation, and a well-developed sternal keel (Senter 2006).

Even some of the more advanced avians, such as *Jeholornis* and *Sapeornis*, are unlikely to have had much better-developed flight capabilities than *Archaeopteryx*, and the very long wings of *Sapeornis* and *Confuciusornis* suggest that gliding or soaring still predominated in the flight of these birds. It has also been shown that, despite a superficial similarity in overall shape, the feathers of non-avian paravians and early avians, such as *Jeholornis*, *Sapeornis*, and *Confuciusornis*, differ from those of later birds and extant Neornithes in the smaller barb angles of the trailing vanes. This observation further strengthens the supposition that powered flapping flight evolved only in later avian evolution, at the base of Ornithothoraces (Feo et al. 2015).

Which of the basal avians and avian-like theropods were able to fly and how often did flight evolve?

As already detailed, the interrelationships of *Archaeopteryx*, *Microraptor*, and other avian and non-avian theropods are still controversial. *Microraptor*, however, is generally reckoned to be more closely related to typical dromaeosaurs than is *Archaeopteryx* (e.g., Turner et al. 2012; see, however, Agnolín and Novas 2013 for an opposing view). The anatomical evidence suggests that *Archaeopteryx* and *Microraptor* possessed at least the capabilities of gliding flight, whereas large dromaeosaurs with short forelimbs, such as *Deinonychus* and *Velociraptor*, undoubtedly were flightless. This raises the question about how often paravians became airborne and whether some taxa secondarily lost their flight capabilities.

Even before the discovery of *Microraptor*, it was assumed that large dromaeosaurs were secondarily flightless (Paul 2002), but this hypothesis received only slight acceptance. Instead, various authors explicitly or implicitly assumed a multiple origin of flight in paravians (e.g., Foth et al. 2014). In this case, the evolution of flight capabilities would have convergently led to a very similar forelimb morphology and wing feather arrangement in *Archaeopteryx* and *Microraptor*.

Within Neornithes, flight capabilities were lost multiple times in numerous taxa. That the application of parsimony criteria for an explanation of flightlessness may be misleading is especially documented by palaeognathous birds, where the volant tinamous were shown to be phylogenetically nested within various flightless taxa (see Chapter 6). A descent of tinamous from a flightless ancestor and a secondary regaining of flight capabilities in these birds form the most parsimonious explanation for the distribution of flightlessness among palaeognathous birds (Hackett et al. 2008). The more likely and generally favored hypothesis, however, is a multiple convergent loss of flight capabilities (Harshman et al. 2008).

It is likewise more plausible from an evolutionary point of view that flight evolved only once in theropods and that the last common ancestor of *Archaeopteryx* and *Microraptor* was capable of some sort of aerial locomotion (Feo et al. 2015). Under this assumption, large dromaeosaurs and presumably other paravians, too, must have secondarily lost flight capabilities. This hypothesis is in agreement with the small size of *Microraptor*

and other basal dromaeosaurs (Turner et al. 2012), and it was proposed that the characteristic tail of dromaeosaurs, which is stiffened by the elongation of the praezygapophyses and chevrons, originally evolved as a flight stabilizer (Persons and Currie 2012).

Another taxon for which the question of primary or secondary flightlessness was debated is *Caudipteryx* (e.g., Feduccia 2012). Its feathered forelimbs and those of the closely related *Similicaudipteryx* are clearly too short to have provided enough lift for these animals to have become airborne or to have covered even a short distance in gliding flight. *Caudipteryx* furthermore has symmetric forelimb feathers, a feature usually taken as indicative of the non-aerodynamic functions of pennaceous feathers. However, secondarily flightless birds also often lose vane asymmetry of the wing feathers, and *Caudipteryx* appears to have had a propatagium, the wing membrane between humerus and ulna, which forms the cranial margin of the avian wing and whose origin is difficult to understand in a non-aerodynamic context (Feduccia and Czerkas 2015). The fact that the minor wing digit of *Caudipteryx* is reduced as in volant birds is likewise in need of a functional explanation, and the assumption of a secondary loss of (gliding) flight capabilities in caudipterygids and other early paravians (Paul 2002) remains a hypothesis to be considered in future studies.

3 The Mesozoic Flight Way towards Modern Birds

The Late Jurassic *Archaeopteryx* and the toothed Late Cretaceous *Ichthyornis* and *Hesperornis* were for a long time the only well-documented Mesozoic birds. This situation has now dramatically changed and new Cretaceous avian taxa are constantly being described. The bulk of these fossils stems from the Jehol Biota in northeastern China. Remarkably, even birds from the earliest strata of these localities already exhibit very disparate morphologies. The skeletal morphology of most is very different from that of *Archaeopteryx*, therefore indicating a rapid avian diversification in the latest Jurassic or earliest Cretaceous.

Apart from the sheer number of avian fossils and the exceptional preservation of many of these, the Jehol Biota is also remarkable for the circumstance that it yielded many early diverging birds, which are unknown from elsewhere. This preponderance of basal avian taxa in East Asia may be an artifact of the fossil record, since no other Late Jurassic or Early Cretaceous fossil deposits are known, which yielded fossil avifaunas nearly as diversified and rich as those of the Jehol Biota. Alternatively, however, this pattern of the fossil record may convey a true biogeographic signal; that is, an early avian radiation in Asia.

Whereas *Archaeopteryx* and other Jurassic paravians appear to have been carnivorous, the diet of many of the Early Cretaceous birds discussed in the present chapter evidently included seeds. The Late Jurassic/Early Cretaceous radiation of more advanced birds temporally coincides with that of angiosperm plants (Friis et al. 2011), and it is likely that coevolutionary processes were involved. One of the key attributes of angiosperms is their dispersal via seeds and fruits, and the formation of flowers as reproductive organs. Without doubt, angiosperm diversification was therefore shaped by a coevolution with potential flower pollinators

Avian Evolution: The Fossil Record of Birds and its Paleobiological Significance,
First Edition. Gerald Mayr.
© 2017 John Wiley & Sons, Ltd. Published 2017 by John Wiley & Sons, Ltd.

and seed dispersers, and fossil stomach contents suggest that birds played a role in seed dispersal early on. Avian evolution, on the other hand, especially the refinement of aerial maneuverability, may have been triggered not only by plant evolution itself, but also by the greater abundance and availability of pollinating insects around the newly emerged flowers.

Jeholornithids: Early Cretaceous Long-Tailed Birds

Only a little more than a decade ago, our knowledge of long-tailed birds was solely based on the anatomy of *Archaeopteryx*. This has changed with the description of *Jeholornis* (Jeholornithidae) from the Early Cretaceous Jiufotang Formation of the Jehol Biota, of which some 100 skeletons are now known (Plate 3a; Zhou and Zhang 2002a, 2003a; Zheng et al. 2014a). Other *Jeholornis*-like fossils from the Jiufotang Formation, such as *Shenzouraptor* or *Dalianraptor*, are of doubtful validity, and some may be synonyms of *Jeholornis*, which is possibly also true for *Jixiangornis* from the Yixian Formation (Zhou and Zhang 2007; Turner et al. 2012; O'Connor et al. 2012b, 2013a).

The pheasant-sized *Jeholornis* (Figure 3.1) differs from *Archaeopteryx* in its shorter and higher skull, as well as in various other cranial and **postcranial** features (Zhou and Zhang 2003a). The dentition is greatly reduced and only three teeth are present in the tips of each of the robust lower jaws (a single tooth has also been reported in the maxillary bone of one fossil; O'Connor et al. 2012b). Unlike in *Archaeopteryx*, there is a mandibular symphysis. Also unlike in *Archaeopteryx*, *Jeholornis* has an ossified sternum consisting of two fused sternal plates, which are laterally bordered by a pair of perforated structures (see Figure 4.6; Zheng et al. 2014a). The coracoid of *Jeholornis* is narrower than that of *Archaeopteryx*, but still not as strut-like

as in more advanced birds. The minor digit consists of three phalanges and a claw. In contrast to *Archaeopteryx*, however, the metacarpals of the minor and major digits of *Jeholornis* are proximally fused and the minor metacarpal is distinctly bowed. The ribs lack uncinate processes and the number of fused sacral bones (six) is higher than in *Archaeopteryx* (five). The metatarsals are not co-ossified and a rudimentary fifth metatarsal is present. As in *Archaeopteryx*, the first toe is not reversed (Zhou and Zhang 2007). Being composed of 27 vertebrae, the tail is even longer than in *Archaeopteryx* and the tail vertebrae have long chevrons similar to those of dromaeosaurs; unlike in the latter, the praezygapophyses do not, however, form greatly elongated rods. Some specimens of *Jeholornis* exhibit well-preserved tail feathers, which are restricted to the tip of the tail and form a peculiar fan-shaped frond; another feather bundle was identified at the tail base (Plate 3a; O'Connor et al. 2012b, 2013a). This unusual "two-fanned" tail morphology is not matched by any other feathered paravian and was considered to be primarily for display.

Some *Jeholornis* fossils exhibit seeds as stomach contents, which documents an at least partially **granivorous** diet; the location of ingested seeds in one specimen even indicates the presence of a crop (Zhou and Zhang 2002a, 2003a, 2007; O'Connor et al. 2012b; Zheng et al. 2014b). Purported fossilized ovaries in another *Jeholornis* fossil were contested and are more likely also to represent stomach contents (Mayr and Manegold 2013).

Figure 3.1 Skeleton of the long-tailed avian *Jeholornis* (Jeholornithidae) from the Early Cretaceous Jehol Biota. Reconstruction © Scott Hartman.

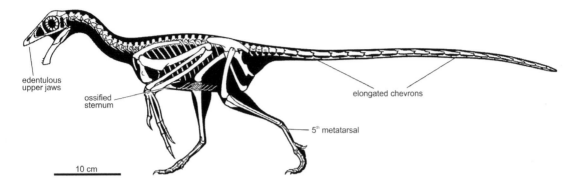

The locomotory habits of *Jeholornis* have not yet been studied in detail. Although these animals may have fed on fruits and seeds that fell on the ground, the presumed omni- or granivorous diet suggests some arboreality. *Jeholornis* was certainly capable of at least gliding flight. However, differences in skeletal anatomy and plumage structure indicate that its aerodynamic characteristics departed from those of both *Archaeopteryx*, which has serially arranged feathers along the entire tail, and *Microraptor*, which has long, pennaceous hindlimb feathers.

Confuciusornis, *Sapeornis*, and Kin: Basal Birds with a Pygostyle

Avian evolution is characterized by a high degree of homoplasy and it is difficult to characterize particular clades. One morphological character complex, however, which was long unchallenged as an apomorphy of a major avian clade, is the reduction of the tail and the formation of a pygostyle. There is a clear-cut distinction between avians that lack a pygostyle and those that have it, and the latter are usually classified in the Pygostylia. Surprisingly, monophyly of this latter taxon is now challenged by several analyses, which placed the long-tailed Jeholornithidae within pygostylians (Zhou et al. 2010; Turner et al. 2012). If these results are corroborated in future studies, they would suggest the convergent origin of a pygostyle in more than one avian clade.

Confuciusornithids: The first beaked birds

Confuciusornis from the Early Cretaceous Yixian and Jiufotang formations of the Jehol Biota is one of the most abundant fossil birds, of which hundreds of skeletons were discovered (Figure 3.2; Plate 4; Chiappe et al. 1999). Although several *Confuciusornis* species were described, only one is currently considered well established (Chiappe 2007; O'Connor et al. 2011a). Two further confuciusornithid taxa are known from strata that are older than those yielding the *Confuciusornis* fossils; that is, *Changchengornis* from the lowermost part of the Yixian Formation ("Chaomidianzi Formation"; Chiappe et al. 1999) and *Eoconfuciusornis* from the Huajiying Formation of Hebei Province (Zhang et al. 2008b). A putative confuciusornithid was also reported from the Early Cretaceous of Siberia (*Evgenavis*; O'Connor et al. 2014).

Confuciusornithids lack teeth and are the phylogenetically most basal lineage of edentulous birds, but tooth reduction undoubtedly occurred independently of that in more advanced (ornithuromorph) birds.

Figure 3.2 Skeleton of the pygostylian *Confuciusornis* (Confuciusornithidae) from the Early Cretaceous Jehol Biota (after Chiappe et al. 1999).

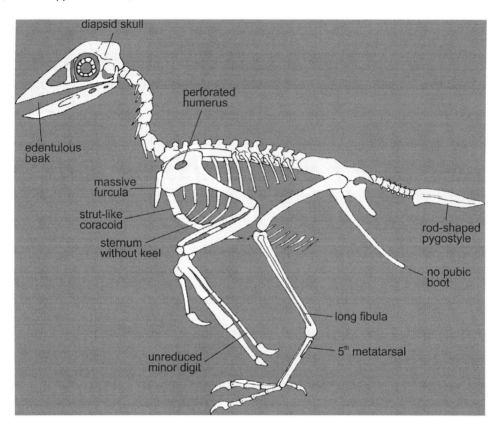

Teeth are also reduced, albeit not completely lost, in other early avians, but confuciusornithids are distinctive in that they have an essentially modern-type, pointed beak. The lower jaws are fused at their tips and form a mandibular symphysis, even though this fusion appears to be incomplete and the tip of the symphysis exhibits a deep cleft (Figure 3.3a). Foramina and grooves for blood vessels indicate that the beak was covered with a horny rhamphotheca.

Confuciusornithids are the only avian group for which a fully diapsid skull with a well-developed upper temporal bar – like in most non-avian archosaurs – was reported (Plate 4; Peters and Ji 1998). Another skull characteristic is the large and caudally situated nostril, which at first sight may be mistaken for the antorbital fenestra (the latter being very small in confuciusornithids; see Figure 4.3).

The postcranial skeleton of confuciusornithids shows a peculiar mix of derived and primitive features, which readily distinguishes them from other pygostylians. The pectoral girdle exhibits an ossified sternum, which in some specimens bears a very low midline ridge. The strut-like coracoid is fused with the scapula to form a scapulocoracoid (Chiappe et al. 1999). The boomerang-shaped furcula is even more robust than that of *Archaeopteryx*. The proximal end of the humerus is very

Figure 3.3 (a) Mandibular symphysis of *Confuciusornis*; the cleft on the tip of the symphysis (arrow) is visible in many specimens. (b) X-ray photograph of a *Confuciusornis* wing.

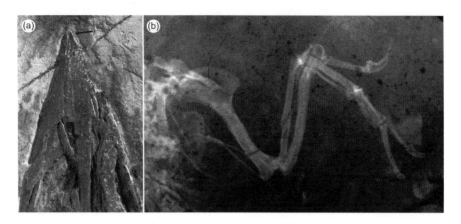

large (Figure 3.3b); in *Confuciusornis* and *Changchengornis*, it is pierced by a large foramen of unknown functional significance, which is absent in *Eoconfuciusornis* (the only known fossil of which is, however, from a juvenile individual). The minor wing digit is still well developed and composed of three phalanges and a claw (Figure 3.3b). Unlike in *Archaeopteryx*, however, the phalanges of the major digit show some craniocaudal widening for an improved attachment of the primary feathers, which are very long in *Confuciusornis*. Furthermore, unlike the "Urvogel," the metacarpals of the minor and major digits of confuciusornithids are proximally fused with each other and the distal carpals, thereby forming a carpal trochlea. The metacarpal of the alular digit, however, remains distinct. Whereas the claw of the alular digit is very large, the major digit bears an unusually small one. Uncinate processes were reported to be present on the ribs of some *Confuciusornis* fossils (Chiappe et al. 1999), but they are absent in the majority of the specimens.

In contrast to the derived features of the wing and pectoral girdle, the pelvic girdle and the hindlimbs of confuciusornithids still exhibit a relatively primitive morphology. A pubic symphysis is present, as are gastralia. The metatarsals are only fused at their proximal ends and there is a rudimentary fifth metatarsal. The hind toe is already reversed, albeit – judging from its medial position – perhaps not to the same degree as in extant birds.

The pygostyle is elongated and rod shaped, and many specimens exhibit a pair of long, streamer-like tail feathers. These occur in about one-fifth of the *Confuciusornis* fossils (O'Connor et al. 2012a) and are also present in the single individuals known of *Changchengornis* and *Eoconfuciusornis*. The morphology of these paired tail streamers is peculiar in that the feathers largely consist of a wide shaft, with the vane being restricted to the tip (see Figure 4.11c). Whether they represent ornamental feathers of male individuals was controversially discussed (Chiappe et al. 2008; Peters and Peters 2010). The identification of **medullary bone** allowed the determination of the gender of some fossils and suggests that at least some of the specimens without long tail streamers indeed represent females (Chinsamy et al. 2013; but see Chapter 4). The *Confuciusornis* fossils show a considerable disparity in size, which is attributed to different individual ages of the fossils and a possible sexual dimorphism (Chiappe et al. 1999).

The toe proportions indicate that confuciusornithids were arboreal to a certain degree, but probably they also spent some time foraging on the ground (Chiappe 2007). Mainly because of the large wing claws, some earlier authors assumed that they had tree-climbing habits. The claw of the major digit is, however, reduced in confuciusornithids and the long primary feathers would have constituted a hindrance in climbing (Chiappe et al. 1999; Peters and Ji 1999). Proposed poor flight capabilities were based on incorrect assumptions of body mass and feather shaft widths (Zheng et al. 2010), and the unusually long wing feathers, as well as the large humerus head and ossified sternum, instead indicate that confuciusornithids were aerial animals (Peters and Ji 1999; Chiappe 2007). However, and as already noted, the primitive morphology of the pectoral girdle (e.g., the ventral position of the glenoid fossa, which did not allow an elevation of the humerus much above the back, as well as the absence of a well-developed sternal keel) suggests that confuciusornithids were predominantly soaring or gliding and not capable of sustained flapping flight (Peters and Ji 1999; Senter 2006).

The reasons for tooth reduction in confuciusornithids remain elusive, not least because the diet of these birds is unknown. Interestingly, for none of the numerous *Confuciusornis* fossils were definitive stomach contents reported, whereas these are known from several less abundant avian taxa from the Jehol Biota (one *Confuciusornis* specimen has the remains of a fish preserved next to the neck vertebrae, which may well, however, be an accidental association). Likewise, none of the confuciusornithid skeletons exhibits gastroliths, which are usually ingested by granivorous birds and were described for jeholornithids and other birds from the Jehol Biota. Because the fossil record does not provide positive evidence for the diet of *Confuciusornis*, it likely consisted of items that do not easily fossilize, such as fruits or leaves, and specialized feeding habits may explain the formation of a beak with sharp cutting edges in confuciusornithids.

Omnivoropterygids ("Sapeornithids"): Some of the largest volant Mesozoic birds

Sapeornis is another distinctive and abundantly represented avian taxon from the Early Cretaceous Jiufotang and Yixian formations of the Jehol Biota, and more than 100 skeletons of this taxon have so far been found (Figure 3.4, Plate 5; Zhou and Zhang 2002b, 2003b; Gao et al. 2012; Zheng et al. 2014a). Even though a number of putatively close taxa were described in the past few years, these are now all considered junior synonyms of *Sapeornis*, which probably includes only a single species (Gao et al. 2012). These birds were often assigned to the taxon "Sapeornithidae," but the less often used Omnivoropterygidae was published before (Czerkas and Ji 2002) and therefore has nomenclatural priority.

With an estimated wing span of about one meter, *Sapeornis* is one of the largest volant Mesozoic birds known to date. The snout is short and high. Teeth are only present in the praemaxilla and the rostral portion of the maxilla; the lower jaws are edentulous. Like confuciusornithids and other early avians, *Sapeornis* retains a well-developed postorbital bar (Hu et al. 2010).

In the postcranial skeleton of *Sapeornis*, differences to confuciusornithids are especially found in the wing and pectoral girdle. One of these concerns the absence of a sternum, or even ossified sternal plates, in *Sapeornis* (Zheng et al. 2014a). The coracoid is squarish and has a similar shape to that of *Archaeopteryx*. Unlike in *Confuciusornis* and *Archaeopteryx*, the furcula exhibits an apophysis, which may have served as an alternative attachment site of some pectoral

Figure 3.4 Skeleton of the pygostylian *Sapeornis* (Omnivoropterygidae) from the Early Cretaceous Jehol Biota. Reconstruction © Scott Hartman.

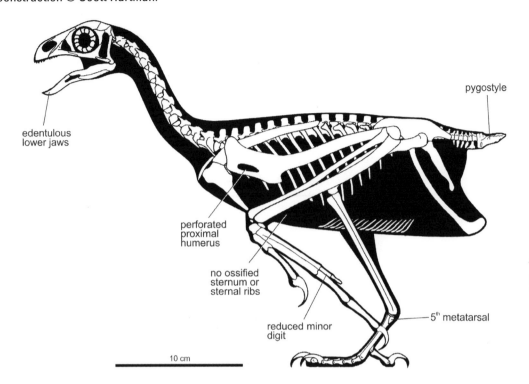

pygostyle

edentulous
lower jaws

perforated
proximal
humerus

no ossified
sternum or
sternal ribs

reduced minor
digit

5ᵗʰ metatarsal

10 cm

muscles (Chiappe 2007). The wings are very long, and as in confuciusornithids there is a perforation in the proximal end of humerus. The minor and major metacarpals are proximally fused, but unlike in confuciusornithids the minor digit is strongly reduced and consists of only the metacarpal and two very thin phalanges without a claw. The ribs lack uncinate processes, and *Sapeornis* has an exceptionally high number of 15–16 pairs of gastralia (O'Connor et al. 2015a). The wing is formed by at least 11 primary feathers, which are long and very narrow (Gao et al. 2012).

The hindlimb skeleton of *Sapeornis* agrees well with that of confuciusornithids, but unlike in the latter, at least some *Sapeornis* specimens exhibit a tuft of long feathers on the metatarsals. Instead of two streamers as in confuciusornithids, the tail of *Sapeornis* consists of multiple feathers (Zheng et al.

2013a), and the pygostyle is not as long as in confuciusornithids.

In several *Sapeornis* fossils gastroliths are preserved, which indicate a herbivorous diet (Czerkas and Ji 2002; Zhou and Zhang 2003b). That is also supported by the presence of seeds as crop contents, and it was assumed that these specialized feeding habits were functionally correlated with the reduction of the teeth (Zheng et al. 2011).

The long wings of *Sapeornis* and its primitive pectoral girdle suggest that it was probably not capable of powerful flapping flight and that gliding was the predominant way of aerial locomotion (see also Feo et al. 2015). Its flight characteristics may have been similar to those of *Confuciusornis*, with which *Sapeornis* shares a few unusual skeletal features, such as the presence of a large opening in the proximal end of the humerus.

Ornithothoraces and the Origin of Sustained Flapping Flight Capabilities

Ornithothoraces comprises more advanced pygostylians, which are characterized by derived traits of the pectoral girdle, including a well-developed sternal keel and a strut-like coracoid, and it is in this clade where the origin of powerful flapping flight is localized. Ornithothoraces falls into two major subclades, which are both already known from the earliest Cretaceous: Enantiornithes and Ornithuromorpha, the clade including extant birds (Figure 3.5). At least the early members of these two clades appear to have had different habitat preferences and ecologies: whereas enantiornithines are generally considered to have been arboreal, the skeletal morphology of early ornithuromorphs indicates that these birds were more terrestrial or aquatic (X. Wang et al. 2013; Liu et al. 2014).

The enantiornithine radiation

Enantiornithes, or "opposite birds," form the clade of Mesozoic birds that exceeds all others in species richness as well as in its temporal and geographic range. These birds occurred throughout the Cretaceous and achieved a global distribution by the end of the Mesozoic (Chiappe and Walker 2002; O'Connor et al. 2011a).

Enantiornithes were small to medium-sized birds and show a formidable morphological diversity. Among the key characteristics is a peculiar configuration of the articulation between scapula and coracoid, with the coracoid exhibiting a convex facet and the scapula a concave one. In most neornithine birds, the articulation facets of these two bones show the reverse condition, and it is this "opposite" morphology of the pectoral girdle of Enantiornithes that prompted their scientific name. Other derived enantiornithine features are a minor metacarpal that protrudes farther distally than the major metacarpal, a very long furcular apophysis, and a slender fourth metatarsal. In most Enantiornithes, the pygostyle is very long and the tail is formed by one or two pairs of long, streamer-like feathers in which the rachis is very wide and the vaned part restricted to the tip, therefore forming a "rachis-dominated" feather.

Enantiornithine fossils from the earliest Cretaceous were only found in Eurasia, where the group may have originated. The oldest specimens are from the Huajiying Formation of the Chinese Jehol Biota and have an age of about 130 million years (e.g., *Protopteryx*, *Eopengornis*; Plate 7c; Zhang and Zhou 2000; Jin et al. 2008; X. Wang et al. 2010, 2014). Only slightly younger are skeletons from the Early Cretaceous of Spain (*Iberomesornis*, *Concornis*, *Noguerornis*, *Eoalulavis*; Chiappe and Walker 2002). The latest enantiornithine records stem from just before the K/Pg boundary, from the Maastrichtian of North America, some 67–65.5 mya (Longrich et al. 2011).

The Yixian and Jiufotang formations of the Jehol Biota yielded particularly large numbers of enantiornithine fossils, and about half of the 60 known enantiornithine species stem from these Chinese localities (Figure 3.6; Wang and Zhang 2011; Wang et al. 2015a). One of the most distinctive enantiornithine taxa from the Jehol Biota is Longipterygidae (Plates 6 and 7), whose representatives are characterized by a very long snout that has teeth only at its tip (*Longipteryx*, *Longirostravis*, *Rapaxavis*, *Shanweiniao*, *Boluochia*; O'Connor et al. 2009, 2011b). Bohaiornithidae, on the other hand, which is a further enantiornithine taxon from the Jehol Biota (Plate 6a), includes birds with a short snout, large teeth, and feet with long and slender claws (*Bohaiornis*, *Parabohaiornis*, *Longusunguis*, *Shenqiornis*, *Sulcavis*, *Zhouornis*; O'Connor et al. 2013b; M. Wang et al. 2014a; Y. Zhang et al. 2014). Another

Figure 3.5 Temporal occurrences of major groups of Mesozoic birds and their interrelationships as obtained in recent analyses (e.g., M. Wang et al. 2015b). Some key apomorphies are indicated.

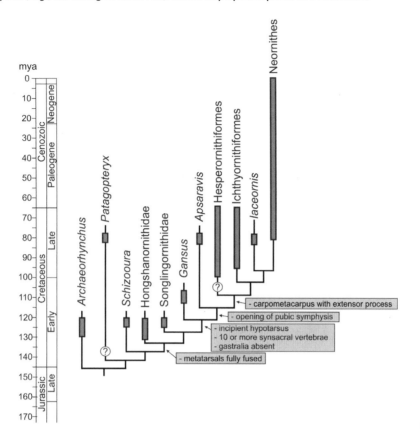

rather short-snouted enantiornithine from the Jehol Biota is *Eoenantiornis* (Plate 6c; Zhou et al. 2005). Pengornithidae comprises *Eopengornis* from the Huajiying Formation as well as *Pengornis* and *Parapengornis* from the Jiufotang Formation (Plate 6b), which are among the largest enantiornithines from the Jehol Biota and have unusually low teeth and a particularly large and globose humerus head (Zhou et al. 2008; Hu et al. 2014, 2015; X. Wang et al. 2014a). Unlike other enantiornithines, pengornithids exhibit a fifth metatarsal and lack a greatly elongated pygostyle as well as rachis-dominated tail feathers.

Early Cretaceous enantiornithines had a wide distribution and were also reported from Australia (*Nanantius*; see Close et al. 2009). The earliest New World remains stem from the Early Cretaceous (Aptian) Crato Formation in Brazil (de Souza Carvalho et al. 2015), and numerous Late Cretaceous enantiornithines were described from both South and North America. A particularly rich and diverse record exists from the Late Cretaceous of Argentina and includes morphologically disparate taxa, such as *Neuquenornis*, *Enantiornis*, *Yungavolucris*, *Lectavis*, and *Soroavisaurus* (Walker and Dyke 2009). Some of these New World enantiornithines belong to the Avisauridae, which occurred in the Late Cretaceous of both Americas (*Halimornis*, *Avisaurus*,

Figure 3.6 Skeleton of the enantiornithine *Sinornis* from the Early Cretaceous Jehol Biota.

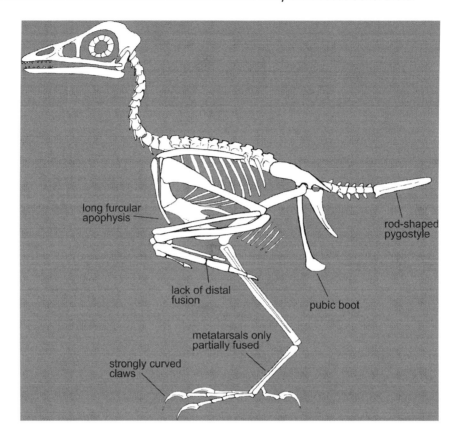

Soroavisaurus, Intiornis; e.g., Novas et al. 2010).

The phylogenetic interrelationships of Enantiornithes are not well resolved. In part this is due to the circumstance that many species are represented by only a few rather poorly preserved specimens (O'Connor et al. 2011a). Critical skeletal details therefore remain unknown for many taxa, particularly as several species are based on a few bones or juvenile individuals. The better-resolved current analyses, however, supported an early divergence of *Protopteryx*, which is among the geologically oldest enantiornithine taxa (e.g., Li et al. 2014a; M. Wang et al. 2014b, 2015b). Some analyses also suggested that Pengornithidae are the earliest diverging Enantiornithes (e.g., X. Wang et al. 2014), which conforms with the fact that pengornithids exhibit a fifth metatarsal and lack the rod-like pygostyle and rachis-dominated tail feathers characterizing other Enantiornithes; *Eopengornis* is likewise among the earliest known enantiornithines.

Most Enantiornithes have toothed jaws, but the extent of their dentition as well as the size and shape of the teeth show much variation (O'Connor and Chiappe 2011). Some taxa bear teeth in the maxillary, praemaxillary, and dentary bones (e.g., the Early Cretaceous *Iberomesornis* and *Sinornis*), but in others they are restricted to the tips of the jaws (e.g., Longipterygidae). An edentulous beak occurs in *Gobipteryx* from the Late Cretaceous of Mongolia (Chiappe et al. 2001).

Gobipipus, another edentulous and putatively enantiornithine taxon from the Late Cretaceous of Mongolia, is only known from embryonic remains (Kurochkin et al. 2013), and its phylogenetic affinities and taxonomic distinctness still have to be established with specimens of adult individuals. In *Holbotia* from the Early Cretaceous of Mongolia, the teeth are unusually widely spaced (Zelenkov and Averianov 2016).

Enantiornithes also display a remarkable diversity in terms of skull proportions and the fusion of the cranial bones (Figure 3.7; O'Connor and Chiappe 2011), with an *Archaeopteryx*-like skull shape likely being primitive. The extremes in snout lengths are exemplified by *Eoenantiornis* and bohaiornithids, where the snout is very short, and longipterygids, in which it is greatly elongated and may have even been rhynchokinetic (O'Connor and Chiappe 2011). Some taxa, such as *Protopteryx*, *Shenqiornis*, and *Sulcavis* (Plate 7a), exhibit a long postorbital bar (Zhang and Zhou 2000; Wang et al. 2010; O'Connor et al. 2013b), but whether that is characteristic for all enantiornithines remains uncertain owing to the poor preservation of the skulls of many fossils. The bones of the braincase are separate in Early Cretaceous Enantiornithes, but they are fused in the Late Cretaceous *Neuquenornis* (O'Connor and Chiappe 2011).

The enantiornithine coracoid lacks a well-developed acrocoracoid process and the dorsal surface of the sternal extremity bears a deep fossa (see Figure 4.6). The Y-shaped furcula has a long apophysis, and the well-developed keel of the sternum is restricted to the caudal portion of the bone. A superficially similar morphology of the furcula and sternum occurs in the extant South American Hoatzin (Opisthocomiformes), in which it is correlated with the presence of a large crop, for which there is, however, no evidence in Enantiornithes (Zheng et al. 2014b). More likely, the restriction of the keel of the enantiornithine sternum to the caudal part of the bone is because a part of the pectoral muscles inserted on the furcular apophysis. Otherwise, the enantiornithine sternum has a fairly modern morphology, and its caudal margin bears a pair of deep incisions (Hu et al. 2014). Enantiornithine sterna exhibit some variation in their shapes, and the bone is very elongated and narrow in *Eoalulavis*. Unlike in extant birds, the sternal ribs of Enantiornithes do not show a marked caudal lengthening, which may indicate that their lung ventilation differed from that of extant birds (Z. Zhang et al. 2014). The occurrence of uncinate processes appears to be variable, and gastralia are present. The carpometacarpus still lacks a distal fusion of the metacarpals. The minor digit is reduced and some variation exists regarding the size of the wing claws of the alular and major digits.

The enantiornithine synsacrum consists of eight fused sacral vertebrae (Hu et al. 2014). A pubic symphysis is present, albeit this is short. Usually, the tarsals are fused (Chiappe and Walker 2002), but a lack of fusion was reported for several taxa (O'Connor et al. 2011a), which in some cases may, however, be due to a juvenile or subadult age of the fossils. A fifth metatarsal is present in *Eopengornis* (X. Wang et al. 2014a) and, possibly, *Pengornis* (Hu et al. 2014), but it is absent in other enantiornithine taxa. The metatarsal bones are only proximally fused and exhibit some disparity in proportions, from being very slender in, for instance, *Neuquenornis* and *Lectavis* to very wide in *Yungavolucris*. In *Mystiornis* from the Early Cretaceous of Siberia, the second metatarsal is very short (Kurochkin et al. 2011). These disparate metatarsal shapes probably correlate with different uses of the feet and indicate a high ecological diversification of Enantiornithes.

Unlike more basal pygostylians, most Enantiornithes were small to very small, sparrow-sized birds, although some taxa,

Figure 3.7 Disparate skull shapes of enantiornithines from the Jehol Biota. (a) The short-snouted *Bohaiornis* (Bohaiornithidae; photograph by Zhonghe Zhou). (b) The long-snouted *Rapaxavis* (Longipterygidae; photograph by Jingmai O'Connor).

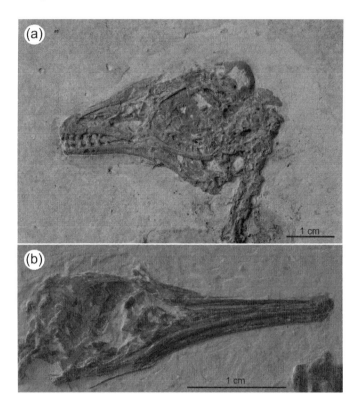

such as *Pengornis* and *Bohaiornis*, reached the size of a crow, and *Enantiornis* from the Late Cretaceous of Argentina had an estimated wingspan of 1.2 meters (Chiappe and Walker 2002). Furthermore, unlike phylogenetically more basal pygostylians, enantiornithines were probably well capable of sustained flapping flight and complicated aerial maneuvers. In concordance with their presumably more advanced flight capabilities, Enantiornithes are the earliest birds for which an alula was reported (*Eoalulavis*, *Protopteryx*; Sanz et al. 1996; Zhang and Zhou 2000). Their superior flight capabilities are also reflected by the wide distribution of these birds, which so far constitute the only known Mesozoic avian group that reached a global distribution. One Late Cretaceous taxon, *Martinavis*, is considered to have had a particularly wide distribution and includes fossils from Argentina, North America, and France (Walker et al. 2007).

Like confuciusornithids, most enantiornithines have a long, rod-shaped pygostyle and, some diversity of tail morphologies notwithstanding, most taxa exhibit greatly elongated, streamer-like, and distally expanded tail feathers. One pair of these tail feathers is present in *Protopteryx*, *Eopengornis*, and *Dapingfangornis* (Zhang and Zhou 2000; O'Connor et al. 2012a; X. Wang et al. 2014), two pairs occur in *Paraprotopteryx* (Zheng et al. 2007), whereas the longipterygid *Shanweiniao* has a fan-shaped tail (O'Connor et al. 2009). A fan-shaped tail also occurs in a recently described enantiornithine from the Chinese Xiagou Formation (*Feitianus*; O'Connor et al. 2016). In most enantiornithines with streamer-like tail feathers, the vaned portion is largely restricted to the distal part of the feather. In pengornithids, however, the pennaceous portions reach to the base of the tail feathers, even though the rachis is unusually wide as in other enantiornithines (X. Wang et al. 2014; de Souza Carvalho et al. 2015; Hu et al. 2015).

The majority of enantiornithine fossils stem from lacustrine or fluviatile sediments, but a few records also exist from marine offshore deposits (*Nanantius*, *Halimornis*). It is assumed that most species either fed on invertebrates or were **piscivorous**, but stomach contents appear to be remarkably rare in enantiornithine fossils (see, however, Chapter 4) and ingested arthropods were so far only documented for *Eoalulavis* (Chiappe and Walker 2002).

The marked differences in snout shapes, dentitions, and foot morphologies suggest that Enantiornithes underwent significant ecological diversification. Most species are likely to have been arboreal (O'Connor et al. 2011a), but others, such as the long-legged *Lectavis*, may have been more terrestrial (Chiappe and Walker 2002). For Pengornithidae and *Fortunguavis* from the early Cetaceous of China even trunk-climbing habits were assumed (M. Wang et al. 2014b; Hu et al. 2015). The long pedal claws of *Bohaiornis* as well as gastroliths preserved in one specimen – exceptional for an enantiornithine bird – were interpreted as evidence of raptorial habits (Li et al. 2014a). However, even the diversity of this largest Mesozoic avian radiation was still far below that of extant birds, and, apart from specializations of the snout and feet, most Enantiornithes exhibit similar limb proportions and skeletal characteristics. Neither were there extremely long-legged waders, nor aerial specialists like swifts or frigatebirds (Mitchell and Makovicky 2014).

The Ornithuromorpha: Refinement of Modern Characteristics

More advanced birds than those discussed so far are classified in the taxon Ornithuromorpha (alternatively sometimes also termed Euornithes, which, if current phylogenies are correct, has nearly the same content but differs slightly in its definition). Ornithuromorpha includes crown group birds, and one of the key characteristics of this clade is the high degree of fusion of the metacarpal and metatarsal bones. This increased ossification of both fore- and hindlimbs may go back to correlated developmental processes and is likely to have formed the basis for the great diversity of locomotory specializations in ornithuromorph birds, particularly within the crown group. In addition, typical ornithuromorphs share a modern-type coracoid with procoracoid and lateral processes, an U-shaped furcula, a sternum with a keel extending over the entire length of the bone, as well as a synsacrum composed of at least eight vertebrae. Most of these features represent adaptations for a refinement of the flight apparatus, and it is within Ornithuromorpha that a modern-type fan-shaped tail evolved, together with a ploughshare-shaped pygostyle.

The early evolution of ornithuromorph birds has long been poorly known, with some of the first described fossils representing highly specialized taxa. Numerous recent discoveries, however, have shed new light on Cretaceous ornithuromorphs.

Songlingornithids, hongshanornithids, and other Asian early ornithuromorphs with and without teeth

The Chinese Jehol Biota yielded an unparalleled diversity of basal ornithuromorphs, some of which show an unexpected mosaic distribution of derived and primitive characters. The best and longest-known of these are the Songlingornithidae, which include *Songlingornis*, *Yanornis*, and *Yixianornis* from the Jiufotang Formation (Figure 3.8; Plate 7d; Zhou and Zhang 2001; Clarke et al. 2006). Songlingornithids are characterized by a pair of medial fenestrae in the caudal margin of the sternum (Zhou and Zhang 2001; Clarke et al. 2006). The praemaxilla has a reduced dentition and there is an intersymphyseal ("predentary") bone (Clarke et al. 2006; Zhou and Martin 2011). This small ossicle of unknown functional significance is situated on the tips of the lower jaws and also occurs in other early ornithuromorphs (see later discussion). Some of the ribs bear uncinate processes and gastralia are still present, albeit few in number. The coracoid has an essentially modern shape, with a

well-developed procoracoid process. The pelvis retains a pubic symphysis and there are nine synsacral vertebrae. The metatarsals are fully fused into a tarsometatarsus. Gastroliths are present in some specimens (Zhou et al. 2004), which is notable because stomach contents indicate that at least *Yanornis* was piscivorous (Zhou et al. 2004; Zheng et al. 2014b).

Piscivoravis, a comparatively large taxon from the Jiufotang Formation, shares several features with songlingornithids, including medial fenestrae in the caudal margin of the sternum. As for *Yanornis*, a piscivorous diet was assumed based on the presence of a presumed gastric pellet consisting of fish remains (Zhou et al. 2014a). The skull of *Piscivoravis* is unknown, but feather remains

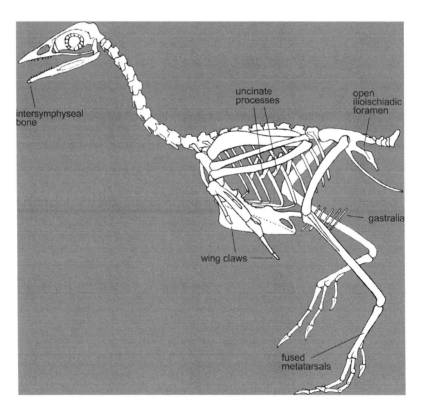

Figure 3.8 Skeleton of the ornithuromorph *Yixianornis* (Songlingornithidae) from the Early Cretaceous Jehol Biota. Adapted from Clarke et al. (2006).

document a long, fan-shaped tail. Whereas earlier analyses did not support close affinities of *Piscivoravis* and songlingornithids (O'Connor and Zelenkov 2013; Zhou et al. 2014a), it was obtained as the sister taxon of *Yanornis* in a recent study (M. Wang et al. 2015b). Another songlingornithid-like taxon from the Jiufotang Formation, *Jianchangornis*, is characterized by a high number of very small teeth (Zhou et al. 2009).

A further ornithuromorph group from the Early Cretaceous of China are the Hongshanornithidae. These include *Archaeornithura* from the Huajiying Formation, which is the oldest known ornithuromorph bird, as well as *Hongshanornis*, *Longicrusavis*, and *Tianyuornis* from the Yixian Formation, and *Parahongshanornis* from the Jiufotang Formation (Plate 8c; Zhou and Zhang 2005; O'Connor et al. 2010; Chiappe et al. 2014; Zheng et al. 2014c; M. Wang et al. 2015b). For *Longicrusavis*, teeth were reported as absent, although alveoli – the cavities of the jaws in which teeth are anchored – were observed (O'Connor et al. 2010). Contrary to initial assumptions that *Hongshanornis* is edentulous (Zhou and Zhang 2005), a few teeth are present in both its maxillae and dentaries (Chiappe et al. 2014); teeth are also present in the upper and lower jaws of *Tianyuornis* (Zheng et al. 2014c). At least *Hongshanornis* has an intersymphyseal bone (Zhou and Martin 2011). Hongshanornithids are much smaller than songlingornithids and have longer legs, which indicate a more terrestrial way of life. The pubic symphysis is short and the tail fan-shaped (Chiappe et al. 2014). For *Hongshanornis*, crop contents consisting of seeds were reported (Zheng et al. 2011). Based on the proportions of the long hindlimbs, the long toes, and the weak pedal claws, hongshanornithids were considered wading birds (Zhou and Zhang 2005; Chiappe et al. 2014). The skeletal morphology of these birds would, however, equally conform to a terrestrial, ground-dwelling way of life.

Besides these toothed taxa, the Jehol Biota also yielded the earliest edentulous ornithuromorphs. These are *Archaeorhynchus* from the Yixian and Jiufotang formations, and *Zhongjianornis* and *Schizooura* from the Jiufotang Formation (Plate 8a, b; Zhou and Zhang 2006; Zhou et al. 2010; Zhou et al. 2012, 2013). The presence of neurovascular grooves and foramina suggests that the jaws of these taxa were covered with a horny rhamphotheca. Unlike in *Confuciusornis* and more advanced ornithuromorphs, however, the tips of the lower jaws are not fused into a mandibular symphysis; an intersymphyseal bone is absent. *Archaeorhynchus* and *Zhongjianornis* have a very robust, U-shaped furcula, whereas the equally robust furcula of *Schizooura* is V-shaped. The sternum of *Archaeorhynchus* exhibits two pairs of deep caudal incisions. The sternal body is, however, shorter than in other ornithuromorphs, which may be correlated with the very long wings of this taxon. The ribs of *Zhongjianornis*, *Archaeorhynchus*, and *Schizooura* bear uncinate processes. Gastralia were only reported for *Archaeorhynchus* and *Schizooura*. The humeri of *Zhongjianornis* and *Schizooura* are characterized by a large deltopectoral crest, which gives the bone a similar outline to that of *Confuciusornis*. *Schizooura* has distally fused metacarpals, but a distal fusion was reported as absent for *Zhongjianornis* and the major and minor metacarpals are not co-ossified in the subadult specimens of *Archaeorhynchus* (Zhou et al. 2010; Zhou et al. 2012, 2013). *Zhongjianornis* appears to have a short pubic symphysis, but whether this is present in *Schizooura* and *Archaeorhynchus* is uncertain, not least because all known specimens of *Archaeorhynchus* are subadult individuals. The metatarsals are incompletely fused in all three taxa. Numerous gastroliths are preserved in each of the three known specimens of *Archaeorhynchus* and suggest a herbivorous diet consisting of coarse plant matter (Zhou et al. 2013).

Initially, *Zhongjianornis* was considered to be one of the most basal pygostylians, but subsequent studies found it to be nested within Pygostylia (O'Connor and Zhou 2013, and in some analyses *Zhongjianornis* and *Schizooura* form a clade with *Chaoyangia*, a poorly known ornithuromorph from the Jiufotang Formation (O'Connor and Zelenkov 2013; O'Connor and Zhou 2013). *Archaeorhynchus* resulted as the earliest diverging ornithuromorph taxon in some analyses (e.g., O'Connor and Zelenkov 2013; M. Wang et al. 2015b), but an assessment of its affinities is hampered by the fact that only subadult individuals are known. At least judging from the published descriptions, these edentulous taxa may be more closely related than is apparent from current phylogenies.

Xinghaiornis from the Yixian Formation superficially resembles *Zhongjianornis* in skeletal morphology. This edentulous bird has a similar bill shape to *Zhongjianornis*, but unlike the latter it features an Enantiornithes-like furcula with a long apophysis and the hind toe inserts very high on the tarsometatarsus (X. Wang et al. 2013). In the original description, the affinities of *Xinghaiornis* within Ornithothoraces were considered unresolved, and the taxon could not be convincingly assigned to either Ornithuromorpha or Enantiornithes. In a more recent analysis, *Xinghaiornis* was placed within Ornithuromorpha (O'Connor et al. in press). In this latter study, another particularly long-beaked edentulous ornithuromorph (*Dingavis*) was reported from the Yixian Formation.

Other early ornithuromorphs are too incompletely known to determine their affinities precisely. This is particularly true for various taxa from the Early Cretaceous Chinese Xiagou Formation (*Changmaornis, Yumenornis, Jiuquanornis*; Y.-M. Wang

et al. 2013); because these are largely based on non-overlapping skeletal elements, their taxonomic distinctness also needs to be established by future finds. Currently likewise poorly resolved is the phylogenetic position of *Ambiortus*, a teal-sized ornithuromorph from the Early Cretaceous of Mongolia (Kurochkin 1999; O'Connor and Zelenkov 2013).

The aquatic *Gansus*

In the Early Cretaceous, some birds had already become adapted to an aquatic or semi-aquatic way of life. However, as is the case in some of the piscivorous ornithuromorphs from the Jehol Biota discussed earlier, the extent of these specializations is often difficult to determine. An Early Cretaceous taxon for which a more aquatic lifestyle is safely established is *Gansus* (Gansuidae) from the Early Cretaceous (lower Aptian, about 120 mya; Liu et al. 2014) lacustrine deposits of the Xiagou Formation in Gansu Province in China (You et al. 2006). The skull of this tern-sized bird is unknown, but its postcranial skeleton is well documented by numerous remains. The ribs bear uncinate processes (Y.-M. Wang et al. 2016), gastralia are absent, and the pygostyle is very small. As in songlingornithids, the caudal margin of the sternum exhibits medial fenestrae. The metatarsals are fully fused and the tarsometatarsus bears an incipient hypotarsus (Y.-M. Wang et al. 2016). The tips of the pubes of *Gansus* are distally expanded and contact each other, but a true pubic symphysis is absent or very short. A suite of skeletal characteristics, such as a narrow pelvis, an elongated cnemial crest of the tibiotarsus, a very short tarsometatarsal trochlea for the second toe, and long feet, suggest that *Gansus* was an aquatic bird, which used its hindlimbs for propulsion (You et al. 2006). The shape of the pedal claws, which have far distally located tubercles for the attachment of the tendons of the flexor muscles of the foot, may indicate the former presence of webbed feet.

A putative gansuid, *Iteravis*, was also reported from the Jehol Biota, some 2000 kilometers away from the type locality of *Gansus* (Plate 8d; Liu et al. 2014; Zhou et al. 2014b; see Mortimer 2014 concerning the muddled taxonomic history of this bird, which in one of the aforementioned studies was assigned to *Gansus*). Of *Iteravis* the skull is known from multiple specimens and bears small teeth in the lower jaws. The praemaxilla is edentulous and an intersymphyseal bone is present. Some skeletons are preserved with gastroliths. Unlike in *Gansus* there is a pubic symphysis in *Iteravis* and the cnemial crests are less projected, and whether *Iteravis* is indeed a gansuid needs to be evaluated further (Mortimer 2014). The exact stratigraphic provenance of the taxon also seems to be controversial (Jiufotang Formation according to Liu et al. 2014, whereas Zhou et al. 2014b considered the fossils to be from the Yixian Formation).

It is tempting to assume that *Gansus*-like birds gave rise to an iconic and long-known group of foot-propelled diving birds, the hesperornithiforms (see next section). The analyses performed so far did not support such close affinities, however, and suggested that *Gansus* is the sister taxon of Ornithurae, the clade including hesperornithiforms and extant birds (e.g., Liu et al. 2014; Y.-M. Wang et al. 2016; see also Chapter 4).

Ornithurae and the Origin of Modern Birds

Ornithurae was established in the 19th century as the taxon including crown group birds as well as *Hesperornis* and *Ichthyornis*, then the two best-known Mesozoic avian taxa other than *Archaeopteryx*. Today, it is defined as the least inclusive clade comprising Hesperornithiformes and crown group birds. The members of Ornithurae exhibit an essentially modern-type skeletal morphology, and one of the features that may have been critical for the evolutionary success of this clade is the opening of the pubic symphysis of the pelvis. Possibly functionally correlated therewith, ornithurine birds also lost gastralia. Many Mesozoic Ornithurae, however, still retain plesiomorphic characteristics, such as teeth, which readily distinguish them from crown group (neornithine) birds.

The foot-propelled hesperornithiforms

Hesperornithiforms were highly specialized diving birds and mainly occurred in the northern latitudes of the Northern Hemisphere. More than 25 species were described, most of which were flightless and lived in marine environments (Figure 3.9; O'Connor et al. 2011a). New finds and reinterpretations of some fossils, however, have shown that the ecological diversity of these birds was greater than was assumed only a few years ago.

The earliest currently recognized hesperornithiform is *Enaliornis* from Early Cretaceous (Albian, 100 mya) deposits in England, which already was a specialized diving bird (Galton and Martin 2002). This taxon is represented by a number of cranial and postcranial remains, but wing and pectoral girdle bones have not yet been identified. Of the skull only the neurocranium is known, so that the presence and extent of a dentition remain elusive. *Enaliornis* is one of the smallest hesperornithiforms, with an estimated weight of only about 400–600 grams. Because its size is below that of other flightless aquatic birds, it may well have been capable of flight (Elzanowski and Galton 1991). The fossil material of the next oldest hesperornithiform, *Pasquiaornis* from the earliest Late Cretaceous of Canada, includes humerus remains, which greatly

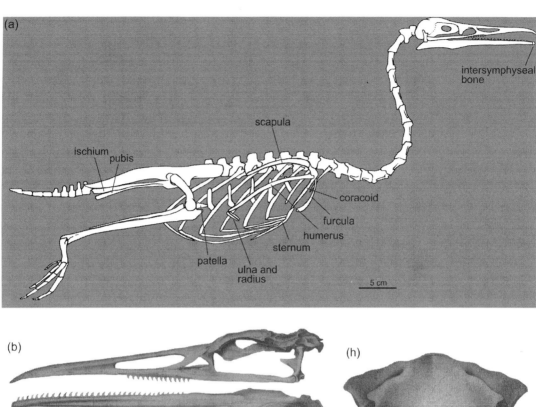

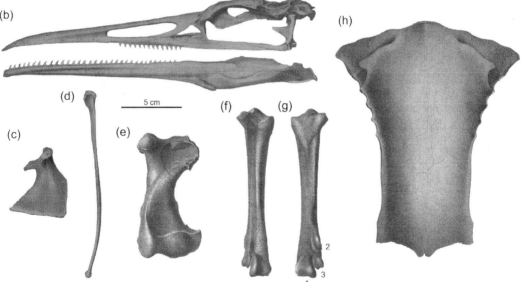

Figure 3.9 Late Cretaceous North American hesperornithiforms. (a) Skeleton of *Baptornis* (after Martin and Tate 1976). (b–h) Bones of *Hesperornis* (from Marsh 1880; b: skull, c: coracoid, d: humerus, e: femur, f, g: tarsometatarsus in dorsal and plantar view, h: sternum). Note the feeble humerus, highly modified leg bones, and absence of a sternal keel in these flightless, foot-propelled birds.

differ from the feeble humeri of later hesperornithiforms and suggest that *Pasquiaornis* may have been capable of flight too (Tokaryk et al. 1997).

All other hesperornithiforms likewise occur in Late Cretaceous rocks. In particular the Western Interior Seaway – an **epicontinental** sea that divided North America in the late Mesozoic – featured a high diversity of these birds. Sediments of this seaway formed the Niobrara Chalk of Kansas, from which various hesperornithiforms were reported. The best represented of these is *Hesperornis*, which was described in the classic monograph by Marsh (1880). *Hesperornis* encompasses several species and includes the largest hesperornithiforms (Figure 3.9b–h; Martin and Lim 2002; Rees and Lindgren 2005). Other Late Cretaceous North American hesperornithiforms are classified into the taxa *Fumicollis*, *Parahesperornis*, and *Baptornis*, which include a single species each (Martin and Tate 1976; Rees and Lindgren 2005; Bell and Chiappe 2015, 2016). *Baptornis* (Figure 3.9a) exhibits a less specialized morphology than *Hesperornis* and *Parahesperornis*, and is usually assigned its own higher-level taxon, Baptornithidae. The northernmost occurrences of hesperornithiforms are specimens from the mid-Maastrichtian of Canada (Hills et al. 1999; Hou 1999).

With the exception of the Early Cretaceous *Enaliornis*, the fossil record of hesperornithiforms was for a long time an exclusively North American one. Hesperornithiforms are now, however, known to have had a wide distribution in the Late Cretaceous of the Old World, and fossils have been found in Sweden, Kazakhstan, Russia, and Mongolia. These specimens were assigned to the taxa *Hesperornis* and *Baptornis*, as well as to the Mongolian baptornithid *Judinornis* and to *Asiahesperornis*, which is well represented by numerous fossils from Kazakhstan (Panteleyev et al. 2004; Rees and Lindgren 2005; Dyke et al. 2006; Martin et al. 2012; Bell and Chiappe 2016).

Brodavis (Brodavidae), one of the most recently described hesperornithiform taxa, comprises four species from the Late Cretaceous of North America and Asia and constitutes the only lacustrine occurrence of Hesperornithiformes (Martin et al. 2012). The *Brodavis* species differ distinctly in size and some existed in the latest Maastrichtian, just before the end-Cretaceous mass extinction events (see also Longrich et al. 2011). Apart from being one of the latest fossil records of hesperornithiforms, *Brodavis* also includes some of the smallest known species.

Within hesperornithiforms a large size was gained several times independently, and in particular some species of *Hesperornis* were very large, reaching a length of 1.5 meters (Bell and Chiappe 2016). As an adaptation for their diving habits, the limb bones of hesperornithiforms are pachyostotic; that is, they have very thick bone walls. Pachyostotic bones also occur in other flightless diving birds, such as penguins, and reduce buoyancy by increasing the weight of the bird. The complete dentition is only known from *Hesperornis*, in which teeth are absent from the tips of the upper jaws, with the praemaxillary bones therefore being edentulous. The mandibles are not co-ossified, and there is an intersymphyseal bone at their tips (Zhou and Martin 2011). The absence of a mandibular symphysis and the presence of an intraramal joint – that is, a joint within the bones of the lower jaw – increased the flexibility and movability of the lower jaws and allowed swallowing of large prey items. Hesperornithiforms are considered to have been predominantly piscivorous birds, and direct evidence for the diet of *Baptornis* comes from coprolites, which contain fish remains (Martin and Tate 1976).

The wings of the large *Hesperornis* are greatly reduced and appear to have consisted only of the very thin humerus. Wing reduction is less pronounced in *Baptornis*, for which an ulna is known, and the wings of this taxon may have acted as steering devices during diving (Martin and Tate 1976). The sternum of *Hesperornis* and *Baptornis* lacks a keel. The pelvis is very narrow and the femoral articulation is constrained in such a way that the femora were laterally splayed in the living animal. As in some extant birds, which use the legs for aquatic propulsion, the patella (kneecap) is greatly enlarged.

Owing to the unique sprawling posture of the legs, at least the highly derived hesperornithiforms, such as *Hesperornis* and *Parahesperornis*, were probably not able to walk well on land (Rees and Lindgren 2005). In light of these apparently very limited terrestrial locomotory capabilities, it was concluded that hesperornithiforms may have been viviparous (Feduccia 1999), but in birds this reproductive mode is prevented by developmental and other constraints, including the advanced lung system and solid eggshell (Blackburn and Evans 1986). Accordingly, even the most specialized hesperornithiforms must have visited firm breeding grounds, which were likely on islands or in other predator-free areas. Because hesperornithiforms were found in far northern latitudes with cold polar winters, it was surmised that some species were migratory, although conclusive evidence for this assumption has not yet been presented (Wilson and Chin 2014).

Other secondarily flightless Mesozoic ornithuromorphs

Aside from hesperornithiforms, flightlessness occurred in at least one further Mesozoic ornithuromorph taxon, *Patagopteryx* (Patagopterygiformes) from the Late Cretaceous (Campanian, about 80 mya) of Argentina. This turkey-sized bird is known from multiple partial skeletons and is characterized by an unusual combination of primitive and derived features. Of the skull, only portions of the neurocranium were found, so that it is unknown whether *Patagopteryx* had teeth. Unlike in all other birds, the quadrate appears to have been fused with the pterygoid (Chiappe 2002). The wings are very short and the sternal keel is reduced. The loss of flight capabilities in this taxon are of particular interest, because it lived in a continental environment with potential predators, but achieved neither a large size nor cursorial habits (Chiappe 2002). *Patagopteryx* is currently considered to be one of the earliest diverging ornithuromorphs (e.g., Turner et al. 2012; M. Wang et al. 2015b), but some derived characteristics (e.g., the absence of a pubic symphysis) may suggest that it is more deeply nested within Ornithuromorpha (see the next chapter).

A tiny putative relative of *Patagopteryx*, *Alamitornis* from the Late Cretaceous of Argentina (Agnolín and Martinelli 2009) is based on a humerus that appears to be non-avian and is more likely from a squamate (lizards and allies). A further enigmatic taxon that was likened to *Patagopteryx* is the cassowary-sized *Gargantuavis* from the Late Cretaceous of France, of which pelvis remains and a few other bones were found (Buffetaut 2010; Buffetaut and Angst 2013). Especially the wide pelvis and the cranial position of its socket for the femur (acetabular foramen) are unparalleled by other large flightless birds, but doubts concerning the avian identity of *Gargantuavis* (Mayr 2009a) were countered by Buffetaut (2010).

Ichthyornis and _Apsaravis_: Was the ancestor of modern birds aquatic?

Together with *Hesperornis*, Marsh (1880) described a volant toothed bird from the Late Cretaceous Niobrara Chalk of Kansas – the small and long-snouted *Ichthyornis* (Ichthyornithiformes). Several species of this taxon were named from Late Cretaceous (late Cenomanian to Campanian, 90-80 mya) fossil sites in North America, but all are likely to be synonyms of the type species (Clarke 2004). Some of the youngest fossils stem from Mexico and were thus found farther south than the majority of the geologically older specimens, which are from localities in Kansas and Canada (Porras-Múzquiz et al. 2014). Fragmentary remains of *Ichthyornis*-like birds were also reported from the Maastrichtian of North America and Eurasia, but some of these records need further verification (Clarke 2004; Longrich et al. 2011).

Ichthyornis has teeth in the upper and lower jaws, but as in *Hesperornis* the praemaxillary bone is edentulous. Also as in *Hesperornis*, the mandibles lack a symphysis and exhibit an intraramal joint. The articular facets of the thoracic vertebrae are biconcave (amphicoelous), whereas they are saddle-shaped (heterocoelous) in most extant birds. *Ichthyornis* has a deeply keeled sternum, which together with the large proximal end of the humerus and the well-developed deltopectoral crest indicates proficient flight capabilities. These are also suggested by a large distal process on the first phalanx of the major wing digit which is unknown from other non-neornithine birds, and which in extant birds occurs in taxa with very long primary feathers. The hindlimb bones exhibit essentially modern-type morphologies, and even though a supratendinal bridge on the distal tibiotarsus is still absent, the tarsometatarsus bears an incipient hypotarsus.

Ichthyornis is presumed to have been piscivorous, which is also true for hesperornithiforms, gansuids, and other early ornithuromorphs (e.g., songlingorniths). This prompted the hypothesis that the ancestor of modern birds had similar habits and lived in an aquatic environment (e.g., You et al. 2006). The phylogenetically most basal extant neornithine birds – that is, Palaeognathae and Galloanseres – are, however, predominantly herbivorous or omnivorous and mainly occur in terrestrial habitats, so that the ecological preferences of *Ichthyornis* and kin are rather unlikely to have been ancestral for the avian crown group.

That there also existed terrestrial taxa at the base of Ornithurae is exemplified by *Apsaravis* from the Late Cretaceous (late Campanian to early Maastrichtian) of Mongolia, which stems from an arid, continental paleoenvironment (Clarke and Norell 2002). *Apsaravis* exhibits a mandibular symphysis, but unfortunately little further anatomical information on the skull can be obtained from the single known skeleton. There are, however, no teeth in the lower jaws, and because tooth reduction in ornithuromorph birds starts from the tips of the upper jaws (see the next chapter), it is likely that *Apsaravis* was edentulous. The pelvis lacks a pubic symphysis and as in *Ichthyornis*, but unlike in more basal ornithuromorphs, the tarsometatarsus exhibits an incipient hypotarsus. In contrast to Early Cretaceous ornithuromorphs, the carpometacarpus of *Apsaravis* bears an extensor process, which serves for the attachment of a muscle extending the hand section of the wing (Clarke and Norell 2002). *Apsaravis* was hypothesized to be the sister taxon of a clade including *Ichthyornis*, *Hesperornis*, and neornithine birds (Clarke 2004; Turner et al. 2012), and especially the distal ends of the humerus and tibiotarsus exhibit a more plesiomorphic morphology than the corresponding bones of *Ichthyornis*. *Apsaravis*-like birds may have had a wider distribution in the Late Cretaceous, and with regard to a similar overall shape and the presence of a deep fossa on the ventral surface of the shaft, the coracoid of *Apsaravis* resembles that of *Palintropus* from the Late Cretaceous (Maastrichtian) North American Lance Formation (Longrich et al. 2011).

4

Mesozoic Birds: Interrelationships and Character Evolution

Only two decades ago, the phylogenetic interrelationships of the few then known Mesozoic birds were reasonably well understood, and clear-cut character evidence existed in support of widely accepted phylogenetic hypotheses. Most of these early phylogenies had a ladder-like shape, and the sequential branching of more advanced groups suggested a straightforward evolution of birds towards the neornithine crown clade. With the numerous new taxa described in the past few years, however, a reconstruction of the interrelationships of Mesozoic birds has become more difficult.

By and large, the temporal sequence of Mesozoic avian higher-level taxa broadly conforms to their phylogenetic positions, and earlier fossils usually belong to more basally diverging taxa. Most long-tailed, *Archaeopteryx*-like avians occur in Late Jurassic deposits, from which no pygostylians are known. In the Early Cretaceous of China various basal avians – including both long-tailed taxa and pygostylians – coexisted, but from the Late Cretaceous only the more advanced ornithothoracine birds are known. Taken as it is, the current fossil record therefore indicates that the evolution of short-tailed birds commenced in the latest Jurassic or earliest Cretaceous and that there were strong selective pressures in these early phases of avian evolution. Basal forms were soon replaced by more derived ones, and only Enantiornithes and some specialized marine birds existed over long periods during the Mesozoic.

Avian Evolution: The Fossil Record of Birds and its Paleobiological Significance,
First Edition. Gerald Mayr.
© 2017 John Wiley & Sons, Ltd. Published 2017 by John Wiley & Sons, Ltd.

In the more than 150 million years of avian evolution, the characteristics of modern birds were sequentially acquired. Major evolutionary changes first concerned the refinement of the flight apparatus and mainly involved the forelimbs, pectoral girdle, and tail, whereas key innovations of the pelvic girdle and the hindlimbs occurred later. Because early Mesozoic birds display a high degree of homoplasy in character distribution, a reconstruction of the early phases of avian evolution is, however, not straightforward.

The Interrelationships of Mesozoic Birds: Controversial Phylogenetic Placements and Well-Supported Clades

Although a basic framework for the interrelationships of non-neornithine birds exists (see Figure 3.5), major parts of current phylogenies are poorly resolved and bush-like, and the position of some critical taxa varies from analysis to analysis. The difficulties start with a well-founded placement of long-tailed feathered paravians like *Aurornis*, *Xiaotingia*, and *Anchiornis* relative to *Archaeopteryx* and deinonychosaurs. The conflicting hypotheses regarding the affinities of these basal taxa have already been outlined, and in the following the focus lies on the interrelationships of more advanced avians.

Most authors considered the long-tailed *Jeholornis* to be the sister taxon of a monophyletic Pygostylia; that is, all avians with a greatly shortened tail and a pygostyle (e.g., O'Connor et al. 2012b). However, some analyses resulted in a more basal position of *Sapeornis* (Zhou et al. 2010; Turner et al. 2012), and a clade including *Jeholornis* and the remaining pygostylians to the exclusion of *Sapeornis* was regarded as "extremely well supported" (Turner et al. 2012: 115). The proposed apomorphies of this latter clade include strut-like coracoids and other characters associated with the coracoid morphology, as well as a bowed metacarpal

of the minor digit and edentulous upper jaws; unlike other avians except *Archaeopteryx*, *Sapeornis* also lacks an ossified sternum. A non-monophyletic Pygostylia is, however, not as well supported as has been stated. Tooth reduction in the upper jaws represents unconvincing character evidence, because the jaws of basal enantiornithines show a full dentition, which indicates the convergent loss of teeth in *Sapeornis* and more advanced pygostylians. Apart from being elongated, the coracoids of *Jeholornis* and *Confuciusornis* exhibit very different morphologies, each being distinguished from *Sapeornis* and later avians, so that homoplasy in coracoid evolution is likely, too (so much the more as the coracoid of some early ornithuromorphs, such as *Archaeorhynchus*, is fairly wide, rather than strut-like). Unlike in more derived avians but as in *Archaeopteryx*, *Jeholornis* furthermore has a non-reversed first toe (Zhou and Zhang 2007), which supports its placement as a sister taxon of all pygostylians including *Sapeornis*.

Even disregarding the relationships of *Jeholornis*, the interrelationships of basal pygostylians are difficult to resolve. Again, the character mosaic of *Sapeornis* proves to be problematic, and it is controversial whether this taxon (e.g., Zhou et al. 2008; Y.-M. Wang et al. 2013) or the Confuciusornithidae (e.g., Gao et al. 2008; O'Connor et al. 2009; M. Wang et al. 2014b) diverges first (Figure 4.1). As just noted, *Sapeornis* differs from other pygostylians in the

Figure 4.1 Three alternative hypotheses on the interrelationships of early diverging avians, with some key apomorphies (see text for further discussion). (a) The "*Jeholornis-Sapeornis*-sequence" (e.g., Zhou et al. 2008; Y.-M. Wang et al. 2013). (b) The "*Jeholornis*-Confuciusornithidae-sequence" (e.g., O'Connor et al. 2009; Y. Zhang et al. 2014; M. Wang et al. 2015b). (c) The "*Sapeornis-Jeholornis*-sequence" (e.g., Zhou et al. 2010; Turner et al. 2012).

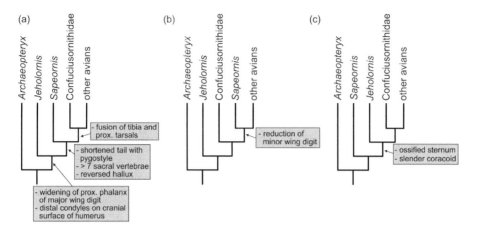

plesiomorphic, *Archaeopteryx*-like shape of its coracoid and the absence of ossified sternal plates. Confuciusornithids, on the other hand, are the only pygostylians, which retain an unreduced minor wing digit, whereas some derived features that are absent in *Sapeornis*, such as uncinate processes of the ribs (but see Chapter 3), suggest closer affinities to Ornithothoraces. Because *Sapeornis* and *Confuciusornis* are among the best represented Early Cretaceous pygostylians, it is to be expected that future analyses will shed more light on their exact interrelationships and lead to robust phylogenies.

One of the less controversial clades is Ornithothoraces, which includes Enantiornithes and Ornithuromorpha and is obtained in all current analyses. With the ever-increasing diversity of basal ornithuromorphs, the numbers of enantiornithine apomorphies significantly decreased in the past few years, however, and some taxa exhibit a character mosaic that aggravates their assignment to either enantiornithines or ornithuromorphs (Zhou et al. 2008; X. Wang et al. 2013).

Of particular interest for an understanding of tooth reduction in birds are the affinities of the edentulous ornithuromorph taxa from the Early Cretaceous of China, such as *Archaeorhynchus*, *Schizooura*, and *Zhongjianornis*. The incomplete fusion of the metatarsals indicates that these taxa occupy a more basal phylogenetic position than the toothed songlingornithids and hongshanornithids (Figure 4.2), which is also suggested by various other plesiomorphic traits, such as the low count of fused synsacral vertebrae of *Archaeorhynchus* and the absence of a globose proximal humerus end in *Schizooura* (Zhou et al. 2012, 2013). However, the exact position of *Archaeorhynchus*, *Schizooura*, and *Zhongjianornis* is controversially resolved in current analyses, so that it remains difficult to assess how often teeth were lost in avian evolution. Because some of these edentulous Early Cretaceous ornithuromorphs are based on juvenile specimens, seemingly primitive traits may represent early ontogenetic ones, and future discoveries of adult specimens will have to show whether these taxa are indeed as

Figure 4.2 Interrelationships of Mesozoic ornithuromorphs as resulting from current analyses of comprehensive data sets (e.g., M. Wang et al. 2015b). Some key apomorphies are indicated; see the text concerning the affinities of *Patagopteryx* and hesperornithiforms.

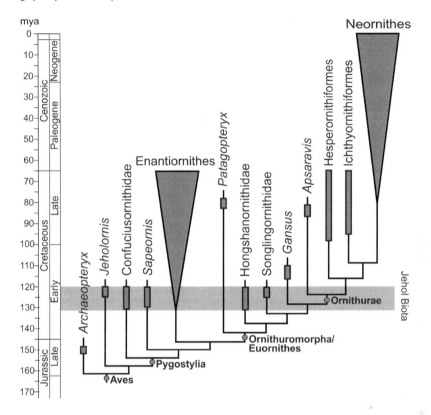

widely separated as they are in some current analyses (e.g., Zhou et al. 2012).

In addition to the plethora of new and therefore often still insufficiently known basal ornithuromorphs, there are also several taxa that were described a while ago, but exhibit unusual character mosaics that aggravate an unambiguous phylogenetic placement. The Late Cretaceous *Patagopteryx*, for example, resulted as one of the earliest branching ornithuromorphs in some analyses (e.g., Zhou et al. 2010; Turner et al. 2012). This position not only conflicts with the comparatively young geologic age of the taxon, but also with some derived characters shown by *Patagopteryx*, such as the absence of a pubic symphysis and the presence of a tarsometatarsus with an incipient hypotarsus and proximal vascular foramina. It has already been noted (Chapter 2) that the origin of flightlessness is often due to paedomorphosis, and some of the plesiomorphic features displayed by *Patagopteryx* therefore may well represent secondary reversals into the primitive condition.

A confusing character mosaic is also displayed by the enigmatic *Vorona* from the Late Cretaceous of Madagascar, the metatarsals of which are largely fused, whereas a rudimentary fifth metatarsal is still present (Forster et al. 2002). Likewise poorly resolved are the affinities of other incompletely known and more recently described taxa, such as

Hollanda from the Late Cretaceous of Mongolia, which is represented by a few hindlimb bones and may have been a cursorial bird (Bell et al. 2010).

In light of the fact that the origin of hesperornithiforms remains elusive and aquatic habits were rare in Cretaceous birds, it may be readily assumed that the geologically somewhat older and more primitive gansuids gave rise to the highly specialized hesperornithiforms. In current analyses, however, hesperornithiforms form a clade together with ichthyornithiforms and crown group birds, and *Gansus* is not obtained as part of this clade (e.g., Clarke 2004; O'Connor et al. 2011a; Turner et al. 2012). Some of the derived characteristics related to hindlimb-propelled aquatic locomotion shared by hesperornithiforms and *Gansus*, such as the narrow pelvis and prominent cnemial crests of the tibiotarsi, are, however, not included in these studies, in which proximal vascular foramina of the tarsometatarsus are furthermore erroneously considered to be present in hesperornithiforms (Bell and Chiappe 2016 emphasize the presence of these foramina on the plantar surface of the tarsometatarsus, but note that they do not perforate the bone as in more advanced ornithurans). The possibility that hesperornithiforms derive from an Early Cretaceous *Gansus*-like ancestor and occupy a more basal position within Ornithurae therefore has yet to be scrutinized with revised data sets.

Given the pace of recent discoveries, it is likely that many uncertainties concerning the interrelationships of Mesozoic birds will be settled in the near future. It should, however, be noted that a drawback of many current analyses of large morphological data sets is the lack of adequate discussion of the character evidence for the clades obtained, and the evolutionary insights that can be gained from phylogenies are quite limited if the apomorphies of critical clades are not identified and discussed.

Character Evolution in Mesozoic Birds

Some of the characteristic traits of birds already evolved in non-avian theropods. This is true for bipedal locomotion, extensive skeletal pneumatization and hollow limb bones, the fusion of the clavicles into a furcula, the reduction of the wing digits and toes, as well as the emergence of feather homologues and possibly even true pennaceous feathers (Xu et al. 2014). Aside from this evolutionary heritage, however, numerous morphological transformations occurred in the long evolutionary history of birds, and some of these are outlined in the following sections.

Skull evolution

It has been hypothesized that birds have paedomorphic dinosaur skulls, with short snouts and large orbits (Bhullar et al. 2012). The skull of *Archaeopteryx*, however, still resembles that of some troodontids in its proportions, whereas early birds show a considerable diversity of skull shapes (Figure 4.3). Basally diverging avians, such as *Jeholornis* and *Sapeornis*, are characterized by robust skulls with tall snouts and a reduced dentition that is restricted to the tips of the upper or lower jaws. This cranial morphology may be indicative of herbivory (Zanno and Makovicky 2011). It is also found in oviraptorosaurs and scansoriopterygids, and was considered to be plesiomorphic for birds (Xu et al. 2011). In that case, some enantiornithines must have secondarily acquired *Archaeopteryx*-like skull proportions and a full dentition (Fig. 4.3), which is not a likely assumption. Alternatively, a deep snout with a reduced dentition may have evolved multiple times independently in the above

Figure 4.3 Schematic depiction of the skull of early avians and close avian relatives. Note the similar shapes of the skulls of the oviraptorosaur *Similicaudipteryx*, the scansoriopterygid *Scansoriopteryx*, and the basal avians *Jeholornis* and *Sapeornis* on the one hand, and those of *Archaeopteryx* and the enantiornithine *Shenqiornis* on the other. Adapted from Xu et al. (2011) and O'Connor and Chiappe (2011). Not to scale.

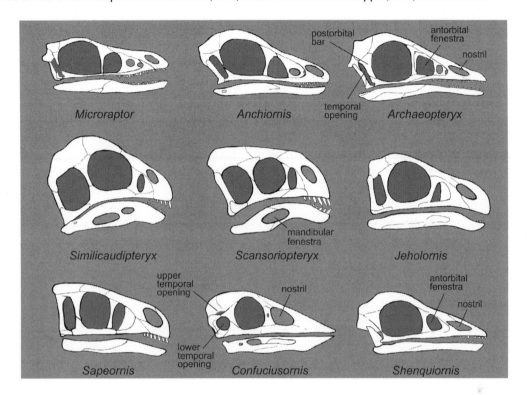

lineages in response to a herbivorous diet, or some of the taxa with this skull morphology may be more closely related than is apparent from current phylogenies.

Birds primitively had a fully developed postorbital bar, and at least in *Confuciusornis* there is also an upper temporal bar, which forms a second temporal opening (Figure 4.3; Peters and Ji 1998; Chiappe et al. 1999). The condition in *Archaeopteryx* is uncertain owing to the poor preservation of the temporal region in the known specimens, but at least a postorbital bone with a short ventral process is present (Wellnhofer 2009). A postorbital, which is absent in extant birds, was also reported for *Sapeornis* and some Enantiornithes (Hu et al. 2010;

Wang et al. 2010; O'Connor and Chiappe 2011).

Archaeopteryx already has large orbits and a large, bird-like brain (Domínguez Alonso et al. 2004) and these attributes are also present in the Troodontidae (Balanoff et al. 2013). The bones of the neurocranium and skull roof are not co-ossified in *Archaeopteryx*, and extensive fusion of the cranial bones occurred convergently in some Late Cretaceous Enantiornithes (e.g., *Neuquenornis*) and in Ornithuromorpha. In the hesperornithiform *Enaliornis* there remains a suture between the frontal and parietal bones, which among Neornithes also occurs in the palaeognathous

Lithornithiformes and in Tinamiformes (Elzanowski and Galton 1991).

The palate of most Mesozoic birds is poorly known, but reconstructions or descriptions exist for *Archaeopteryx* (Elzanowski and Wellnhofer 1996; Mayr et al. 2007), *Confuciusornis* (Chiappe et al. 1999), the enantiornithine *Gobipteryx* (Chiappe et al. 2001), and *Hesperornis* (Elzanowski 1991). The bones of the palate of *Archaeopteryx* are best known and resemble those of the troodontid *Gobivenator* (Figure 4.4; Tsuihiji et al. 2014). Unlike in more advanced birds, the palatine of *Archaeopteryx* is tetraradiate – that is, it has four processes, as in non-avian theropods – whereas this bone is triradiate (with only three processes) in extant birds (initial observations of a triradiate palatine in *Archaeopteryx* were based on a misinterpretation of the damaged Munich specimen; Mayr et al. 2005, 2007). The palate of *Archaeopteryx* furthermore exhibits an ectopterygoid (see Figure 4.4). This bone occurs in non-avian theropods but is absent in neornithine birds, although a possible homology was suggested to the uncinate bone, a small accessory ossicle of varying shape and position found in some Neornithes (Elzanowski 1999). A possible ectopterygoid was also reported for *Confuciusornis* (Chiappe et al. 1999), but it appears to be absent in the enantiornithine *Gobipteryx*.

In agreement with the disparate skull architectures of early Mesozoic birds, different types of cranial kinesis were suggested. The skull of *Archaeopteryx* was considered to have been prokinetic (Wellnhofer 2009), whereas that of *Confuciusornis* probably

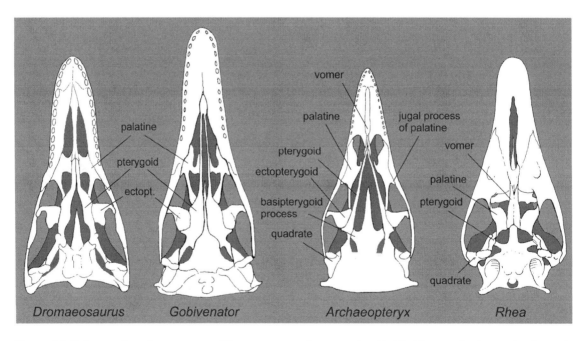

Figure 4.4 Palates of a dromaeosaur (*Dromaeosaurus*), a troodontid (*Gobivenator*), the early Jurassic *Archaeopteryx*, and an extant palaeognathous bird (*Rhea*, Rheiformes). Although the palatal morphology of *Rhea* is superficially similar to that of the Mesozoic taxa, there are distinct differences in detail, with the palatines of *Rhea* being in a more caudal position and the pterygoids being much shorter. Fossil taxa adapted from Tsuihiji et al. (2014). Not to scale.

lacked cranial kinesis owing to the diapsid temporal construction, and therefore was akinetic (Chiappe et al. 1999). Some of the long-snouted, schizorhinal enantiornithines, by contrast, may have had rhynchokinetic skulls (O'Connor and Chiappe 2011).

Teeth and their reduction

Birds primitively had fully toothed jaws with teeth in the maxillary, praemaxillary, and dentary bones. Unlike in many non-avian theropods, the crowns of avian teeth are not serrated. Avian teeth exhibit broadly similar morphologies, but some variation exists in the details of their number, size, and shape (Figure 4.5). The teeth of *Archaeopteryx* are short and peg-like, whereas *Jeholornis* and *Sapeornis* have procumbent teeth at the tip of the snout. A particularly high number of teeth is found in some early ornithuromorphs, such as *Jianchangornis* and the long-snouted *Yanornis* and *Hesperornis*. Dental specializations occur in Enantiornithes, with *Sulcavis* showing distinct grooves on the lingual sides of the teeth and *Shenqiornis* having unusually bulbous teeth (Wang et al. 2010; O'Connor et al. 2013b). The teeth of Mesozoic birds were regularly shed and cases of tooth replacement have been documented (e.g., Clarke 2004).

Tooth reduction occurred in many avian lineages and led to complete edentulism in Confuciusornithidae, the enantiornithine *Gobipteryx*, the basal ornithuromorphs *Archaeorhynchus*, *Zhongjianornis*, and *Schizooura*, as well as in Neornithes (Louchart and Viriot 2011). Because tooth loss in ornithuromorphs started from the praemaxilla (see the next paragraph), it is likely that the Late Cretaceous *Apsaravis* was also edentulous; of this taxon the upper beak is unknown, but the lower jaws lack teeth. In fact, a full dentition is only present in *Archaeopteryx* and some Enantiornithes (e.g., *Iberomesornis*), with all avian lineages that branch between these two taxa showing

a reduction of teeth. The full enantiornithine dentition is optimized as a secondary reversal in **parsimony-based phylogenetic analyses** (Turner et al. 2012), but a secondary gain of a full dentition is certainly a less plausible evolutionary scenario than a multiple independent loss.

That avian teeth were reduced numerous times convergently is not only indicated by the phylogenetic positions of the taxa involved, but also by the different patterns of tooth reduction (Figure 4.5; Louchart and Viriot 2011). Tooth loss started caudally in basal non-ornithuromorphs and resulted in the retention of teeth only at the tip of the snout, a pattern also observed in taxa allied with oviraptorosaurs (e.g., *Caudipteryx*, *Protarchaeopteryx*). In *Jeholornis*, teeth are only present in the lower jaws (except for the occurrence of a single tooth in the maxillary bone of one specimen; O'Connor et al. 2012b), whereas the dentition of *Sapeornis* is restricted to the rostral portions of the upper jaws.

In Ornithuromorpha, by contrast, the dentition was first lost in the praemaxillaries – that is, in the tips of the upper jaws – which is the case in, for example, hongshanornithids (*Hongshanornis*) and hesperornithiforms (*Hesperornis*, *Baptornis*; Louchart and Viriot 2011). Owing to the poorly resolved ornithuromorph interrelationships, it remains an open question how often the dentition was lost in the lineage leading to Neornithes. Tooth reduction in the praemaxilla already started in songlingornithids and hongshanornithids, but according to current phylogenetic reconstructions (e.g., O'Connor et al. 2013b), some fully edentulous ornithuromorph taxa, such as *Zhongjianornis* and *Archaeorhynchus*, lost their teeth independently of Neornithes.

It has been assumed that the reduction of avian teeth was due to weight reduction in these flying animals (e.g., Zhou et al. 2010), but dietary specializations are a more

Figure 4.5 Different patterns of tooth reduction in Mesozoic birds. (a) *Archaeopteryx* with a full dentition. (b) *Sapeornis*, in which teeth are restricted to the praemaxillae and the rostral portions of the maxillae. (c) *Jeholornis*, where teeth are only present at the tips of the lower jaws. (d) The enantiornithine *Bohaiornis*, which has teeth in the maxillary, praemaxillary, and dentary bones. In the enantiornithines (e) *Rapaxavis* and (f) *Longipteryx*, the dentition is restricted to the tip of the snout. In (g) *Hesperornis*, the praemaxillae lack teeth and an intersymphyseal bone is situated on the tips of the lower jaws. Not to scale.

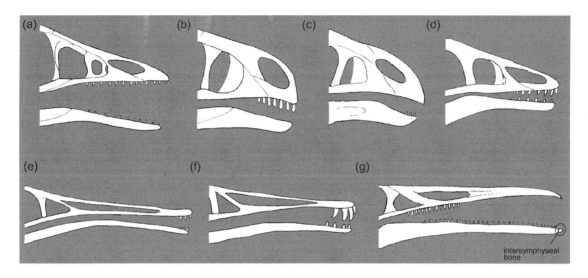

likely cause. At least in the neornithine lineage, tooth reduction appears to have been correlated with the formation of a horny rhamphotheca and a muscular **gizzard**, which took over the main role in food processing (Louchart and Viriot 2011). Several early non-ornithuromorphs with reduced dentitions have unusual peg-like teeth, which were interpreted as being indicative of a herbivorous diet (Zanno and Makovicky 2011). Crop and stomach contents do indeed show some of the early avians with reduced dentitions to have been seed eating, and a herbivorous diet may have played a role in the loss of teeth in these taxa (Zheng et al. 2011).

However, there were probably other reasons that also led to the loss of teeth in birds. In Enantiornithes, for example, a reduced dentition with teeth restricted to the very tip of the snout is found in the long-snouted Longipterygidae, for which a piscivorous diet is assumed (O'Connor and Chiappe 2011). Other early ornithuromorph lineages with a reduced dentition, such as Hesperornithiformes, were likewise piscivorous. Tooth reduction in the praemaxilla of ornithuromorphs seems to be correlated with the occurrence of an intersymphyseal bone on the tips of the lower jaws, which was reported for songlingornithids (*Yanornis*, *Yixianornis*), *Jianchangornis*, *Iteravis*, hongshanornithids (*Hongshanornis*), and hesperornithiforms (*Hesperornis*, *Baptornis*; Zhou and Martin 2011; Zhou et al. 2014b). According to current phylogenies, an intersymphyseal bone either evolved several times independently or represents a primitive feature of ornithuromorphs, which was secondarily lost in some taxa closer to the crown group. The functional significance of this ossicle remains elusive.

In any case, the loss of teeth was a significant step in avian evolution and led to the formation of a multitude of different beak shapes in Neornithes (Louchart and Viriot 2011). In addition to enabling the diversity of dietary specializations found in extant birds, tooth loss and the formation of a horny rhamphotheca were also of significance for the use of the avian beak as a tool, which is likely to have been an important factor in the evolutionary success of the crown group. A beak may have been one of the **preadaptations** for advanced nest constructions, and extant birds also spend much time on feather maintenance (preening), for which they draw feathers through the beak to remove ectoparasites and to restore the integrity of the vanes. Whether effective preening of pennaceous feathers is possible with fully toothed jaws has yet to be evaluated, and tooth loss may have constituted a selective advantage for preening.

Morphological transformations of the pectoral girdle

In non-avian theropods, scapula and coracoid are tightly joined or even fused into a scapulocoracoid, and individualization of these two bones was one of the major prerequisites for the evolution of advanced flight capabilities; that is, sustained flapping flight. *Archaeopteryx* and *Sapeornis* still exhibit a squarish coracoid, similar to that of non-avian theropods. Both taxa also lack an ossified sternum, and their broad coracoids as well as the robust furcula probably served as the main attachment sites of the flight musculature (Olson and Feduccia 1979; Chiappe 2007; Zheng et al. 2014a). Confuciusornithids are unique among early avians in that coracoid and scapula are fused and form a scapulocoracoid.

The arms of the furcula of non-avian theropods are very robust and form a wide angle. The furcula of early avians still has robust arms and is either boomerang shaped or exhibits a long apophysis (Figure 4.6; Close and Rayfield 2012). The scapulae are widely spaced in non-avian theropods and lie lateral of the ribcage, whereas they are oriented dorsally thereof in extant birds and other

Ornithothoraces. This dorsal shift elevated the glenoid fossae and allowed these birds to perform recovery strokes in flapping flight (Senter 2006). Elongation of the coracoid and formation of acrocoracoid and procoracoid processes are further prerequisites of sustained flapping flight capabilities. These processes contribute to guiding structures for the tendon of the ventrally situated supracoracoideus muscle, and therefore enable a more powerful wing elevation. Another evolutionary novelty of more advanced birds was the development of a ligament that prevents ventral dislocation of the humerus during flight strokes (Baier et al. 2006). All of these derived features may have evolved only once, at the base of Ornithuromorpha.

Evolution of the sternum

Ossified sternal plates are widely distributed among maniraptorans, and in some oviraptorosaurs (*Ajacingenia*) and dromaeosaurs (*Microraptor*) they are even fused and form a large sternal plate (Figure 4.6). Still, there appears to have been much homoplasy, and ossified sternal plates are absent in the Troodontidae, *Archaeopteryx*, and *Sapeornis* (Zheng et al. 2014a). A plausible hypothesis has yet to be proposed as to why the volant latter two taxa lack an ossified sternum, whereas this bone is present in some flightless paravians. In Jeholornithidae and Confuciusornithidae, the sternal plates are fused into a single bone, although in *Jeholornis* a suture is still visible (Zheng et al. 2012; Hu et al. 2014). *Jeholornis* lacks any evidence of a sternal keel. In *Confuciusornis*, by contrast, there is a low ridge in at least some specimens. A well-developed keel first occurs in Ornithothoraces, the clade including Enantiornithes and neornithine birds. In Enantiornithes it is, however, restricted to the caudal part of the sternum (Zheng et al. 2012; Hu et al. 2014) and the cranial portion of the pectoral musculature may instead have

Figure 4.6 Ossified sternal plates and sternum (upper two rows), as well as coracoid and furcula (lower row) of oviraptorosaurs (*Caudipteryx*, *Citipati*), dromaeosaurs (*Bambiraptor*, *Microraptor*), and various early avians. *Philomachus* (Charadriiformes, Scolopacidae) and *Bucco* (Piciformes, Bucconidae) exemplify two different sternum morphologies of extant birds. Fossil sterna after Zheng et al. (2012), furcula of *Sapeornis* after Gao et al. (2012). Not to scale.

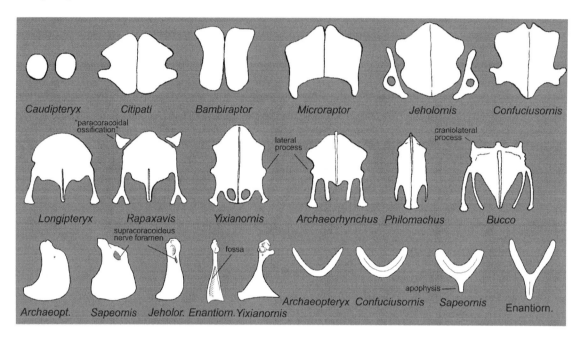

inserted on the very long furcular apophysis of these birds.

The sternum of basal avians has a characteristic shape and exhibits a pair of lateral projections ("ziphoid processes") on each side, which are also found in some oviraptorosaurs (Figure 4.6). In the course of avian evolution, the caudal margin of the bone developed an increasingly complex shape. Whereas it bears a pair of deep incisions in Enantiornithes, there are two pairs of incisions in more advanced Mesozoic ornithuromorphs, with the medial ones being closed to fenestrae in some taxa.

Ontogenetically, the sternum of Enantiornithes developed from four to six ossification centers, which is a similar number as in neornithine birds, where there are two to seven ossification centers (Zheng et al. 2012). However, ossification of the enantiornithine sternum proceeded from the caudal to the cranial end of the bone, whereas the opposite is the case in Neornithes (Zheng et al. 2012). "Paracoracoidal ossifications" identified in the enantiornithine sternum (Figure 4.6; O'Connor et al. 2011c) may represent homologues of the craniolateral processes of the sternum of more advanced birds.

Ribs and gastralia

Uncinate processes of the ribs occur in *Caudipteryx* and some oviraptorosaurs and dromaeosaurs, whereas they are absent in *Archaeopteryx*, *Jeholornis*, and *Sapeornis*. These processes were reported for *Confuciusornis* (Chiappe et al. 1999), but their occurrence in Enantiornithes seems to be variable. Within ornithuromorphs, uncinate processes generally seem to be present, although they are secondarily lost in a few crown group taxa (see Chapter 1).

Whereas there is some homoplasy in the distribution of uncinate processes, gastral ribs (gastralia) may have been lost only once, in Ornithuromorpha. They are present in Songlingornithidae, Hongshanornithidae, *Piscivoravis*, and *Archaeorhynchus*, but absent in *Zhongjianornis*, *Gansus*, and taxa more closely related to Neornithes. Their loss appears to coincide with the opening of the pubic symphysis of the pelvic girdle, but further studies are needed to show whether there was indeed a functional correlation.

The wing skeleton

The major transformations of the avian wing skeleton took place in its distal section – that is, the hand skeleton – and one of the most conspicuous evolutionary modifications concerns the reduction of the minor digit (Figure 4.7). A well-developed minor digit

with three phalanges and a large claw, which represents the plesiomorphic condition for theropods, is present in all long-tailed avians as well as in confuciusornithids among the pygostylians. In *Sapeornis*, by contrast, the minor digit is already reduced and consists of only two phalanges without a claw. The minor digit of some palaeognathous birds still exhibits two free phalanges, but in all other neornithine birds only a single free phalanx is left.

The loss of manual digits is a recurrent theme in theropod evolution and occurs in various maniraptorans that are currently placed outside Aves. In *Caudipteryx*, for example, the minor digit consists of only two phalanges, whereas *Similicaudipteryx* and other oviraptorosaurs have the full phalangeal count. In the enigmatic *Zhongornis* it has two phalanges and a claw, and a reduced minor digit furthermore occurs

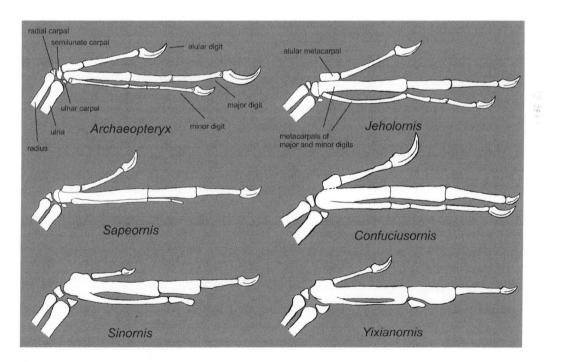

Figure 4.7 Semi-schematic reconstructions of the hand skeleton of early avians. Note the different degree of the reduction of the minor digit. Not to scale.

in the dromaeosaur-like *Balaur*, the affinities of which are not well understood (see Chapter 2).

The reasons for the reduction of the minor digit in birds remain poorly understood, and this is also true for the functional significance of this digit in early avians. Often, a reduced minor digit occurs in taxa with a shortened tail, and a correlation with advanced flight capabilities may therefore be assumed. However, not only is this digit also reduced in the long-tailed *Caudipteryx* and *Balaur*, but the short-tailed Confuciusornithidae retain a well-developed minor digit.

The formation of a carpometacarpus through fusion of the distal carpals and the metacarpals was another important evolutionary step towards the modern avian wing. In non-avian theropods, the carpal and metacarpal bones are not co-ossified, which is also true for *Archaeopteryx*. Fusion starts proximally in avian evolution, and the carpal trochlea is formed through co-ossification of the semilunate carpal with the proximal ends of the metacarpals of the major and minor digits. This stage is already present in confuciusornithids and *Sapeornis*, in which the alular metacarpal does however remain distinct. In Enantiornithes, the metacarpals are not fused distally. A fully developed carpometacarpus, with a co-ossified alular metacarpal and fused distal ends of the metacarpals of the major and minor digits, first occurs in some Early Cretaceous Ornithuromorpha, such as the Hongshanornithidae. A prominent extensor process of the alular metacarpal developed later and is first found in the Late Cretaceous *Apsaravis* and *Ichthyornis*.

In order to provide a larger attachment area for the primary feathers, the proximal phalanx of the major wing digit became widened in more advanced birds. This widening is absent in *Archaeopteryx*, but

already occurs in *Jeholornis*, confuciusornithids, and *Sapeornis*, and it is found in all Ornithothoraces.

Birds inherited large manual claws on all three wing digits from their theropod ancestors, and many extant birds retain vestigial claws on the tips of the alular and major digits. Why wing claws became greatly reduced is an open question, mainly due to their poorly understood functional significance in early avians. Some authors assumed that they assisted in tree climbing, but such habits are considered unlikely for early birds by the majority of current authors. Alternatively, wing claws may have had a preening function in toothed Mesozoic birds (Rietschel 1985), especially those with a long, feathered tail.

Pelvic girdle

In most tetrapods, the movement of the tail is coupled with that of the hindlimbs. Whereas the tail of long-tailed theropods therefore mainly served to maintain balance, the avian tail primarily fulfills aerodynamic functions (Pittman et al. 2013). Surprisingly, the reduction of the long bony tail of early avians seems to have had little influence on the morphology of the pelvic girdle itself. Overall, the pelvic bones of early pygostylians like *Sapeornis* and *Confuciusornis* are similar to those of the long-tailed *Archaeopteryx* and *Jeholornis* (Figure 4.8), which themselves resemble the pelvic bones of dromaeosaurs and troodontids.

A major consequence of tail reduction in avian evolution was a cranial shift of the center of gravity, which resulted in a more upright posture of the trunk. Possibly in correlation therewith, the pelvis developed a marked antitrochanter. This bony projection of the ilium is situated caudodorsal of the acetabular foramen, the socket of the femur, and forms an abutment for the femoral head (see Figure 1.10). Although an incipient antitrochanter is present in non-avian theropods and early avians, such

Figure 4.8 Tail morphologies of early avians and close avian relatives. Adapted from O'Connor & Sullivan (2014).

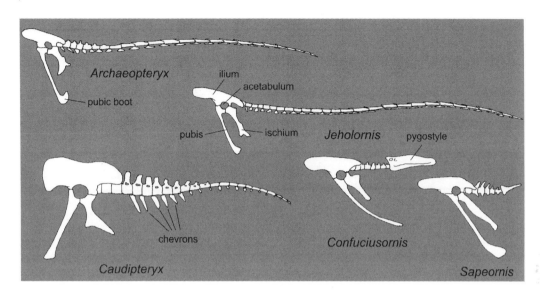

as *Archaeopteryx* and *Sapeornis* (Hutchinson and Allen 2009), it is much less marked than in extant birds. When exactly a pronounced antitrochanter evolved remains elusive. It is absent in *Archaeopteryx* (Hertel and Campbell 2007), but was reported for Enantiornithes (Chiappe and Walker 2002).

In the course of avian evolution, an increasing number of sacral vertebrae became integrated into the synsacrum. Whereas *Archaeopteryx* has only five synsacral vertebrae, there are six in *Jeholornis*, seven or eight in *Confuciusornis*, *Sapeornis*, and enantiornithines, nine in basal ornithuromorphs, and ten or more in neornithine birds.

Another major transformation of the avian pelvic girdle concerns the loss of a pubic symphysis. In *Archaeopteryx* and non-avian theropods close to the ancestry of birds, the tips of the pubes exhibit an expanded "pubic boot" and are co-ossified, forming a pubic symphysis (Figure 4.9; Hutchinson 2001a). The pubic boot is still present in various early avians, including *Sapeornis*, some Enantiornithes (e.g.,

Sinornis), and some early ornithuromorphs, such as songlingornithids and hongshanornithids. Opening of the pubic symphysis first occurred within Ornithuromorpha. Although in some Early Cretaceous ornithuromorphs a well-developed symphysis is present (e.g., Songlingornithidae), only the tips of the pubes meet in others (Hongshanornithidae, *Gansus*). A full opening of the pubic symphysis characterizes taxa of the Ornithurae and is likely to have been correlated with an increased egg size (see later discussion). According to current phylogenies, however, this opening appears to have taken place more than once within that clade, because a pubic symphysis is absent in the Late Cretaceous *Patagopteryx*, which occupies a basal position within ornithuromorphs in most analyses (e.g., Liu et al. 2014; M. Wang et al. 2015b).

Hindlimbs

The paravian femur underwent comparatively few morphological changes on the evolutionary line to birds (Hutchinson

Figure 4.9 The pelvis of (a) the early pygostylian *Confuciusornis* in comparison to that of (b) a neornithine bird (*Clamator*, Cuculiformes). In *Confuciusornis* and other non-ornithurine Mesozoic birds, the tips of the pubic bones are fused and form a pubic symphysis. Not to scale.

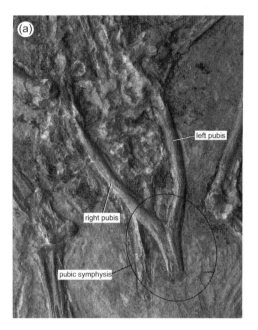

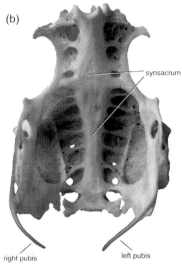

2001b). As in *Archaeopteryx*, the fibula primitively reaches to the distal end of the tibia, and its shortening is likely to be correlated with a reduction of the distal leg muscles in the course of avian evolution.

Non-avian theropods possess a fifth metatarsal, which is retained in *Archaeopteryx*, Jeholornithidae, Confuciusornithidae, and *Sapeornis*. When exactly a fifth metatarsal is lost is difficult to determine because of uncertainties in the identification of this bone in some taxa and the poorly resolved affinities of others in which a fifth metatarsal is present. Most Ornithothoraces lack a fifth metatarsal, but it was reported to be present in the enantiornithines *Eopengornis* and *Parapengornis* (Hu et al. 2015) and in the putative ornithuromorph *Vorona*.

The evolution of the hindlimbs paralleled that of the forelimbs in the formation of compound bones, and this trait may again have been functionally correlated with a reduction of the distal hindlimb muscles in the lineage leading to modern birds. In more advanced birds, the distal end of the tibia fuses with the proximal tarsal bones, the astragalus and the calcaneus, which are still separate in *Archaeopteryx* and other early avians (Figure 4.10). Possibly owing to correlated developmental processes, fusion of the three major metatarsal bones broadly coincides with that of the metacarpals in the wing. Extensive fusion of the metatarsals occurred within Early Cretaceous Ornithuromorpha, and the metatarsals of songlingornithids and hongshanornithids are already completely co-ossified. Formation of a fully formed tarsometatarsus is, however, difficult to trace, because some taxa with only partially fused metatarsals (e.g., *Archaeorhynchus*) are only known from subadult individuals, and incomplete fusion may therefore be an ontogenetic rather than a phylogenetic trait.

Figure 4.10 Distal end of the tibiotarsus of (a) *Archaeopteryx* (Thermopolis specimen) and (b) a juvenile palaeognathous bird (*Rhea*, Rheiformes). The astragalus is phylogenetically (*Archaeopteryx*) or ontogenetically (*Rhea*) not yet fused with the tibia and exhibits a long ascending process.

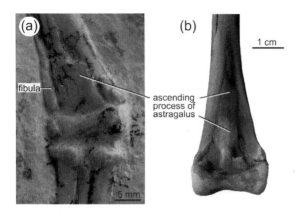

In Neornithes, the ontogenetic fusion of the metatarsals starts distally, whereas their phylogenetic co-ossification begins proximally in non-ornithuromorph avians. Some authors considered these different fusion modes of the metatarsals to be indicative of a basal split of Aves into a clade including *Archaeopteryx*, confuciusornithids, and enantiornithines on the one hand, and one including ornithuromorphs on the other. It is more likely, however, that the mode of ontogenetic fusion of these bones in neornithine birds does not reflect the fusion pattern in the phylogenetic history of birds.

One of the key innovations in the foot of birds is the reversal of the hind toe, which is prerequisite for effective grasping capabilities, and is accompanied by a twisting of the shaft of the first metatarsal. In *Archaeopteryx* and *Jeholornis*, the first toe is still medially positioned and not fully reversed. Its full reversal occurred in Pygostylia and there may have been a functional correlation with the reduction of the tail, perhaps in consequence of a different posture or increased arboreality

of early pygostylians in comparison to more basal long-tailed birds.

Evolution of the avian tail as an aerodynamic device

Archaeopteryx has a long bony tail with two rows of serially arranged feathers of roughly equal length. These formed a cohesive airfoil, which is likely to have both generated lift and acted as a steering device. Having long been the only Mesozoic avian with a well-preserved feathering, *Archaeopteryx* served as a template for early hypotheses on the evolution of the avian tail. Fossil finds of the past few years, however, revealed a surprising diversity of the tail shapes of long-tailed Mesozoic birds and early paravians that potentially opens a new perspective on the evolution of the avian tail.

The tail feathers of *Anchiornis* are similar to those of *Archaeopteryx*, but in the dromaeosaurs *Microraptor* and *Changyuraptor* long pennaceous feathers are restricted to the tip of the tail (Li et al. 2012; Han et al. 2014). The bony tail of dromaeosaurs is stiffened against dorsoventral flexion by extremely elongated chevrons and praezygapophyseal processes. A strikingly similar tail morphology occurs in rhamphorhynchid pterosaurs, in which the tail vertebra likewise form long, rod-like processes, and the tip of the tail bears a rhombic skin flap. These resemblances were interpreted as being indicative of aerodynamic functions of the tails of small, volant dromaeosaurs, which might have been used as flight stabilizers or aerial rudders (Persons and Currie 2012). In *Caudipteryx* the feathers are also restricted to the tip of the tail, and a particularly unusual morphology occurs in *Jeholornis*, in which the tail tip bears frond-like feathers and a second feather bundle is situated at the tail base (Plate 3a; O'Connor et al. 2012b, 2013a). With all of these new discoveries, it now appears possible that a tail with long pennaceous feathers restricted to its tip is plesiomorphic for Pygostylia, rather than the serially arranged tail feathers of *Archaeopteryx* (Chiappe 2007).

The transition from a long tail to a short one with a pygostyle is poorly understood, owing to the lack of phylogenetically well-constrained fossils with intermediate tail lengths. The only taxon to come into question for such an intermediate stage is *Zhongornis*, which has a short tail but lacks a pygostyle. Disregarding the uncertain phylogenetic affinities of *Zhongornis*, the only known skeleton of this taxon is from a juvenile individual. Therefore, its tail morphology is equally likely to represent an intermediate phylogenetic or an early ontogenetic stage, so much the more since a similar tail morphology occurs in juvenile Enantiornithes (e.g., Chiappe et al. 2007a). As already detailed, and because of the uncertain positions of *Sapeornis* and *Jeholornis*, it is also uncertain whether tail reduction and formation of a pygostyle formed a singular event in the evolutionary history of birds, or whether these occurred more than once.

Confuciusornithids and non-pengornithid Enantiornithes feature a rod-shaped and very long pygostyle, which is much more elongated than the ploughshare-shaped pygostyle of the Ornithuromorpha. These disparate pygostyle shapes seem to go along with differences in the morphology of the tail feathers, which in Confuciusornithidae and most Enantiornithes form a pair of very long, rachis-dominated streamers (Figure 4.11; two pairs of tail streamers occur in the enantiornithine *Paraprotopteryx*: Zheng et al. 2007). In the pengornithids *Eopengornis* and *Parapengornis* the entire shaft of the tail feathers is vaned, which coincides with the absence of an elongated pygostyle in these taxa (X. Wang et al. 2014; Hu et al. 2015). It has been hypothesized that the presence of two greatly elongated tail feathers represents the primitive condition for Aves (Zhang and Zhou 2004; Clarke et al. 2006). That

Figure 4.11 Different tail feather morphologies of Mesozoic birds. (a) Fan-shaped tail of ornithuromorphs (e.g., Hongshanornithidae and most extant birds). (b) Vaned tail streamers of the enantiornithine Pengornithidae. (c) Rachis-dominated tail streamers of most other Enantiornithes. Adapted from X. Wang et al. (2014).

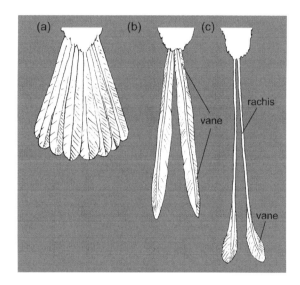

the evolution of avian tail feathering was more complicated, however, is suggested by the fact that *Sapeornis*, which is phylogenetically more basal than enantiornithines, has multiple, moderately long tail feathers (Zheng et al. 2013a; X. Wang et al. 2014).

A display function was proposed to explain the origin of these elongated tail feathers (O'Connor et al. 2012b), and in the case of confuciusornithids they were interpreted as ornamental feathers of adult males (e.g., Feduccia 1999). However, in *Confuciusornis* the presence of tail streamers does not correlate with the observed size differences between the fossils, which are often attributed to sexual dimorphism (Chiappe et al. 2008; Peters and Peters 2010). Moreover, the long, rachis-dominated tail streamers exhibit very similar morphologies in only distantly related taxa, whereas a multitude of different structures would be expected if these feathers evolved for display reasons. Their association with a greatly elongated pygostyle suggests a strong attachment, which also argues for an aerodynamic or bracing function. If pygostyle shape and feather morphology are correlated, the presence of rod-shaped pygostyles in all individuals of confuciusornithids and non-pengornithid enantiornithines likewise conflicts with a sexually dimorphic distribution of long tail streamers (among extant birds, the pygostyle of male peacocks, in which the tail coverts are greatly elongated, is much larger than that of the females, which lack a tail fan).

On the other hand, the occurrence of elongated tail streamers in both confuciusornithids and enantiornithines is not easily explained in an aerodynamic context, because these two taxa probably had very different flight characteristics – whereas confuciusornithids more likely were gliders or soarers, enantiornithines were capable of flapping flight. For enantiornithines a bracing function of the tail was assumed (Hu et al. 2015), even though functional analyses still have to show whether two narrow and greatly elongated feathers do indeed serve this purpose.

In contrast to the elongated, rod-shaped pygostyle of confuciusornithids and most enantiornithines, the pygostyle of ornithuromorphs is ploughshare shaped. In extant birds there are adipose structures on each side of the bone, which are termed rectricial bulbs and, together with the surrounding musculature, play an important role in the control of tail movements. The origin of the ploughshare-shaped pygostyle of ornithuromorphs was considered to be correlated with the evolution of these rectricial bulbs and the formation of a fan-shaped tail, which allows increased flight control in aerial maneuvers, especially during take-off and landing (Clarke et al. 2006).

Ontogenetic Development of Mesozoic Birds

Birds and mammals are the only extant endothermic tetrapods; that is, they are capable of maintaining a constant body temperature independent of that of the surrounding environment. Endothermic animals usually exhibit a higher metabolism and faster growth rates than ectothermic ones, and there are distinctive differences in the growth modes of non-avian dinosaurs, which grew continuously, and extant birds, in which growth terminates after sexual maturity is reached. That non-avian theropods nevertheless had some ability to regulate their body temperature is indicated by various matters of fact, such as the Late Cretaceous occurrence of troodontids in cold Polar regions (e.g., Godefroit et al. 2009). Because early birds inherited their physiological characteristics from theropod ancestors, they were probably likewise able to regulate their body temperature to some degree.

It is, however, not straightforward to assess the growth rates of Mesozoic birds. Concerning *Archaeopteryx*, for example, a critical point is whether all of the eleven known skeletons, which are of very different sizes, belong to a single species. If so, they would indicate slow growth rates, since such different growth stages would be unlikely to be found in a small specimen sample if growth rates were fast (Wellnhofer 2009). As detailed in Chapter 2, the various *Archaeopteryx* specimens most likely represent more than one species, and their different sizes therefore do not provide immediate insights into the growth rates of the "Urvogel". Still, *Archaeopteryx* differs from non-avian dinosaurs in bone histology, and the parallel-fibered rather than woven-fibered bone matrix suggests that its growth rates were indeed unusually slow for an endothermic vertebrate (Erickson et al. 2009).

Definite ontogenetic growth series are known from other basal Mesozoic avians, and especially in *Sapeornis* from the Chinese Jehol Biota large numbers of differently sized individuals have been found, whose analysis likewise indicated slow growth rates (Pu et al. 2013). Slow adult growth rates of volant birds are furthermore to be expected, because they ensure functionality of the wings; that is, the maintenance of a coherent airfoil over sufficiently long time periods.

The long limb bones of many Mesozoic birds exhibit "growth rings" – that is, multiple lines of arrested growth (LAGs) – which are indicative of cyclic periods of growth (e.g., Chinsamy et al. 1994). These LAGs were reported for *Archaeopteryx*, *Sapeornis*, *Jeholornis*, *Confuciusornis*, and Enantiornithes, but they are absent in more derived ornithuromorphs, such as *Iteravis*, *Hesperornis*, and *Ichthyornis* (Wilson and Chin 2014; O'Connor et al. 2015b). In neornithine birds growth rings are very rare, but they do occur in some palaeognathous birds (kiwis and moas), as well as in the Eocene *Gastornis* and a few extant neornithine taxa, which take several years to achieve adult body size (Turvey et al. 2005; Bourdon et al. 2009a). Caution is, however, warranted in the interpretation of LAGs, as these may constitute a response to both a marked seasonality and unusual environmental conditions, and have also been reported from some fast-growing taxa, such as parrots (Bourdon et al. 2009a).

The evidence for the reproductive behavior of non-avian maniraptorans was reviewed by Xu et al. (2014). These animals had two functional ovaries and sequentially laid paired eggs, which were proportionally smaller and more elongate than those of extant birds. Non-avian maniraptorans were nesting on the ground, and their eggs were incubated within substrate and probably lacked a chalaza, the albumen chord that keeps the yolk of extant bird eggs in position when the egg is turned. Rare fossil skeletons preserved on clutches may document parental care in some non-avian maniraptorans.

The earliest avians must have inherited these reproductive traits and strategies from their theropod ancestors. However, it has been hypothesized that Early Cretaceous birds only retained a single functional ovary, like extant birds (Zheng et al. 2013b). This assumption was based on the purported presence of fossilized ovarian **follicles** in *Jeholornis* and some enantiornithines from the Jehol Biota, but for various reasons it is likely that these structures were misidentified and instead represent stomach contents (Mayr and Manegold 2013). On the one hand, their identification as mature ovarian follicles conflicts with the fact that such easily perishable cell structures are unlikely to be preserved in multiple fossils with little or no other soft tissue preservation. Moreover, these fossil "follicles" have similar dimensions in very differently sized animals, and a simultaneous maturing of many follicles in early birds would conflict with fossil evidence that oviraptorosaurs laid only one pair of eggs at the same time (hence, even non-avian maniraptorans already had the ovulation mode of extant birds – that is, the consecutive maturing of follicles – although they retained two functional ovaries).

Extant birds exhibit two basic modes of hatchling development, with various intermediate forms between the extremes. In precocial birds, the young exhibit advanced development at the time of hatching and leave the nest soon thereafter. In altricial birds, by contrast, the young hatch naked and helpless, and need to be raised by the adult birds until they are able to leave the nest. The fossil record of Mesozoic hatchlings is sparse, but embryonic skeletons exist for *Gobipteryx* and other enantiornithines, and show that these birds were highly precocial ("super-precocial"), with even unhatched birds in the eggs already featuring fully feathered wings (Zhou and Zhang 2004). Evidence from bone histology also indicates that Enantiornithes underwent very rapid embryonic development (Chinsamy and Elzanowski 2001), and the offspring of these birds was probably capable of flight soon after hatching. In extant birds, similar advanced embryonic development is only known from the galliform Megapodiidae, the eggs of which are incubated by the use of external heat sources, so that the young have to dig their way through substrate after hatching.

The fossil record indicates that the nesting habits and breeding strategies of the earliest avians were substantially different from those of most modern birds. For instance, the eggs of non-ornithuromorph birds are narrower, more elongated, and more cylindrical than those of extant birds (Deeming and Ruta 2014). This is due to the fact that the narrow passage formed by the closed pubic symphysis (see Figure 4.9) limited egg size and shape, and the larger, yolk-rich eggs of ornithurine birds could only evolve after opening of the pubic symphysis occurred (Dyke and Kaiser 2010). Several enantiornithine eggs are known, and their elongated, cylindrical shape suggests that they were planted upright within sediment which is also evident from the in situ arrangement of putative enantiornithine clutches (Varricchio and Barta 2015). It has furthermore been hypothesized that the eggs of non-ornithurine birds still lacked a chalaza and were therefore not turned in the nest (Varricchio and Barta 2015).

Hence, if a single most important feature for the evolutionary success of ornithurine birds is to be named, it is probably the opening of the pubic symphysis. An open symphysis characterizes all members of the Ornithurae and appears to have been accompanied by increased egg size, larger amounts of yolk, and subsequent differences in breeding strategies and the ontogenetic development of the hatchlings. The advanced reproduction behavior of extant birds – that is, nesting free of sediment, the construction of complex nests in elevated places, the regular turning of eggs by the incubating adult birds, and the long incubation and nesting periods – probably evolved late in the ornithuromorph lineage. These derived traits seem to be a characteristic of the Ornithurae and may have significantly reduced the predation on eggs and nestlings (Dial 2003b), which in turn may have been key for the successful radiation of crown group birds.

5 | The Interrelationships and Origin of Crown Group Birds (Neornithes)

In the past few years, important progress has been made in unraveling the phylogenetic interrelationships of neornithine birds. Modern phylogenetic analyses have not only corroborated many traditional hypotheses on avian interrelationships, but also provided evidence for unexpected groupings that were not assumed by earlier scientists. In the present chapter, current hypotheses on the interrelationships of the major neornithine clades are outlined, and the fossil evidence for the Mesozoic diversification of neornithine birds is summarized.

As will be discussed in later sections, advanced stem group representatives of penguins (Sphenisciformes) and owls (Strigiformes) already occurred in the early Paleocene. These two taxa are phylogenetically widely separated and deeply nested within Neornithes, which suggests that crown group birds as a whole diversified much earlier, in the Cretaceous. This is also indicated by divergence dates derived from virtually all calibrated molecular phylogenies. What remains controversial, however, is the exact timing of the origin of crown group birds and the extent of the neornithine diversification before the K/Pg boundary.

A precise dating of the early neornithine divergences is not just of paleornithological interest, but is also critical for an understanding of Late Cretaceous ecosystems and an assessment of the impact of the Late Cretaceous mass extinction events. Unfortunately, and in contrast to the much better understood avian diversity in the Early Cretaceous, our knowledge of Late Cretaceous avifaunas is still fairly incomplete. Not only are the affinities of most of the fragmentary neornithine-like fossils

Avian Evolution: The Fossil Record of Birds and its Paleobiological Significance,
First Edition. Gerald Mayr.
© 2017 John Wiley & Sons, Ltd. Published 2017 by John Wiley & Sons, Ltd.

contentious, but the transition between neornithine and non-neornithine birds must furthermore have been a gradual one. It will therefore hardly be possible to determine exactly the temporal origin of the crown group based on the fossil record, because the farther one goes back in time, the more difficult it becomes to assign avian fossils without characteristic morphologies to crown group Neornithes. Nevertheless, the fossil record is sufficient for a rough estimation of the timing of the initial neornithine divergences, and these dates are broadly congruent with those resulting from some of the more recent molecular calibrations.

Phylogenetic Interrelationships of Neornithine Birds

Phylogenies: When can they be considered well supported?

Nowadays most avian phylogenies are based on molecular data, and much effort is put into the analysis of large data sets. Analyses of ever-increasing numbers of nuclear gene sequences did indeed converge on a well-resolved and strongly supported phylogenetic framework (Figure 5.1; Ericson et al. 2006; Hackett et al. 2008; Pacheco et al. 2011; Yuri et al. 2013; Jarvis et al. 2014; Prum et al. 2015). How can we assess whether these sequence-based phylogenies do indeed reflect the true interrelationships of birds?

Analyses of large data sets produce trees that are robust against various statistical tests applied to the data, but adding more data to an analysis by itself does not have to result in more correct phylogenies. A stronger case for the accuracy of a phylogenetic hypothesis can be made if different kinds of data convey the same phylogenetic signal.

Of course, not all of the clades that were obtained in sequence-based analyses are novel. Many of these were proposed before in studies of anatomical characters, in which case molecular data provide independent evidence of earlier morphology-based hypotheses. The number of avian higher-level clades for which strong support exists from both morphological and molecular data is notable, but there are also many clades for which this is not the case.

However, congruent support for some clades also comes from different kinds of molecular data. Individual nucleotide substitutions occur randomly, and the genomes of the cell nucleus and the mitochondria, for example, are likely to evolve independently of each other. Clades, which are obtained in analyses of both nuclear and mitochondrial sequences, may therefore be considered to be more strongly supported than those that resulted from only one kind of gene sequence data. A phylogenetic hypothesis based on molecular data can also be corroborated if congruent trees are found in analyses of different individual genes, especially if these are located on different chromosomes.

In addition, there are increasing efforts to identify **transposable elements** (sometimes also ineptly called "jumping genes"), which are randomly inserted into the DNA strands and exhibit a very low degree of homoplasy (e.g., Suh et al. 2011). Unfortunately, relatively few avian higher-level clades are characterized by these elements and some conflict in transposon distribution has been identified (Han et al. 2011; Suh et al. 2015).

However, sequence-based analyses have congruently yielded a number of clades that were not recognized by morphologists before and are likewise not obtained in current analyses of large morphological data sets. In

Figure 5.1 Phylogenetic interrelationships of neornithine (crown group) birds as obtained in analyses of nuclear gene sequences. (a) Phylogenetic tree resulting from an analysis of 19 gene loci (Hackett et al. 2008). (b) Tree obtained from an analysis of complete nuclear genomes (Jarvis et al. 2014). The asterisks in (a) indicate nodes that are also retained in (b). The exclamation marks in (b) denote taxa with a very different position in the two phylogenies. Taxa of the "metavian" clade in (a) are highlighted in bold in (b).

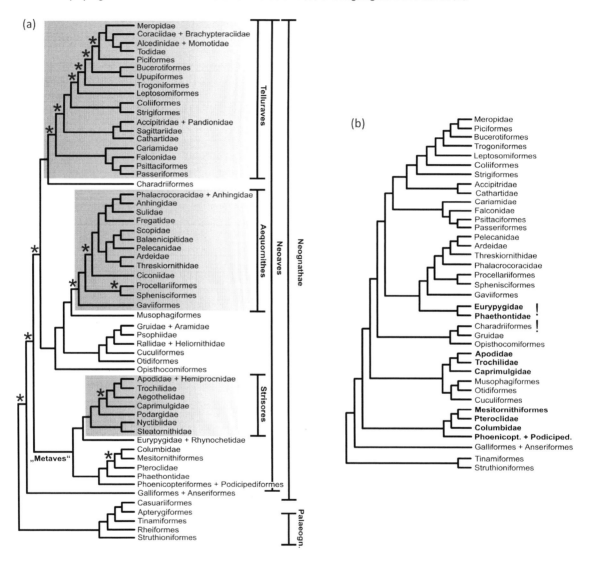

retrospect, some of these novel clades can be corroborated by previously unappreciated morphological apomorphies. Even though such approaches may easily be exposed to allegations of arbitrary "character picking," they add to the corroboration of certain clades, and are further justified because it is only possible to put fossil taxa into a phylogenetic context if clades can be characterized with morphological apomorphies.

In general, large morphological data sets may be more prone to yielding incorrect phylogenetic signals than those based on molecular sequence data, because characters in morphology-based analyses are usually attributed the same weight. Unlike the four nucleotides of DNA sequences, however, morphological characters exhibit a high degree of varying complexity. In analyses of large morphological data sets, a few characters of phylogenetic significance are therefore more likely to be overruled by a greater number of characters of minor importance. In addition, functionally interdependent character complexes can bias the results of such analyses. Analyses of smaller sets of well-defined characters may therefore yield more accurate phylogenies than those in which as many characters as possible are included (Mayr 2008a, 2011a). The use of combined molecular and morphological data sets (e.g., Clarke et al. 2009) may overcome some of these problems, although in such analyses the more comprehensive data are likely just to dominate the others.

The interrelationships of neornithine birds

All current analyses of morphological and molecular data support the long-assumed basal split of neornithine birds in the sister taxa Palaeognathae (tinamous and the flightless "ratites") and Neognathae, with the latter including the great majority of living birds (e.g., Livezey and Zusi 2007; Hackett et al. 2008; Prum et al. 2015). Also well supported is a division of Neognathae in the sister taxa Galloanseres (land- and waterfowl) and Neoaves, the taxon including all other neognathous birds (Figure 5.1). In addition to the previously mentioned derived features of the palate (Chapter 1), neognathous birds are characterized by a derived pelvis morphology in which ilium and ischium are caudally connected by a bony bridge, which encloses an ilioischiadic foramen (Figure 1.10). The tarsometatarsus of most neognathous birds furthermore exhibits a complex hypotarsus,

with furrows and canals for the tendons of the flexor muscles of the toes (Figure 1.12).

The great majority of extant birds belong to Neoaves. The members of this clade are characterized by the lack of a phallus, which was, however, also independently reduced in some of the palaeognathous Tinamiformes and in some Galliformes (Brennan et al. 2008). It is difficult to correlate the evolutionary success of neoavians with a particular attribute, but compared to palaeognathous and galloanserine birds they show a high diversity of beak shapes, which may have been promoted by derived properties of the palate bones and allowed these birds to occupy a multitude of feeding niches.

Neoavian interrelationships have become much better understood in the past few decades, and several of the traditional higher-level taxa have been shown to be **paraphyletic** or **polyphyletic** in analyses of molecular sequence data. In particular, two major neoavian clades, Aequornithes and Telluraves, informally termed the "waterbird" and "arboreal landbird" clades, emerged from various analyses of different data. Aequornithes, the "waterbird" clade (Mayr 2011a), includes most aquatic or semi-aquatic extant avian groups; that is, loons (Gaviiformes), tubenoses and allies (Procellariiformes), penguins (Sphenisciformes), the polyphyletic "Pelecaniformes" except tropicbirds (Phaethontiformes), as well as the taxa of the likewise polyphyletic "Ciconiiformes" (storks, herons, ibises, and allies). Aequornithes is congruently obtained in analyses of nuclear and mitochondrial gene sequences (e.g., Ericson et al. 2006; Hackett et al. 2008; Pacheco et al. 2011; Yuri et al. 2013; Jarvis et al. 2014; Prum et al. 2015), and a comparable clade also resulted from analyses of morphological data (Livezey and Zusi 2007).

Telluraves, the "arboreal landbird" clade (Yuri et al. 2013), encompasses most small, arboreal birds, such as mousebirds (Coliiformes), passerines (Passeriformes), trogons (Trogoniformes), as well as woodpeckers, rollers, and allies (Piciformes, Upupiformes, Alcediniformes, and Coraciiformes). These birds were already recognized as closely related by 19th-century morphologists. However, Telluraves also includes seriemas (Cariamiformes), parrots (Psittaciformes), and the non-monophyletic diurnal birds of prey ("Falconiformes"), which were not considered closely related to the "arboreal landbirds" by earlier authors. Regarding the interrelationships of the taxa included in the clade, sequence-based phylogenetic hypotheses furthermore strongly depart from what has been traditionally assumed, with one of the most surprising results being the recognition of a clade including passerines, parrots, the falconiform falcons (Falconidae), and seriemas (Hackett et al. 2008; Suh et al. 2011; Jarvis et al. 2014; Prum et al. 2015).

For some groups that proved difficult to place based on morphological data, strong molecular support now backs previously undetected phylogenetic placements. This is particularly true for flamingos (Phoenicopteriformes) and the morphologically very different grebes (Podicipediformes), which were shown to be sister taxa in virtually all recent analyses of molecular data (Ericson et al. 2006; Hackett et al. 2008; Pacheco et al. 2011; Jarvis et al. 2014; Prum et al. 2015). Other taxa, however, still result in highly unstable positions in molecular analyses, and the phylogenetic affinities of many neoavians that cannot be assigned to either Aequornithes or Telluraves remain controversial.

Earlier analyses of smaller sets of nuclear gene sequences suggested a basally diverging neoavian clade termed "Metaves," which encompassed morphologically disparate groups such as flamingos and swifts (Apodiformes; Fain and Houde 2004; Ericson et al. 2006; Hackett et al. 2008). This clade is not supported by more recent genome-scale analyses, which found the "metavian" taxa to be sequentially branching at the base of Neoaves, albeit in varying positions (Jarvis et al. 2014; Prum et al. 2015; Figure 5.1). The many open phylogenetic questions notwithstanding, however, a fairly robust framework now exists for the interrelationships of most of the major neornithine groups, which forms the basis of this book (Figure 5.2).

The Mesozoic Fossil Record of Neornithine-Like and Neornithine Birds

The fossil record of undisputed neornithine birds is essentially a Cenozoic one, and the few Mesozoic fossils that can be definitely assigned to crown group taxa stem from the latest Cretaceous deposits (Figure 5.3). There are, however, various older specimens with a neornithine-like morphology, and these indicate that the evolutionary history of Neornithes goes farther back in time.

In his classic monograph on *Ichthyornis and Hesperornis*, Marsh (1880) also described *Apatornis*, another avian taxon from the Late Cretaceous (late Santonian/early Campanian; ~80–85 mya) of Kansas. He referred to *Apatornis* as a synsacrum – the holotype – as well as various other skeletal elements of a different individual. Because the latter bones show no overlap with the holotype, they were subsequently assigned to a new taxon, *Iaceornis* (Clarke 2004). Whether this approach was justified remains to be seen, but in any case, the *Iaceornis* remains exhibit a much more "modern" morphology than the corresponding bones of *Ichthyornis* and constitute the earliest record of a neornithine-like Mesozoic bird. In particular, these bones resemble those of the early Cenozoic palaeognathous Lithornithiformes, with one of the few major differences being the presence of a supratendinal bridge on the

Figure 5.2 The consensus phylogeny of neornithine (crown group) birds, which forms the taxonomic framework of this book. Major clade names are indicated.

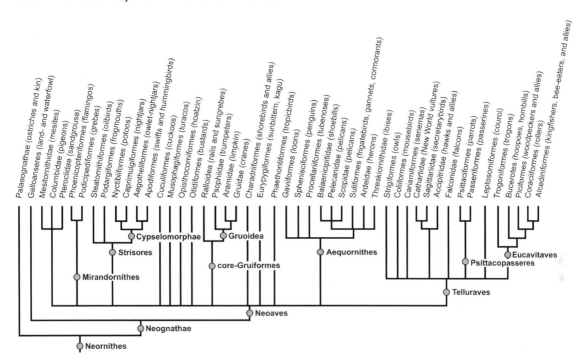

distal tibiotarsus of *Iaceornis* and its absence in lithornithiforms. A phylogenetic analysis supported a position of *Iaceornis* outside a clade including Lithornithiformes and crown group Neornithes, but this placement hinges on a single character, a lower count of synsacral vertebrae, which furthermore had to be estimated for *Iaceornis* (Clarke 2004). The similarities between *Iaceornis* and lithornithiforms may well be plesiomorphic, but even if they are, they may indicate that palaeognathous and neognathous birds diverged well back in the Late Cretaceous.

A few remains with neornithine-like morphologies were also found in the Late Cretaceous (Campanian–Maastrichtian) Allen Formation of northeastern Patagonia, Argentina. In addition to fragmentary wing bones and a carpometacarpus of indeterminate affinities (Agnolín and Novas 2012), these fossils include a more diagnostic

coracoid, which was described as *Lamarqueavis* (Agnolín 2010). This latter fossil was originally assigned to the extinct taxon Cimolopterygidae, but actually it is quite different from the coracoid of this Late Cretaceous North American taxon. Instead, the coracoid of *Lamarqueavis* more closely resembles that of some gruiform birds, such as trumpeters (Psophiidae) and the early Cenozoic Messelornithidae, although the fossil is too fragmentary for a well-founded classification.

The great majority of Mesozoic avian fossils for which neornithine affinities were considered stem from the latest Cretaceous fossil sites, and many are from the Maastrichtian of the North American Lance, Hell Creek, and Freeman formations (Hope 2002; Mayr 2009a; Longrich et al. 2011). Nevertheless, most of these fossils are very fragmentary and defy a reliable phylogenetic assignment. The proposed

Figure 5.3 The earliest temporal occurrences of neornithine birds (see text and Mayr 2014a for further details). The gray bars indicate temporal ranges; open asterisks demarcate the earliest occurrences of modern-type representatives, filled ones those of crown group representatives. The shaded area highlights the stratigraphic range of *Iaceornis* and *Apatornis*, the earliest birds with neornithine-like morphologies.

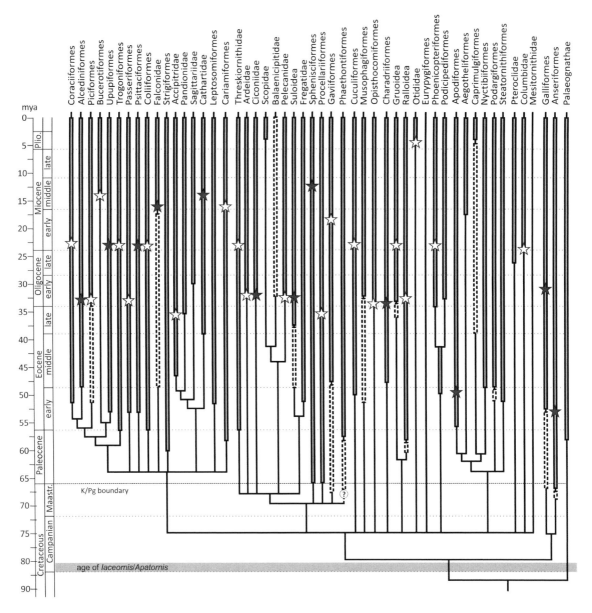

identifications cover a wide range of taxa, including Galliformes (landfowl), Anseriformes (waterfowl), Gaviiformes (loons), and Procellariiformes (tubenoses and allies), but it is notable that none of these fossils belongs to arboreal groups (a tip of a lower jaw from the Lance Formation was considered to be from a parrot, but this identification cannot be upheld: Dyke and Mayr 1999; Mayr 2009a).

Some fossils from the Lance Formation show galloanserine-like morphologies. This is particularly true for a small quadrate, which was previously assigned to the Cimolopterygidae (Elzanowski and Stidham 2011), the known bones of which are themselves similar to those of the anseriform Presbyornithidae (see Chapter 7). It has also been suggested that the taxon *Palintropus*, which is mainly known from fragmentary coracoids from the Lance Formation, may be a galliform bird (Hope 2002). The morphology of these bones does not, however, support the presumed galliform affinities of *Palintropus* (Mayr 2009a), and even its assignment to Neornithes was questioned by Longrich (2009), who identified specimens of the taxon from the Campanian of Canada. Galliform affinities of an even earlier coracoid from the Turonian–Coniacian of Argentina (Agnolín et al. 2006) are likewise far from being well established, owing to the very fragmentary nature of the specimen. Another putative record of a galliform bird, from the Late Cretaceous Austin Chalk of Texas (*Austinornis*), is based on a fossil that is too incomplete for a confident referral, so much the more since the stratigraphic age of the taxon is not well constrained (Clarke 2004).

One of the most substantial records of Mesozoic Neornithes is the partial skeleton of *Vegavis* from the Maastrichtian of Antarctica (Clarke et al. 2005a). The tarsometatarsus of this bird exhibits a complex hypotarsus morphology with sulci and crests, which identifies *Vegavis* as a neognathous bird. This hypotarsus pattern was regarded as indicative of a position of *Vegavis* within crown group Anseriformes, close to ducks and geese (Anatidae; Clarke et al. 2005a). However, a similar hypotarsus morphology evolved independently in several neognathous groups, and anseriform affinities of *Vegavis* still have to be convincingly established (Mayr 2013a).

Several bones from the Maastrichtian of Antarctica were considered to be from a gaviiform bird (*Polarornis*; Chatterjee 2002; see also Acosta Hospitaleche and Gelfo 2015), but the material does not allow an unambiguous identification (Mayr 2004a, 2009a), and more recently anseriform affinities of *Polarornis* have been proposed (Chatterjee 2015). Another Maastrichtian taxon assigned to Gaviiformes is the Chilean *Neogaeornis* (Olson 1992a). Again, gaviiform affinities are not well established, and the single bone known of *Neogaeornis*, a tarsometatarsus, differs distinctly from the tarsometatarsus of early Cenozoic Gaviiformes.

A small but diverse avifauna is finally known from the Hornerstown Formation of New Jersey. The exact age of most of these fossils is controversial and is either latest Cretaceous or earliest Paleocene (Olson and Parris 1987; Parris and Hope 2002). Some of the specimens resemble bones of the Presbyornithidae, an early Cenozoic anseriform taxon (see Chapter 7). This is especially true for *Graculavus*, of which only incomplete humeri were found, however. *Anatalavis*, which is based on a partial humerus, appears to belong to another anseriform lineage and may be closely related to a species from the early Eocene London Clay (Olson 1999a; see Chapter 7). A partial humerus can be referred to the Procellariiformes (*Tytthostonyx*; Olson and Parris 1987; Mayr 2015a), and other fossils from the Hornerstown Formation show similarities to Phaethontiformes (*Novacaesareala*) and the palaeognathous Lithornithiformes (Parris and Hope 2002; Mayr and Scofield 2016).

In summary, there are a few Mesozoic birds for which neornithine affinities are likely or well established, and some of these stem from deposits that go back to earlier stages of the Late Cretaceous. The earliest neornithine-like taxon, *Iaceornis* from the late Santonian or early Campanian (some 80–85 mya), resembles the palaeognathous Lithornithiformes, whereas some fossils of Campanian to Maastrichtian age, such as *Lamarqueavis*, show similarities to neognathous taxa. Fossils with definite neognathous affinities are known from Maastrichtian strata (e.g., *Vegavis*).

Of particular interest is the absence of arboreal neornithine birds in Cretaceous rocks, because arboreal pygostylians constitute the bulk of the avian fossils in some early Cenozoic localities. This discrepancy may be an artifact of the fossil record, due to either an underrepresentation of suitable Late Cretaceous paleoenvironments or a collecting bias that prevented the discovery of small-sized arboreal neornithines. Small species of the Enantiornithes are, however, known from the latest Cretaceous fossil sites (e.g., Longrich et al. 2011), and no plausible explanation has been put forward as to why these fossils are preserved, whereas those of arboreal Neornithes are not. I consider it very possible, if not likely, that arboreal Neornithes indeed did not diversify before the end Cretaceous extinction of the Enantiornithes (Mayr 2014a), but a more complete fossil record from the latest Cretaceous is required for an ultimate assessment of this hypothesis. Predominantly terrestrial or aquatic habitat preferences of Cretaceous Neornithes would, however, be consistent with the fact that these are also assumed for non-neornithine birds close to the origin of the crown group. Most of the early diverging extant neornithine taxa likewise are predominantly terrestrial, although some Galliformes have limited perching capabilities and Anseriformes live in aquatic habitats.

Fossils and divergence dates based on molecular calibrations

As just detailed, the fossil record is in concordance with a Cretaceous origin of the **stem lineages** of several extant avian clades.

Calibrated molecular phylogenies, however, often dated the divergences within morphologically homogenous crown group taxa far into the Cretaceous, as is the case for, for instance, Charadriiformes (shorebirds; Baker et al. 2007), Psittaciformes (parrots; Wright et al. 2008), and Passeriformes (passerines; Ericson et al. 2014). These molecular divergence estimates therefore do not predict just the existence of stem group representatives of extant clades in the Cretaceous, but that of modern-type representatives of these groups. For various reasons, this is an unlikely hypothesis.

On the one hand, a Late Cretaceous divergence of crown group representatives of extant family-level taxa would imply a much earlier initial neornithine divergence. This is in clear conflict with the fossil record, as all of the numerous Early Cretaceous avian fossils from localities around the globe are unequivocally identified as taxa outside crown group Neornithes.

A Late Cretaceous diversification of the crown group representatives of extant avian family-level taxa furthermore conflicts with the fact that such modern-type Neornithes are still unknown in early Cenozoic fossil sites, which yielded numerous well-preserved avian fossils (Mayr 2009a, 2014a). It has been suggested that early crown group taxa may have had a restricted distribution in the Southern Hemisphere, which has a much poorer fossil record compared to Europe and North America (Cracraft 2001). However, various early Cenozoic fossils are also known from localities in Africa and South America (e.g., Mayr et al. 2011a; Mourer-Chauviré et al. 2015), and these likewise do not belong to the crown groups of extant higher-level taxa. Moreover, the undetected occurrence of such modern-type taxa in the Cretaceous of poorly sampled Southern Hemispheric regions would imply an unlikely geographic restriction for dozens of million years until their first occurrence in the early Cenozoic of the Northern Hemisphere. A Cretaceous diversification of crown group representatives of, for example, passerines or parrots would furthermore imply a remarkable evolutionary stasis over 70–80 million years, even though terrestrial ecosystems underwent dramatic changes during that time (Mayr 2013a).

It remains an open question whether nucleotide substitution rates always follow a predictable pattern, and systemic biases in molecular divergence estimates have been identified (e.g., Ksepka et al. 2014). Irrespective of that, some of the earlier molecular calibrations were based on incorrectly identified fossil taxa (e.g., Mayr 2009a, 2011b), and more recent studies with better constrained fossil calibration points have yielded younger divergence dates (Jarvis et al. 2014; Claramunt and Cracraft 2015).

Impact of the end-Cretaceous mass extinction events on avian diversity

It is generally assumed that the mass extinction events at the K/Pg boundary did not only affect non-avian dinosaurs and numerous other marine and terrestrial organisms, but also terminated the existence of non-neornithine avian groups, including the once so successful Enantiornithes (Feduccia 2003, 2014; Longrich et al. 2011). Non-neornithine birds do indeed appear to have been diversified in the latest Cretaceous, and enantiornithines, hesperornithiforms, and ichthyornithiforms are known from shortly before the K/Pg boundary, but were not reported thereafter. In fact, however, reliable extinction dates of these taxa cannot be determined owing to the poor early Paleocene fossil record of birds, and whether, for example, *Qinornis* from the early Paleocene of China – a taxon based on a foot with incompletely fused metatarsals – represents a non-neornithine avian taxon that crossed

the K/Pg border needs to be examined further (Mayr 2007, 2009a).

As already detailed, there is both molecular and fossil evidence that crown group Neornithes had already diverged by the latest Cretaceous. Why Neornithes survived the K/Pg mass extinctions, whereas non-neornithine birds apparently did not, is not well understood, however. That an asteroid impact, which is widely acknowledged as the main cause of the mass extinctions at the K/Pg boundary, led to the selective survival of only neornithine birds is not a likely assumption. In any case, many non-neornithine lineages were extinct before the Late Cretaceous, and those that disappeared at the very end of the Mesozoic belong to marine (Hesperornithiformes and Ichthyornithiformes) or arboreal (most Enantiornithes) taxa.

A selective extinction of avian lineages at the K/Pg boundary could have been triggered by profound vegetation changes, such as large-scale deforestations, which have been assumed at least for North American biotas (Friis et al. 2011). This would not only have led to the extinction of arboreal enantiornithines, but, through changes of atmospheric carbon dioxide levels and accompanied oceanic acidification (Alegret et al. 2012), could have also affected the food chains of marine ecosystems. Predominantly terrestrial birds that did not live in forested environments, by contrast, may have been less affected. Certainly, however, such ad hoc hypotheses have yet to be critically tested once more data on the composition and distribution of fossil birds around the K/Pg boundary become available, and other factors, such as differences in nesting behavior (see Chapter 4), likewise need to be scrutinized in order to explain why only neornithine birds survived the end-Cretaceous mass extinction events.

6 Palaeognathous Birds (Ostriches, Tinamous, and Allies)

The earliest branching of the phylogenetic tree of neornithine (crown group) birds is a split into the sister taxa Palaeognathae and Neognathae. Palaeognathous birds retain a number of plesiomorphic cranial and postcranial features that distinguish them from all other extant birds. In addition to the eponymously primitive palate morphology (see Chapter 1), the caudal ends of the ilium and ischium of the pelvis are not connected in most species and the tibiotarsus often lacks a supratendinal bridge. Because many of their characteristics are plesiomorphic, the monophyly of palaeognathous birds was contested by earlier authors. A clade including these birds is, however, congruently obtained in current analyses of both molecular and morphological data.

The identification of Late Cretaceous records of neognathous birds indicates that at least stem group representatives of the Palaeognathae must have also existed in the Mesozoic. A Cretaceous divergence of palaeognathous and neognathous birds is suggested by all calibrated molecular phylogenies, but from a fossil perspective the earliest history of palaeognathous birds is virtually unknown. As detailed earlier, *Iaceornis* from the Late Cretaceous of North America may at least be similar to the volant stem species of palaeognathous birds. Palaeognathous affinities were also assumed for *Ambiortus* from the Early Cretaceous of Mongolia (Kurochkin 1999), but this taxon is now considered to be outside the avian crown group (O'Connor and Zelenkov 2013). The Cenozoic fossil record of palaeognathous birds is more comprehensive, and, as detailed in this chapter, it includes various taxa that provide insights into the evolutionary history of these birds.

Avian Evolution: The Fossil Record of Birds and its Paleobiological Significance,
First Edition. Gerald Mayr.
© 2017 John Wiley & Sons, Ltd. Published 2017 by John Wiley & Sons, Ltd.

The Interrelationships of Extant Palaeognathae

Except for the volant South and Central American tinamous (Tinamiformes), all extant palaeognathous birds are flightless. These "ratites" comprise New Zealand kiwis (Apterygiformes), South American rheas (Rheiformes), African ostriches (Struthioniformes), the cassowaries and emus (Casuariiformes) of the Australian region, as well as the recently extinct Madagascan elephant birds (Aepyornithiformes) and the moas (Dinornithiformes) of New Zealand. Much controversy existed concerning the interrelationships of these birds, but most analyses of morphological data supported a sister group relationship between Tinamiformes and the "ratites" (e.g., Livezey and Zusi 2007; Bourdon et al. 2009b; Worthy and Scofield 2012). Quite unexpectedly, however, recent sequence-based analyses found the flightless Palaeognathae to be paraphyletic. In contrast to morphology-based analyses, these studies congruently supported a sister group relationship between Struthioniformes and all other palaeognathous birds, which were termed Notopalaeognathae (Hackett et al. 2008; Harshman et al. 2008; Yuri et al. 2013). With the volant tinamous being deeply nested within Palaeognathae, flightlessness must therefore have evolved multiple times within palaeognathous birds. The current morphological evidence for monophyletic Notopalaeognathae is weak at best, although a few derived morphological features shared by non-struthioniform palaeognathous birds have been identified (Johnston 2011).

The basal divergence of ostriches is not the only aspect where analyses of the interrelationships of palaeognathous birds based on molecular and morphological data show incongruent results, and studies of different kinds of molecular data, for example, strongly supported a clade including Apterygiformes and Casuariiformes (e.g., Harshman et al. 2008; Mitchell et al. 2014). This clade is not retained in analyses of morphological data, even though Apterygiformes and Casuariiformes share a particular muscle of the hyoid apparatus (Johnston 2011), a greatly reduced hand portion of the wing, and reduced feather vanes.

Even more unexpected are the results of analyses of ancient DNA, which congruently identified Dinornithiformes as the closest relatives of Tinamiformes and found Aepyornithiformes to be the sister taxon of Apterygiformes (Phillips et al. 2010; Haddrath and Baker 2012; Baker et al. 2014; Mitchell et al. 2014). Neither of these groupings was proposed by morphologists before, and if the results of the new molecular analyses correctly reflect the interrelationships of palaeognathous birds, these birds exhibit a high degree of morphological homoplasy, which impedes the placement of fossil taxa.

Early Cenozoic Palaeognathous Birds of the Northern Hemisphere

Volant lithornithiforms

In the 1980s, a distinctive group of volant palaeognathous birds was identified in Paleogene localities of North America and Europe (Houde 1988). Remains of these birds, the Lithornithiformes, are not uncommon in some early Eocene fossil sites, and *Lithornis vulturinus* from the British London Clay, which had already been described in 1840, is in fact among the earliest named fossil birds.

Lithornithiform fossils are known from the late Paleocene and early Eocene of North America and the early Eocene of Europe (Plate 9b; Houde 1988; Mayr 2009a; Stidham et al. 2014). Tentative records also exist from the latest Cretaceous/earliest Paleocene of North America and from the Paleocene of Germany (Parris and Hope 2002; Mayr 2007, 2009a). Six of the eight currently recognized species are assigned to *Lithornis* and the other two belong to *Pseudocrypturus* and *Paracathartes*; some species have been reported from both North America and Europe (Houde 1988).

Lithornithiforms are medium-sized birds and their limb proportions are similar to those of tinamous. With Tinamiformes they also share a well-developed sternal keel and separate coracoids and scapulae; that is, the absence of a scapulocoracoid, which occurs in all flightless palaeognathous birds. The morphology of the wing and pectoral girdle bones indicates that lithornithiforms were capable of sustained flight (Houde 1988), and their aerial performance was probably much better than that of the weakly flighted tinamous. The long beak of lithornithiforms has schizorhinal nostrils and may have served for probing along shorelines or other bodies of water (Houde 1988). The hind toe is longer than in all extant palaeognathous birds and, together with the curved claws, suggests perching capabilities (Houde 1988).

The affinities of lithornithiforms are controversial and phylogenetic analyses suggested various placements, including sister group relationships to the Tinamiformes, the flightless "ratites," all palaeognathous birds, or even all other neornithine birds (Houde 1988; Livezey and Zusi 2007; Worthy and Scofield 2012). Clearly derived similarities are only shared with the Apterygiformes and include features of the articular end of the mandible, a marked fossa on the pterygoid, and a large olfactory bulb (Houde 1988; Zelenitsky et al. 2011). Because Apterygiformes form a clade with other Australian "ratites," a close relationship of lithornithiforms to kiwis is, however, not very likely from a biogeographic point of view. In fact, even the "palaeognathous" traits of lithornithiforms may be plesiomorphic for neornithine birds, and their position within

crown group Neornithes is not yet strongly based.

Flightless "ratites" in the Paleogene of Europe

In the Paleocene and early Eocene, Europe was isolated from other continents and free of larger carnivorous mammals, with feliforms (mongooses, cats, and allies) and caniforms (weasels, dogs, bears, and allies) later immigrating from Asia and North America, respectively (van Valkenburgh 1999). These geographic and biotic conditions allowed the evolution of a surprisingly high number of flightless birds (Mayr 2009a), including "ratite"-like Palaeognathae, which are classified in the taxa Remiornithidae and Palaeotididae.

Remiornis (Remiornithidae) is only known from leg bones and vertebrae from the late Paleocene of France. Although palaeognathous affinities are uncontested, its precise relationships cannot be established with the material at hand (Martin 1992; Mayr 2009a; Buffetaut and Angst 2014). Of *Palaeotis* (Palaeotididae) from the early and middle Eocene of Germany, by contrast, several skeletons have been found (Figure 6.1; Houde and Haubold 1987; Peters 1988; Mayr 2015b). *Palaeotis* is somewhat smaller than *Remiornis* and has a more elongated tarsometatarsus. Otherwise, this latter bone exhibits a similar morphology in both taxa, which may be more closely related than is apparent from their current classifications (Mayr 2009a).

The long-legged *Palaeotis* reached a standing height of slightly less than one meter and has a narrower beak than all extant "ratites" except cassowaries and kiwis. It was a flightless bird that lived in a forested environment. The gracile hindlimbs indicate that it was cursorial, which deserves particular notice, because the fast-running taxa of extant Palaeognathae – that is, Struthioniformes and Rheiformes – inhabit open areas. As in other cursorial flightless Palaeognathae, the

Figure 6.1 (a) Skeleton of the flightless palaeognathous bird *Palaeotis* from the early Eocene of Messel in Germany. Carpometacarpi of (b) *Palaeotis* and (c–e) extant Tinamiformes, Struthioniformes, and Rheiformes. Scapulocoracoids of (f) *Palaeotis* and (g, h) extant Struthioniformes and Rheiformes.

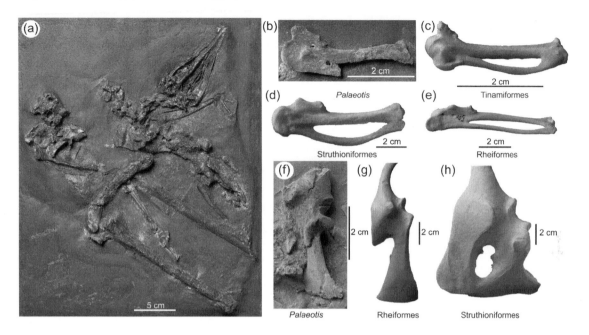

humerus is fairly long. The wing bones are nevertheless greatly reduced, the sternum lacks a keel, and coracoid and scapula are fused into a scapulocoracoid. In these and some other features, such as a narrow pelvis and the lack of a hind toe, *Palaeotis* exhibits a "ratite"-like morphology and differs from the contemporaneous lithornithiforms.

So far, no derived characters have been identified that allow an unequivocal assignment of *Palaeotis* to any of the extant palaeognathous taxa. Overall, its skeleton most closely resembles that of extant Struthioniformes and Rheiformes, with which it was allied by earlier authors (Houde and Haubold 1987; Peters 1988). Even though close affinities to Struthioniformes are plausible for biogeographic reasons, *Palaeotis* likewise shows some similarities to *Diogenornis*, a presumed stem group representative of Rheiformes from the Paleocene of Brazil (Mayr 2009a, 2015b). If a

"ratite"-like morphology did indeed evolve multiple times convergently, as suggested by the results of current molecular analyses, a possible descent of *Palaeotis* from a lithornithiform-like ancestor also has to be taken into consideration (Mayr 2015b).

Long-Winged Ostriches, Rheas, and Tinamous

Ostriches: Birds converging on horses?

The African Ostrich (*Struthio camelus*) occurs in open, savannah-like landscapes, semideserts, and deserts. It is the largest and heaviest extant bird and the only didactyl (two-toed) living avian species. In ostriches not only is the hind toe lost, as in many other palaeognathous birds, but they also lack a second toe and its tarsometatarsal trochlea. This specialized morphology is an adaptation for cursorial locomotion, and ostriches reach speeds of up to 70 kilometers per hour. It has often been noted that the toe reduction of ostriches paralleled that in the evolution of horses, in which only a single functional toe is left in each foot. Unlike in quadrupedal horses, however, an evolution towards a single-toed foot in bipedal animals is impeded by functional constraints, and the fourth toe of ostriches functions as an outrigger in running maneuvers (Schaller et al. 2011). As in other cursorial flightless birds, such as nandus and phorusrhacids (Chapter 11), the wings of ostriches are comparatively long and are used as stabilizing "air-rudders" in fast zigzag runs (Schaller 2008). In flightless birds that are not fast runners, by contrast, the wings are usually greatly reduced.

Apart from being highly cursorial, ostriches are also very self-protective, and these two attributes enable their existence in continental ecosystems with large mammalian predators. The evolutionary origin of ostriches is poorly known, but flight loss in their stem lineage must have taken place in a predator-free environment and therefore under ecological conditions different from those in their extant range.

Extant ostriches only occur in Africa, but their Cenozoic distribution also covered parts of Europe and Asia. As already detailed, the European *Palaeotis* is a candidate taxon for a struthioniform stem group representative, in which case ostriches would have originated outside Africa. The earliest African "ratite"-like fossils stem from the late Eocene of Egypt and belong to the rhea-sized *Eremopezus* (Eremopezidae; Rasmussen et al. 1987, 2001). This taxon is only known from leg bones, the dimensions of which clearly indicate a flightless bird (see Figure 6.4b). Palaeognathous affinities are suggested by the lack of an ossified supratendinal bridge of the tibiotarsus, but the overall morphology of the bones is quite different from that

of extant ostriches and does not indicate a cursorial animal.

The earliest definite skeletal remains of ostriches belong to *Struthio coppensi* from the early Miocene (20 mya) of Namibia (Mourer-Chauviré et al. 1996; Mourer-Chauviré 2008). This species is much smaller than the extant *S. camelus* and still exhibits a vestige of the tarsometatarsal trochlea for the second toe. Otherwise, the morphology of the known bones closely resembles that of extant Struthioniformes, which indicates that the divergence of ostriches from their sister taxon occurred well before the early Miocene. The next oldest African records of ostriches are from the middle Miocene (14 mya) of Kenya and stem from an unnamed larger species (Leonard et al. 2006), therefore suggesting a size increase of ostriches between the early and middle Miocene. Ostrich bones from the Pliocene of South Africa and Morocco were assigned to the large and massive *Struthio asiaticus*, which even exceeded the extant Ostrich in size (Mourer-Chauviré and Geraads 2008; Manegold et al. 2013).

A large part of the African fossil record of ostriches is only based on eggshell fragments, for which various eggshell taxa ("oospecies") were established, which are distinguished by shell thickness and the distribution of the pore openings on the surface (Mourer-Chauviré et al. 1996; Harrison and Msuya 2005; Bibi et al. 2006). Only rarely are these eggshells found together with skeletal remains. Although some of the proposed oospecies may reflect morphological variability across the egg surface rather than different species, the eggshell fossils indicate a higher diversity of large ostrich-like birds in Africa than is apparent from the skeletal remains found so far.

"Aepyornithid" eggshell has been reported from the early Miocene of Namibia, the late Miocene of Kenya and Arabia, the Mio-Pliocene of Turkey, and the Pliocene of India and Mongolia (Bibi et al. 2006). The stratigraphic distribution of this eggshell type corresponds well with that of skeletal remains of ostriches, and *S. coppensi* fossils were found together with "aepyornithid"-type eggshell fragments (Mourer-Chauviré et al. 1996). This "aepyornithid" morphology is therefore likely to represent the primitive eggshell morphology of struthionid eggs, despite the derivation of its name from a similarity to the eggshell of the Madagascan Aepyornithiformes. Eggshells of a *"Psammornis"* type occur in northern Africa and Arabia; their age is uncertain, but possibly they are from Eocene strata (Sauer 1969), in which case they would predate the skeletal fossil record of ostriches and are more likely to come from other large birds, such as *Eremopezus*. In Namibia, several stratigraphically separated eggshell taxa were recognized, and a *"Namornis"* type occurs in the middle Miocene (16–15 mya), a *"Diamantornis"* type in the middle and late Miocene (15–8 mya), and a *Struthio* type from the late Miocene onward (Bibi et al. 2006). Especially the layer of the *"Diamantornis"* type is enigmatic. *"Diamantornis"* eggshells only occur in arid and fluvial environments of southern and eastern Africa as well as Arabia and have a characteristic surface morphology with large circular pore complexes, which are very different from the "aepyornithid" and "struthionid" pore types (Bibi et al. 2006).

Quite puzzling from a biogeographic point of view is the occurrence of "ratite"-type eggs in the late Miocene of Lanzarote, one of the Canary Islands, which is assumed not to have been in contact with continental Africa. These eggs (Figure 6.2) were originally considered to be from ostriches (Sauer and Rothe 1972), but more recently it was hypothesized that they may possibly be from giant seabirds of the taxon Pelagornithidae (García-Talavera 1990). Their shell thickness corresponds, however, with that of

Figure 6.2 Putative ostrich egg from the late Miocene of Lanzarote Island (left), from the collection of Senckenberg Research Institute Frankfurt, in comparison to the egg of an extant ostrich (right).

the eggs of large palaeognathous birds, and it is therefore likely that the birds incubating these eggs were very heavy, which the volant Pelagornithidae were not (Chapter 7). If the egg layer was an ostrich, how its ancestor reached Lanzarote remains mysterious.

Today, the distribution of ostriches is confined to Africa, but numerous fossils are known from the Neogene of Europe and Asia. A fragmentary pedal phalanx and eggshells have been reported from the middle Miocene of Turkey (Sauer 1979). Whereas these fossils may suggest a dispersal of ostriches from Africa during the Miocene Climatic Optimum, even earlier "aepyornithid"-type eggshells have been described from the early Miocene (17.5 mya) of China (Wang et al. 2011). The fact that these Chinese fossils are coeval with the earliest African ostrich remains challenges the current "out-of-Africa" hypothesis for the Cenozoic dispersal of ostriches (Mourer-Chauviré et al. 1996).

In Europe, ostriches appear to have been confined to the southeastern part of the continent (Mlíkovský 1996), and skeletal remains have been found in the late Miocene of Moldova and Ukraine, where ostriches persisted until the end of the Pliocene, as well as in the late Miocene and Pliocene of Greece, Bulgaria, and Hungary (Boev and Spassov 2009). Various species have been described for European and Asian ostriches, but the validity of many of these is questionable, since the fossil record is often very fragmentary and based on non-comparable skeletal elements or on eggshells. In the late Miocene and Pliocene of Eurasia, at least four species of *Struthio* can be distinguished, one of which, *S. karatheodoris*, occurs in Greece, the southern Balkan peninsula, and the Middle East (Boev and Spassov 2009). Other species have been reported from the area north of the Black Sea and from China. Of the latter, *Struthio* ("*Orientornis*") *linxiaensis* from the late Miocene of northwestern China was slightly larger than the extant Ostrich (Hou et al. 2005; Wang 2008; Wang et al. 2011). *S. linxiaensis* is based on a pelvis, and with this element being unknown from most other late Miocene ostriches, its taxonomic status remains doubtful. Further ostrich species have been described from the Pliocene of Asia, with *S. wimani* occurring in the early Pliocene of China and Mongolia, and *S. asiaticus* in the Pliocene of India.

The latest records of ostriches outside Africa stem from the Pleistocene of Georgia (*S. dmanisensis*), China (*S. anderssoni*), Mongolia, and Russia (*S. camelus*; Boev and Spassov 2009). Which of the various fossil species is closest to the extant Ostrich is unknown, and owing to their unresolved interrelationships it also cannot be determined how often ostriches dispersed between Africa and Eurasia. Why ostriches became extinct outside Africa likewise remains elusive, but climatic cooling of the northern continents during the Pleistocene glaciations may have played a role.

South American rheas and tinamous

Today there are two species of rheas in South America, which are classified in the taxa *Rhea* and *Pterocnemia*. The phylogenetic affinities of Rheiformes are controversial and molecular analyses indicate a sister group relationship to either a clade including Casuariiformes, Apterygiformes, and Tinamiformes (Harshman et al. 2008; Prum et al. 2015), or to a clade including Tinamiformes and Dinornithiformes (Smith et al. 2013). Based on shared morphological features, by contrast, a sister group relationship to the South American Tinamiformes was proposed (Johnston 2011), which would be plausible on biogeographic grounds. In any case, the fossil record shows that large, flightless palaeognathous birds have a long evolutionary history in South America.

The earliest fossils assigned to rheas are pedal phalanges from the middle Paleocene of Argentina (Tambussi 1995). Much better represented is *Diogenornis* from the late Paleocene of Brazil, of which various postcranial bones and the tip of the praemaxilla have been discovered (Alvarenga 1983). This flightless bird reached about two-thirds the size of the extant Greater Rhea, *Rhea americana*, from which it differs, among others, in the narrower, more cassowary-like beak. The wing bones of *Diogenornis* are furthermore less reduced than those of extant rheas. Rheiform affinities of *Diogenornis* are not strongly based, but they are certainly highly likely for biogeographic reasons and because of the overall resemblance of the taxon to extant Rheiformes.

More similar to crown group Rheiformes is *Opisthodactylus* (Opisthodactylidae) from the early Miocene of Argentina. A record of *Pterocnemia* stems from the late Miocene of Argentina (Noriega and Agnolín 2008), whereas Rheiformes from the Pliocene of Argentina were considered distinctive enough to merit classification in an extinct taxon (*Hinasuri*; Tambussi 1995).

In South America rheas coexist with tinamous, which are the only volant extant palaeognathous birds. Tinamous are unable to perform long-distance flights. Like galliforms, however, they are capable of explosive burst-off starts, which led to the convergent evolution of similar wing, pectoral girdle, and sternum morphologies in tinamous and landfowl. The fossil record of tinamous dates back to the early-middle Miocene of Argentina, but mainly consists of fragmentary bones. Although some of these can be assigned to extant tinamou lineages (Bertelli et al. 2014), they provide little specific information on the evolutionary history of these birds.

Short-Winged Palaeognathous Birds

Moas and kiwis

The existence of large, flightless birds in New Zealand was first announced by the British anatomist Richard Owen in the early 19th century, and these iconic birds subsequently became known as moas (following Maori grammar, they are sometimes also called moa in the plural – this practice is not adopted here, since it would require grammatical adjustments for many other names derived

from non-English languages). Moa fossils are very abundant and the numerous remains include mummified specimens with preserved soft tissue, as well as eggs, footprints, and coprolites (Worthy and Holdaway 2002).

Some moa species were among the tallest birds that ever lived, and the standing height of the largest one, *Dinornis maximus*, was estimated at more than 3 meters. Moas are the only birds that completely lack wing bones (Figure 6.3a), the genuine absence of which is indicated by fact that the scapulocoracoid has no articulation facet for the humerus.

Nine moa species are currently recognized as valid, and these are assigned to the taxa Dinornithidae, Megalapterygidae, and Emeidae (Worthy and Scofield 2012). Moas spanned quite some size range, and estimated weights of the various species are between 30 and 150 kilograms (Worthy and Holdaway 2002). All species lived contemporaneously and most, albeit not all, are known from both main islands of New Zealand. Although there appears to have been an overlap in habitat preferences and altitudinal occurrences, some species predominantly inhabited the

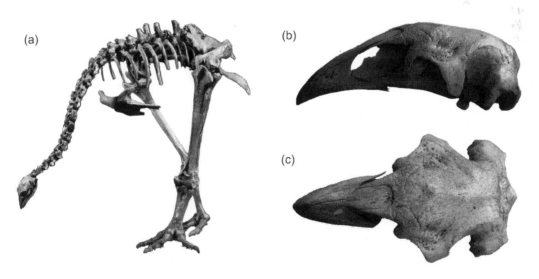

(a)

(b)

(c)

Figure 6.3 Moas (Dinornithiformes) from the Quaternary of New Zealand. (a) Skeleton of *Emeus*. Skull of *Pachyornis* in (b) lateral and (c) dorsal view.

uplands, including the subalpine zones above the tree lines, and others the wet forests in the lowlands (Worthy and Holdaway 2002).

Most moa species belong to the Emeidae, which include the taxa *Emeus, Anomalopteryx, Pachyornis* (Figure 6.3b, c), and *Euryapteryx*, and which differ from other Dinornithiformes in their leg proportions and the simpler structure of the ossified nasal conchae (Worthy and Scofield 2012). In *Megalapteryx*, the sole taxon included in the Megalapterygidae, the tarsus is feathered up to the toes, whereas it is scutellate – that is, covered with scales – in other moas. Unlike the whitish eggs of other moas, *Megalapteryx* furthermore has greenish eggs. The two very large species of *Dinornis*, the only representative of the Dinornithidae, are characterized by a broad and flat skull and a high number of cervical vertebrae. The presence of a vestigial furcula indicates a sister group relationship between *Dinornis* and all other moas, which completely lack this bone.

Many moa fossils were found in cave deposits, which offer favorable climatic conditions for the fossilization of DNA. Moas were therefore among the first fossil birds for which studies of ancient DNA were conducted. These analyses not only contributed to phylogenetic questions, but also showed that some moa species exhibit an extreme reversed sexual size dimorphism, with females being much larger than males (Bunce et al. 2003). Moa bones show lines of arrested growth, and some species probably needed up to a decade to reach sexual maturity (Turvey et al. 2005).

Moas were the main herbivores in the Quaternary of New Zealand and some food plants could be identified through the study of subfossil coprolites, suggesting a diet that consisted of herbs and low shrubs (Wood et al. 2008). Because there were no other large herbivores in New Zealand, the growth of thorns and spines in certain New Zealand plants was interpreted as an evolutionary anachronism, which originally evolved as an adaptation against moa grazing (Worthy and Holdaway 2002).

Almost all moa fossils stem from Holocene and Pleistocene (1.8–2.5 mya) cave, swamp, or dune deposits (Worthy and Holdaway 2002). The youngest bones have been dated to the late 13th century, and it is assumed that moas were hunted to extinction less than 100 years after the first Polynesian settlers arrived on New Zealand (Holdaway and Jacomb 2000). In sharp contrast to their abundant Quaternary fossil record, the early evolutionary history of moas is elusive, and only recently have putative moa eggshells and a few bone fragments been found in the early/middle Miocene (16–19 mya) St Bathans fossil site (Tennyson et al. 2010).

Moas have long been considered to be most closely related to another group of palaeognathous birds from New Zealand, the kiwis, which include five extant species of small, nocturnal, and long-beaked birds. The enormously large eggs of kiwis have been explained by phylogenetic dwarfing of these birds, which affected the size of the body, but not that of the eggs (see Worthy et al. 2013a and references therein). Kiwis and moas share several morphological characteristics that set them apart from other palaeognathous birds, such as a similar morphology of the hypotarsus of the tarsometatarsus, and both resulted as sister taxa in morphology-based analyses (e.g., Bourdon et al. 2009b; but see Worthy and Scofield 2012). However, the first sequence-based studies including moa DNA did not support a sister group relationship between moas and kiwis (Cooper et al. 2001). As noted earlier, more recent analyses congruently resulted in a sister group relationship between moas and tinamous on the one hand, and between kiwis and Madagascan elephant birds on the other (e.g., Mitchell et al. 2014).

Kiwis exhibit some plesiomorphic features that distinguish them from all other neornithine birds, such as two functional ovaries (Kinsky 1971) and an only partially reversed hind toe. These primitive characteristics are, however, likely to be the result of paedomorphosis (Chapter 2). Unfortunately, the early evolutionary history of kiwis is as poorly known as that of moas. The only pre-Pleistocene fossils are a quadrate and an incomplete femur of *Proapteryx* from the early or middle Miocene of New Zealand. *Proapteryx* was smaller than extant Apterygiformes, and if this fossil does indeed represent a stem group representative of the Apterygiformes, it does not support the hypothesis of a phyletic dwarfing of kiwis (Worthy et al. 2013a). Moreover, because of its small size and bone proportions it was hypothesized that *Proapteryx* was a volant bird, in which case the ancestor of kiwis would have reached New Zealand on the wing (Worthy et al. 2013a). The

Proapteryx fossils are, however, too fragmentary for strongly based conclusions about the affinities and way of life of this taxon. The evolutionary history of kiwis therefore remains largely obscure, and this is also true for the factors that triggered the nocturnal way of life of these birds.

Elephant birds

The large island of Madagascar housed another distinctive "ratite" group, elephant birds (Aepyornithiformes), which, like moas, are prominently featured in most textbooks of avian paleontology (Figure 6.4). Not only have these large birds been known for a long time, with the first species, *Aepyornis maximus*, having been described in 1851, *Ae. maximus* was also among the heaviest birds that ever existed, and its eggs are the largest known vertebrate eggs, whose volume

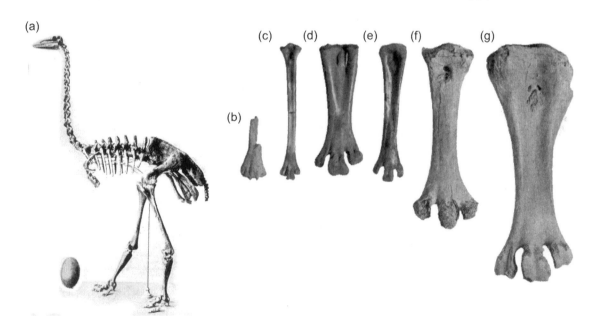

Figure 6.4 (a) Skeleton and egg of the Madagascan elephant bird *Aepyornis* (Aepyornithiformes). Tarsometatarsi of (b) *Eremopezus* (Eremopezidae) from the late Eocene of Egypt, (c) an extant nandu (*Rhea*, Rheiformes), (d, g) two *Aepyornis* species, (e) the aepyornithiform *Mullerornis*, and (f) the moa *Dinornis* (Dinornithidae). Photographs from Lambrecht (1933).

even surpasses that of the largest non-avian dinosaur eggs.

Unlike the cursorial African ostriches, elephant birds have stout leg bones and were probably rather slow-moving birds. Furthermore, unlike in ostriches, the wings of elephant birds are greatly reduced. A marked pitting of the skull roof was interpreted as evidence of ornamental head feathers (Lambrecht 1933).

Two elephant bird taxa, *Aepyornis* and *Mullerornis*, have been distinguished, each including several species that are in need of taxonomic revision. *Ae. maximus*, the largest aepyornithiform, reached a standing height of 2.7 meters, and had an estimated weight of more than 300 kilograms (Worthy and Holdaway 2002). Other *Aepyornis* species and those of *Mullerornis* are smaller, and the latter taxon also has a more slender tarsometatarsus than *Aepyornis* (Figure 6.4).

Like New Zealand moas, elephant birds were herbivorous, and apparently some Madagascan plants also evolved defense structures against elephant bird browsing (Bond and Silander 2007). Compared to moas, however, much less is known about the habitat preferences and way of life of elephant birds.

Aepyornis eggs are fairly common in the south of Madagascar, and a few even contain the remains of embryos (Balanoff and Rowe 2007). Isotope data from eggshell fragments indicate that the diet of elephant birds primarily consisted of coastal wetland plants (Clarke et al. 2006). Numerous eggshell fragments are found in coastal habitats, where they can reach very dense accumulations, which are perhaps indicative of communal nests (Goodman and Jungers 2014). Amazingly, two putative *Aepyornis* eggs have also been reported from dune deposits in Western Australia, and it has been hypothesized that they drifted to this continent on oceanic currents (Long et al. 1998). However, after recognition of the putative eggshells of the large dromornithid *Genyornis* as those of the galliform *Progura* (Grellet-Tinner et al. 2016, see Chapter 7), the possibility should be reconsidered that these Australian "*Aepyornis*" eggs are in fact the true *Genyornis* eggs.

The evolutionary origin of elephant birds is unknown and there are no pre-Pleistocene fossils of these birds from Madagascar. For biogeographic reasons, earlier hypotheses of close affinities to the late Eocene *Eremopezus* from continental Africa are unlikely, because Madagascar and Africa separated too early for the overland dispersal of a flightless bird (Rasmussen et al. 2001). As already noted, "aepyornithid" eggshell fragments in the Cenozoic of Africa probably stem from ostriches and do not prove the existence of aepyornithiforms outside of Madagascar.

Most unexpectedly, and also as already detailed, analyses of mitochondrial genome sequences indicated a sister group relationship between Aepyornithiformes and Apterygiformes (Mitchell et al. 2014). Both taxa differ in numerous aspects of their skeletons, but they share a similar derived morphology of the sternum, which is craniocaudally very short and bears a marked concavity in its cranial margin. As in Apterygiformes, the hand part of the wing is furthermore greatly reduced in Aepyornithiformes. However, these features are clearly related to flightlessness and must therefore have evolved convergently, because Madagascar had already split from the Mesozoic supercontinent Gondwana in the Middle to early Late Jurassic, some 155–160 mya, and no dispersal routes existed for a flightless ancestor of either elephant birds or kiwis (e.g., Smith et al. 1994).

The exact extinction date of aepyornithiforms is unknown. The youngest radiocarbon dates are from bones that stem from the 8th century, but survival into the 13th century has been considered likely (Goodman and Jungers 2014). Even at the earlier date humans were already present on Madagascar, but whether they were involved in the elephant bird extinction, or whether the latter had natural causes such as increasing aridification and changes in vegetation, remains unknown (Goodman and Jungers 2014).

Casuariiformes (emus and cassowaries)

Casuariiformes encompass the Casuariidae (cassowaries), the three extant species of which live in forested environments of northern Australia and New Guinea, and the more cursorial Dromaiidae (emus), whose single extant species, the emu, is widely distributed across open areas of Australia. Despite their disparate life habits and some differences in external appearance, Casuariidae and Dromaiidae are very similar in skeletal morphology and were shown to be sister taxa in virtually all phylogenetic analyses of whatever data.

The earliest fossil representatives of the Casuariiformes are two species of *Emuarius* from the late Oligocene to middle Miocene (24–15 mya) of Australia. Like in the extant emu, the bill of *Emuarius* has a rounded tip, whereas the beak of cassowaries is narrower and more pointed. The eyes of *Emuarius*, however, are proportionally smaller than those of the emu, and the more slender femur indicates that the fossil taxon, which had a weight of around 20 kilograms, was less cursorial than the larger emu (Boles 1997b). *Emuarius* has been identified as a stem group representative of Dromaiidae (Worthy et al. 2014). If so, it indicates that cassowaries and emus had already diverged in the Paleogene, which stands in some contrast to the strong anatomical similarities of both groups. The only other pre-Quaternary fossil record of emus is an extinct Pliocene species, which was assigned to *Dromaius*, but was smaller than the extant species. Two small emu species, which became extinct at the beginning of the 19th century, also occurred on

King Island (*D. ater*) and Kangaroo Island (*D. baudinianus*) off the coast of South Australia.

Biogeography: A Textbook Example of Gondwanan Vicariance Has Been Dismantled

Disjunct distributions of related organisms can be explained by either active dispersal or vicariance; that is, habitat fragmentation through the formation of physical barriers. Under the traditional hypothesis of a clade including all flightless extant Palaeognathae ("ratites") and a descent of that clade from a flightless ancestor, dispersal of palaeognathous birds would have strongly depended on land corridors. Because the extant species are mainly found on the Southern Hemisphere, it was assumed that plate tectonics, and the break-up of the Southern Hemispheric supercontinent Gondwana in particular, shaped the evolution of palaeognathous birds (Cracraft 2001). The flightless "ratites" were therefore for a long time among the textbook examples of a vicariant biogeography.

The fossil record indicates a long evolutionary history of palaeognathous birds, and flightless "ratite"-like taxa are known from the Paleocene of both the Northern (*Remiornis*) and Southern (*Diogenornis*) Hemispheres. However, although palaeognathous and neognathous birds most likely already diverged in the Cretaceous, there exist no Mesozoic records of flightless "ratite"-like forms. Even if these were identified in the future, the late Mesozoic paleogeography would not have allowed the dispersal of flightless birds across all of the southern continents, let alone into the Northern Hemisphere.

From a paleogeographic point of view, a late Mesozoic or earliest Cenozoic dispersal of flightless palaeognathous birds between South America and Australia via Antarctica was only possible before the glaciation of the latter continent started in the late Eocene (Mayr 2009a). There does indeed exist a partial distal tarsometatarsus of a putative "ratite" from the late Eocene of Antarctica (Tambussi and Acosta Hospitaleche 2007), but this fossil is too fragmentary for a well-founded identification (Mayr 2009a).

The dispersal of flightless "ratites" to Africa or Madagascar cannot be explained by vicariance due to the break-up of Gondwana, because Africa was already isolated from other parts of this supercontinent about 110 mya, and Madagascar separated even earlier (Smith et al. 1994). The occurrence of "ratite"-like flightless palaeognathous birds in the early Cenozoic of the Northern Hemisphere poses another biogeographic problem if the descent of these birds from a flightless ancestor is assumed, because, according to current paleogeographic reconstructions (Smith et al. 1994), there were no Late Cretaceous/early Paleogene overland dispersal routes between South America or Australia and the Northern Hemisphere.

In any case, biogeographic scenarios depending on land corridors are no longer required to explain the dispersal of palaeognathous birds, since the new molecular phylogenies supported a placement of the volant Tinamiformes within the flightless "ratites," and therefore indicate a multiple loss of flight capabilities in the evolution of palaeognathous birds (Hackett et al. 2008; Mitchell et al. 2014; Prum et al. 2015). The ancestors of flightless palaeognaths therefore could have reached their current geographic ranges on the wing. As detailed in the preceding sections, the early evolutionary history of the extant "ratite" lineages is poorly known, but at least for Apterygiformes some fossils have indeed been considered indicative of a volant ancestor (Worthy et al. 2013a).

Although recent phylogenetic hypotheses based on molecular data have simplified some biogeographic scenarios, they also raised new, unresolved questions. Earlier sequence-based analyses, which only sampled extant palaeognathous taxa, found the New Zealand Apterygiformes to be most closely related to the Australian Casuariiformes (e.g., Harshman et al. 2008). Such a relationship would make perfect sense in biogeographic terms. However, more recent studies of ancient DNA have suggested that Dinornithiformes are not part of this Australasian clade but are in fact the sister group of the New World Tinamiformes, whereas Apterygiformes resulted as the sister group of the Madagascan Aepyornithiformes (Mitchell et al. 2014). If these phylogenetic placements stand up to future scrutiny, the biogeographic history of palaeognathous birds will become as intricate as it always used to be.

7

Galloanseres: "Fowl" and Kin

The phylogenetic tree of neognathous birds is far from being fully understood, but a basal dichotomy between Galloanseres and Neoaves was obtained in virtually all analyses. The key galloanserine apomorphies pertain to cranial specializations and include basipterygoid processes, which are ovate and sessile rather than being stalked or absent as in other neornithine birds, as well as a bicondylar mandibular process of the quadrate (this process is tricondylar in most other neognathous birds). In addition, most galloanserine birds possess elongated **retroarticular processes** on the caudal ends of the mandibles, which increase the leverage of the jaw muscles.

Extant Galloanseres fall into two clades, Galliformes (landfowl) and Anseriformes (waterfowl), each being further divided in three subclades, which show a remarkable congruence in their geographic distribution. Within crown group Galliformes, the Australasian Megapodiidae (megapodes) are the sister group of a clade including the Neotropic Cracidae (chachalacas, guans, and curassows) and the globally distributed Phasianoidea (grouse, quails, pheasants, and allies). Concerning crown group Anseriformes, the South American Anhimidae (screamers) and the Australasian Anseranatidae (magpie geese) are successive sister taxa of the Anatidae (swans, geese, and ducks), which have a worldwide distribution. Both Galliformes and Anseriformes therefore include two species-poor and early diverging Australasian and Neotropic clades, and a highly diversified third one with a global distribution.

This geographic pattern was considered indicative of a Cretaceous origin of crown group Galloanseres owing to the break-up of Gondwana (Cracraft 2001). This hypothesis is, however, not supported by the fossil record. Not only does the occurrence of successive sister taxa of crown

Avian Evolution: The Fossil Record of Birds and its Paleobiological Significance,
First Edition. Gerald Mayr.
© 2017 John Wiley & Sons, Ltd. Published 2017 by John Wiley & Sons, Ltd.

group Galliformes in the early Cenozoic of the Northern Hemisphere conflict with a Southern Hemispheric origin of crown group galliform birds in the Mesozoic era (Mayr 2009a), but, as will be detailed in this chapter, there are also records of putative anseranatids and anhimids from the early Cenozoic of the Northern Hemisphere. The pervasive Southern Hemispheric distribution pattern of basal galloanserines is therefore more likely the result of a relictual distribution of these lineages, which succumbed to competition elsewhere.

Galloanserine affinities have been proposed for a number of aberrant extinct groups. If these have been correctly identified, the ecological and morphological diversity of Galloanseres was remarkable, spanning large flightless species with greatly reduced wings and extremely long-winged soaring birds. Alternatively, some of the features used to characterize Galloanseres may be primitive for neognathous birds or may have convergently evolved in only distantly related avian taxa.

Galliformes: From Herbivorous Forest Dwellers to Seed Eaters of Open Landscapes

The landfowl are predominantly herbivorous birds, which today occur in a large variety of habitats. Most galliform birds are not capable of long-distance flight, which is why only a few species – mainly of the Megapodiidae – colonized remote oceanic islands. Galliform birds are, however, by no means weak fliers. The members of the Phasianoidea in particular employ powerful burst take-offs, and these locomotory specializations are functionally correlated with derived modifications of the wing and pectoral girdle, including a robust humerus and a sternum with unusually deep caudal incisions (Figure 7.1d).

A plethora of stem group Galliformes
The early Eocene Gallinuloididae, which include the North American *Gallinuloides* and the very similar European *Paraortygoides*, are the earliest and phylogenetically most basal galliform stem group representatives (Mayr 2009a; Figure 7.1). Gallinuloidids differ from later landfowl in several

plesiomorphic traits that indicate profound differences in their life habits (Mayr and Weidig 2004; Mayr 2006b). The slender humerus is more similar to that of anseriform birds in its proportions (Figure 7.1e–g), and gallinuloidids were therefore probably not capable of the powerful flight bursts that characterize modern landfowl. As in other stem group Galliformes, the coracoid exhibits a plesiomorphic, cup-like scapular articulation facet (Figure 7.1h). The furcula is much more robust than in the crown group taxa and the tip of the sternal keel reaches farther cranially (Figure 7.1h, j). All of these features are likely to be functionally correlated. The slender furcula and caudally shifted tip of the sternal keel of extant Galliformes are a consequence of the large crop of these birds, which serves to store and process hard-to-digest food items (Stegmann 1964). The robust furcula and cranially projected sternal keel of gallinuloidids indicate that these early stem group galliforms lacked a voluminous crop (Mayr 2006b). That gallinuloidids did not predominantly feed on seeds and other coarse plant matter is also suggested by the absence of gastroliths in the known skeletons, whereas grit and small pebbles are regularly ingested by extant

Figure 7.1 (a) Skeleton of the stem group galliform *Gallinuloides* (Gallinuloididae) from the early Eocene North American Green River Formation. Sternum of (b) *Gallinuloides*, (c) extant Megapodiidae (*Alectura*), and (d) extant Phasianidae (*Rollulus*). (e) Humerus of an anatid (*Nettapus*), (f) the gallinuloidid *Paraortygoides* from the early Eocene of Messel in Germany, and (g) a cracid (*Nothocrax*). (h) Coracoid and furcula of *Gallinuloides* in comparison to (i, j) the coracoid and furcula of a phasianid (*Rollulus*). Some of the differences in the pectoral girdle bones of stem group and crown group Galliformes are related to the evolution of a large crop in the latter, which is especially true for the caudally shifted tip of the sternal keel and the more slender furcula.

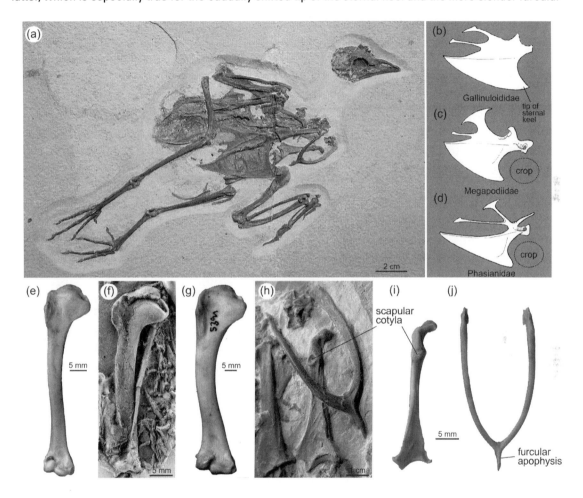

Galliformes to break down plant matter mechanically in the gizzard. Gallinuloidids are likely to have been omnivorous, feeding on invertebrates and fruits, and, as evidenced by the structure of its feet, at least *Paraortygoides* was probably a forest-dwelling bird with better perching capabilities than most crown group Galliformes.

The presence of a well-developed, bipartite pneumotricipital fossa on the proximal end of the humerus is another characteristic of gallinuloidids, albeit one of unknown functional significance. Such a double fossa also occurs in the late Eocene to late Oligocene *Paraortyx*, which belongs to another taxon of stem group

Galliformes, the Paraortygidae. Definitive records of paraortygids are only known from Europe (Mourer-Chauviré 1992a; Mayr 2009a), although two late Eocene and early Oligocene North American taxa, *Procrax* and *Archaealectrornis*, show a close resemblance to the paraortygid *Pirortyx* (Mayr 2009a). Concerning their bone morphology, paraortygids were more similar to extant Galliformes, but several plesiomorphic attributes, such as a cup-like scapular articulation facet of the coracoid, clearly show them to be outside the crown group.

A distinctive further taxon of stem group Galliformes are the Quercymegapodiidae, which occurred in the late Eocene and early Miocene of France and the late Oligocene or early Miocene of Brazil (e.g., *Quercymegapodius*, *Taubacrex*, *Ameripodius*; Mourer-Chauviré 1992a, 2000; Mayr 2009a). As suggested by one skeleton with preserved gastroliths, quercymegapodiids fed on coarse plant matter. Their much narrower carpometacarpus, however, indicates that they differed from extant Galliformes in flight characteristics. Better capabilities for sustained long-distance flight would conform to the fact that quercymegapodiids are the only stem group Galliformes known to have reached the geographically isolated South America. Derived humerus features show quercymegapodiids to be more closely related to crown group Galliformes than are paraortygids (Mourer-Chauviré 1992a; Mayr 2009a), but the interrelationships of these taxa are in need of further scrutiny. Future phylogenetic analyses will also have to address the affinities of *Sobniogallus* from the early Oligocene of Poland, which already exhibits a more modern-type sternal morphology (Tomek et al. 2014).

Outside Europe and North America, stem group Galliformes have been reported from the early Eocene of Mongolia, the middle Eocene of Namibia (*Namaortyx*, *Scopelortyx*), and the early or middle Eocene of Tunisia (*Chambiortyx*; Hwang et al. 2010; Mourer-Chauviré et al. 2011a, 2013a, 2015). All of these fossils represent small to very small species, and, except for *Scopelortyx*, which was assigned to the Paraortygidae, their relationships to the better-represented European and North American taxa are uncertain. These fossils not only document a wide distribution of stem group Galliformes in the early Cenozoic, but also show that the absence of crown group taxa in Europe is not a mere geographic artifact.

An Old World origin of Galliformes is suggested by the occurrence of diverse stem group Galliformes in the early Paleogene of Europe, which are successively more closely related to the crown group (Figure 7.2). The geographic origin of the crown group itself, however, is more difficult to determine.

Early diversification and biogeographic history of crown group Galliformes

Megapodes, the earliest diverging crown group Galliformes, are today mainly confined to the Australasian region, and only a single species occurs west of **Wallace's Line**. These omnivorous birds are best known for their breeding biology, with egg incubation being due to the use of external heat sources. The earliest known megapode is *Ngawupodius* from the late Oligocene of Australia, which reached only two-thirds the size of the smallest extant species (Boles and Ivison 1999). Otherwise, the fossil record of Megapodiidae is confined to the Quaternary of Australia and Oceania and mainly includes extinct species of extant taxa (Boles 2008). A notable exception is the very large *Progura* from the Plio-Pleistocene of Australia, which was considered most similar to the extant taxon *Leipoa* (Boles 2008).

Even sparser is the fossil record of cracids. Cracids are today only found in the Neotropic region, but these birds most likely originated outside South America (see the discussion at the end of this section). The earliest

Figure 7.2 Interrelationships of fossil and extant Galliformes with apomorphies characterizing some key nodes. The gray bars indicate the temporal ranges of some taxa.

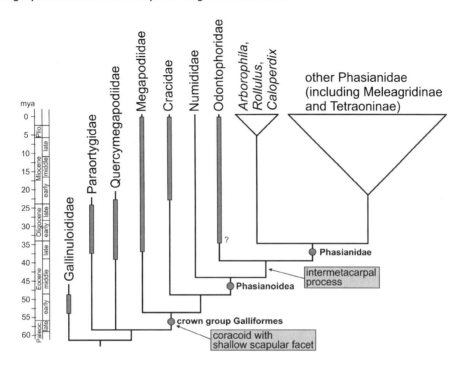

fossils are fragmentary remains from the early Miocene of Florida (Olson 1985a; Mayr 2009a). The later fossil record of cracids is also very sparse and tells us little about the evolutionary history of these birds.

In terms of its wide geographic distribution and high number of species, the third major taxon of crown group Galliformes, the Phasianoidea, is the evolutionarily most successful one, and the extant species are found in a great variety of habitats, from semideserts to tropical forests and subantarctic tundra. Within phasianoideans, the African guinea fowl (Numididae) and the New World quail (Odontophoridae) are successive sister taxa of the remaining species, which are classified in the Phasianidae and mainly occur in the Old World (N. Wang et al. 2013).

Fragmentary remains of putative Odontophoridae have been reported from the late

Eocene and early Oligocene of North America, but their identification is tentative at best (Mayr 2009a). The earliest undisputed fossil Phasianoidea stem from the early Oligocene of France and belong to *Palaeortyx*, which occurred in Europe until the early Pliocene and was also reported from the early Miocene of Namibia (Mourer-Chauviré 1992a, 2003; Göhlich and Mourer-Chauviré 2005; Göhlich and Pavia 2008). *Palaeortyx* is the first galliform taxon with an intermetacarpal process of the carpometacarpus (see Figure 1.9e), which is a derived trait of the Odontophoridae and Phasianidae. The preservation of gastroliths in one skeleton documents a similar diet to extant Phasianoidea (Plate 10a; Mayr et al. 2006). *Palaeortyx* is characterized by a double pneumotricipital fossa of the humerus. Earlier authors took this feature as indicative of closer affinities of *Palaeortyx* to either Odontophoridae or the Southeast

Asian *Arborophila*, one of the earliest diverging taxa of crown group Phasianidae. The exact affinities of *Palaeortyx* are, however, not well constrained and are in need of a modern revision. In any case, *Palaeortyx*-like phasianoideans existed during much of the Cenozoic and had a wide distribution. One of these, *Palaeocryptonyx*, occurred in Europe from the middle Miocene to the Pleistocene (Pavia et al. 2012), and a similar taxon was also described from the middle Miocene of Mongolia (*Tologuica*; Zelenkov and Kurochkin 2009a).

The sudden appearance of phasianoidean Galliformes in the early Oligocene of Europe may indicate an origin of crown group Galliformes in the late Eocene of Asia (Mayr 2009a), which after the early Oligocene closure of the **Turgai Strait** (e.g., Brikiatis 2014) could disperse into Europe. Colonization of Australia by galliform birds probably did not occur until northward drifting brought this continent close to Asia towards the Oligocene, and crown croup Galliformes may have reached South America via North America and the Beringian land bridge, which connected the latter continent with Asia in the early Cenozoic (Mayr 2009a).

In the past few years, analyses of molecular data have improved our understanding of the interrelationships of the Phasianidae (e.g., N. Wang et al. 2013), and the affinities of many fossil taxa need to be revisited on the basis of this phylogenetic framework. This is true for *Miogallus* ("*Miophasianus*"), a large phasianid, which was widely distributed in the Miocene of Europe (Mlíkovský 2002). The tarsometatarsus of the males of *Miogallus* exhibits a distinct spur that appears to have evolved several times independently within Phasianidae, being present in Pavoninae (peafowl and allies), Phasianinae (pheasants and allies), and various other taxa. Some authors considered *Miogallus* to be most closely related to peafowl (Cheneval 2000), but a reassessment of its affinities in

light of the new sequence-based phylogenetic hypotheses would be desirable. Large phasianid galliforms are also known from the middle Miocene of Mongolia and China (e.g., *Linquornis*, *Lophogallus*; Zelenkov and Kurochkin 2010); their relationships to *Miogallus* and the extant taxa are likewise insufficiently understood.

True peafowl have a disjunct extant distribution, with *Pavo* occurring in southern Asia and *Afropavo* in tropical Africa. It is therefore noteworthy that fossil species assigned to *Pavo* occurred in the late Miocene to Pliocene of France and southeastern Europe, as well as in the early Pliocene of Africa (Mourer-Chauviré 1989; Boev 2002; Louchart 2003; Pickford et al. 2004).

The interrelationships of extant phasianids indicate that their last common ancestor lived in the Old World, but the current fossil record contributes only little to an understanding of the biogeographic history of the extant clades. Most extant African phasianids belong to the Numididae or to the phasianid francolins. However, fossils from the early Miocene of Namibia, which resemble some extant Asian phasianid taxa, suggest that the past diversity of African galliforms was higher (Mourer-Chauviré 2003). Phasianids were diversified in the middle and late Miocene of Asia (Zelenkov and Kurochkin 2009b, 2010), but the exact affinities of these fossils to coeval ones from Europe and to the extant taxa are elusive. At least some Neogene phasianids appear to have achieved a wide distribution, such as *Plioperdix*, to which fossils from the Pliocene of Europe, Morocco, and Mongolia were attributed (Mourer-Chauviré and Geraads 2010; Zelenkov and Kurochkin 2009b).

Phasianids did not reach South America, and all native New World species belong to either Meleagridinae (turkeys) or the Palearctic Tetraoninae (grouse), which are sister groups and deeply nested with Phasianidae

(Figure 7.2; N. Wang et al. 2013). The origin of the clade including these two taxa is likely to have been in North America, from where Tetraoninae dispersed into Europe in the late Cenozoic. The earliest fossil assigned to the Meleagridinae is *Rhegminornis* from the early Miocene of Florida (Olson 1985a), but this taxon is only known from a distal tarsometatarsus and its identification can therefore not be considered unambiguous. The earliest uncontroversial turkeys are from the Pliocene of North America (Stidham 2011). Earlier records of Tetraoninae from the Miocene of North America still need to be verified (Olson 1985a), and from Europe no pre-Pliocene grouse fossils have been described (Mlíkovský 2002).

The mid-Cenozoic evolution and diversification of the Phasianidae may have been shaped by the spread of open grasslands, and many of the characteristic extant phasianid taxa seem to be of a comparatively young age. Only after a revision of the numerous fossil phasianids, however, will it be possible to draw well-founded conclusions on the evolutionary history of these birds.

The Waterfowl

Of the extant Galloanseres, only the Anseriformes entered aquatic habitats, and exploitation of this new ecological zone provided the basis for the formidable diversification of these birds. Unlike most of the highly aquatic neoavian taxa, which are carnivores, most anseriforms feed on plants or small aquatic invertebrates. The members of Anatidae, the most species-rich taxon of crown group Anseriformes, are characterized by horny lamellae along the cutting edges of the beak, which represent an adaptation for filter feeding.

The two early diverging anseriform groups, screamers and magpie geese

Anhimidae today only occur in South America and include three species that are distinguished from other anseriform birds in numerous features, including an unspecialized and rather galliform-like beak, the absence of webbed feet, and a highly pneumatized skeleton. There exists fossil evidence for an origin of screamers outside South America, and undescribed remains of early Eocene anhimid-like birds have been reported from North America and Europe (Ericson 1997; Feduccia 1999). One of the oldest published fossil screamers is *Chaunoides* from the late Oligocene or early Miocene of Brazil (Alvarenga 1999). *Chaunoides* was smaller than the Northern Screamer, *Chauna chavaria*, the smallest extant species of the Anhimidae, and it had a less pneumatized skeleton. Possibly also related to screamers is *Paranyroca* from the early Miocene of South Dakota, which is based on a tarsometatarsus that exhibits a notable similarity to the corresponding bone of the Anhimidae (Miller and Compton 1939). A screamer-like quadrate from the early Eocene of Australia (Elzanowski and Boles 2012), on the other hand, is probably too fragmentary for a reliable identification.

The single extant member of the Anseranatidae is the Australasian Magpie Goose (*Anseranas semipalmata*). Despite its goose-like external appearance, this species is distinguished from true geese and other Anatidae in a number of plesiomorphic features, including only partially webbed feet and a beak with poorly developed lamellae. The earliest Australian stem group anseranatid, the late Oligocene or early Miocene *Eoanseranas* (Worthy and Scanlon 2009), is based on a few pectoral girdle bones that resemble those of the somewhat larger Magpie Goose.

That Anseranatidae had an even earlier history, and one outside the Australasian

Figure 7.3 Various Paleogene and extant Anseriformes. (a–d) *Anatalavis* (?Anseranatidae) from the early Eocene London Clay. (e–h) The extant *Anseranas* (Anseranatidae). (i–l) The early Eocene presbyornithids (i–k) *Telmabates* and (l) *Presbyornis*. (m–o) The extant *Anas* (Anatidae) (a, e: skulls; b, f, k, n: coracoids; c, g: furculae; d, i, j, h, m: humeri; l, o: tarsometatarsi). Note the greatly elongated tarsometatarsus of presbyornithids.

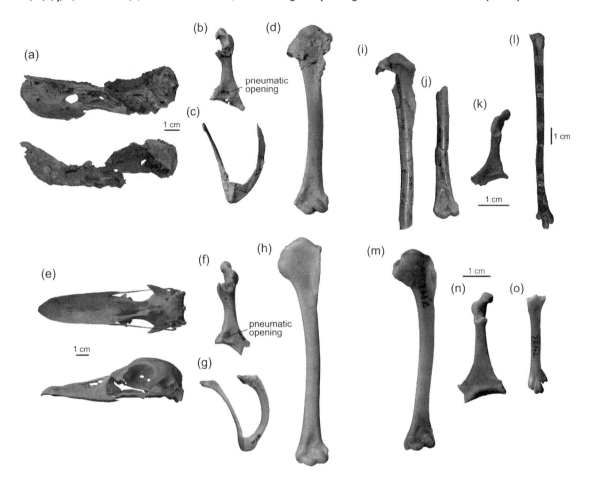

region, is suggested by a remarkable fossil of a goose-sized anseriform from the early Eocene British London Clay (Figure 7.3a–d). This partial skeleton was assigned to the taxon *Anatalavis*, which was first established for a humerus from the Late Cretaceous or early Paleocene of North America (see Chapter 5; Olson 1999a). Whether this classification is correct remains to be seen, but in any case the London Clay fossil represents one of the earliest modern-type anseriforms, therefore being critical for an understanding of the early evolution of waterfowl. Assignment of the London Clay anseriform to the Anseranatidae was based on several derived characters, but in many skeletal traits it differs from *Anseranas* (Figure 7.3e–h) and some even set it apart from all extant Anseriformes (Olson 1999a; Mayr 2009a). Unlike extant Anseranatidae, the fossil has a wide, duck-like beak (Olson 1999a). If it is indeed a stem group representative of the Anseranatidae, it would either indicate that such a beak occurred convergently in taxa outside Anatidae or that it evolved early in anseriform evolution and was lost in crown group Anseranatidae. Another putative anseranatid was described from the late Oligocene of France (*Anserpica*; Mourer-Chauviré et al. 2004), but this taxon is only known from a coracoid and for a definite identification more bones would be desirable.

Presbyornithidae, or what did the anseriform stem species look like?

Certainly among the most peculiar anseriforms are the long-legged Presbyornithidae, which were misclassified as shorebird or flamingo relatives before their duck-like skulls were described. Well-known representatives of these birds are *Telmabates* from the early Eocene of Patagonia and *Presbyornis*, which is abundantly represented in the Paleocene and early Eocene of North America and Mongolia (Figure 7.3i–l; Howard 1955; Olson and Feduccia 1980a; Ericson 2000; Kurochkin and Dyke 2010). Although a putative presbyornithid from the Late Cretaceous of Mongolia (Kurochkin et al. 2002) is based on remains too fragmentary for a well-founded identification, the evolutionary history of presbyornithids may well extend into the Late Cretaceous, and the humerus of *Presbyornis*, for example, is very similar to that of the poorly known taxon *Graculavus* from the Late Cretaceous/early Paleocene of North America (Olson and Parris 1987).

The beak of the particularly well-represented *Presbyornis* resembles that of extant filter-feeding ducks, but it is unknown whether horny lamellae – a requisite of filter feeding – were present. The postcranial skeleton of presbyornithids, by contrast, is very different from that of ducks and all other extant Anseriformes in many respects. Some of these differences may be functionally correlated with the greatly elongated legs of these birds (Figure 7.3l), but

others, such as a non-pneumatized humerus, are more likely due to the retention of plesiomorphic features. In this regard, it is notable that the caudalmost thoracic vertebrae of *Telmabates* are opisthocoelous (with convex cranial and concave caudal articular surfaces), rather than heterocoelous as in extant Anseriformes and most other neornithine birds (Howard 1955).

Some analyses found a sister group relationship between Presbyornithidae and Anatidae, but this hypothesis is only weakly supported in terms of derived characters (Ericson 1997; Livezey 1997; Mayr 2008b). Unfortunately, an assessment of the affinities of presbyornithids is hampered by the poorly known morphology of the skull, which has not yet been studied in detail, despite the availability of numerous well-preserved specimens. The quadrate, the only skull bone of which more detailed descriptions exist, exhibits several presumably plesiomorphic characteristics, which suggest a position of presbyornithids outside crown group Anseriformes (Elzanowski and Stidham 2010).

The phylogenetic affinities of presbyornithids are of significance for an understanding of character evolution in Anseriformes. If they occupy a more basal position within Anseriformes than is currently assumed, their bill shape would provide evidence for the hypothesis that a duck-like bill evolved early in anseriform evolution and was secondarily lost in the Anhimidae (Olson and Feduccia 1980a).

Towards a filter-feeding duck

Most of the about 150 extant species of the Anatidae are short-legged birds and occur in aquatic and semi-aquatic habitats. Swans, geese, and ducks display a wide variety of dietary specializations and include specialized grazers and filter feeders, as well as species that predominantly feed on fish or mollusks. All species exhibit lamellae along the margins of the beak, which indicate that the stem species of Anatidae was a filter-feeding aquatic bird that nourished on small-sized food particles.

With regard to the fossil record, the early origin of Anatidae remains obscure. Disregarding the poorly known and doubtfully anseriform *Eonessa* from the middle Eocene of North America (Olson and Feduccia 1980a; Mayr 2009a), the earliest uncontested "duck-like" anseriform is the small, teal-sized *Romainvillia* from the late Eocene of France. Overall, the skeletal morphology of this taxon is similar to that of modern Anatidae, with which it shares a reduced hind toe. *Romainvillia* is, however, outside crown group Anatidae, from which it is distinguished by, for example, a coracoid with a well-developed foramen for the supracoracoideus nerve; this foramen is reduced in crown group Anatidae (Mayr 2008b). The short tarsometatarsus of *Romainvillia* differs from that of crown group Anatidae in the shape of the trochlea for the second toe, which was interpreted as possible evidence that the taxon was less adapted to swimming than its extant relatives (Mayr 2008b). *Romainvillia*-like anseriforms were also reported from the early Oligocene of England and Belgium (Mayr 2009a), and putative records exist from the late Eocene of China and the late Oligocene of France (Mayr and De Pietri 2013; Stidham and Ni 2014).

The evolution of swans, geese, and ducks

Anatidae is the most widely distributed and most species-rich taxon of crown group Anseriformes. In spite of the ubiquity and abundance of these birds, there are comparatively few studies of their interrelationships, and this is true for both molecular and morphological data. It is generally assumed that Dendrocygninae (whistling ducks) are the sister taxon of all remaining Anatidae. The latter are usually divided into Anserinae (swans and geese) and Anatinae (typical ducks). One analysis of morphological data, however, found Anserinae to be the most basally diverging taxon of Anatidae (Worthy and Lee 2008), which is, among other features supported by the fact that Dendrocygninae share with Anatinae a bulbous enlargement of the trachea (syringeal bulla) that is absent in the Anserinae.

The fossil record of Anatidae is extensive, but only in recent years have attempts been made to put it into a phylogenetic context. The oldest modern-type anatids stem from the earliest Oligocene of Europe, but their affinities cannot be assessed owing to the fragmentary nature of the specimens (Mayr and Smith 2001; Mayr 2009a). Overall, however, these fossils resemble the next-oldest anatids, which belong to the widely distributed taxon *Mionetta*. This medium-sized anatid first occurs in the late Oligocene of France and is very abundant in early Miocene European fossil sites, mainly the Saint-Gérand-le-Puy area in France, where thousands of bones were found (Mlíkovský 2002; Mayr 2009a; Zelenkov 2012a). Outside Europe, *Mionetta* was reported from the early Miocene of Namibia (Mourer-Chauviré 2008), from where another *Mionetta*-like anatid, "*Anas luederitzensis*," was described in the early 20th century. Phylogenetic analyses placed *Mionetta* as the sister taxon of either all other Anatidae exclusive of Dendrocygninae (Livezey and Martin 1988) or of all Anatidae exclusive of Anserinae and Dendrocygninae (Worthy and Lee 2008). No matter what its exact affinities are, *Mionetta* is therefore the earliest anatid taxon for which a position within crown group Anatidae is assumed.

Anserinae, the clade including geese (Anserini) and swans (Cygnini), dates back to the early Miocene, by which time large anserines resembling the Australian *Cereopsis* and the New Zealand *Cnemiornis* occurred in the early Miocene of New Zealand (Worthy et al. 2008). Anserines have also been described from the early and middle Miocene of Europe (Mlíkovský 2002), the late Miocene of Mongolia (*Heteroanser, Bonibernicla*; Zelenkov 2012a), and the middle Miocene of California (*Presbychen*; Howard 1992). However, unraveling the exact phylogenetic relationships of most of these taxa, which are only known from fragmentary remains, requires further studies. Definitive fossils of the Cygnini first occurred in the middle Miocene of Germany, from where a fragmentary, and more questionable, early Miocene record also exists (*Cygnavus*; Mlíkovský 2002). Of particular biogeographic interest is the occurrence of an extinct taxon of Cygnini (*Afrocygnus*) in the latest Miocene of Chad and Libya (Louchart et al. 2005a), because swans do not naturally occur in Africa today.

Anatinae, the clade including most of the "duck-like" anseriforms, is divided into the Oxyurini (stiff-tailed ducks), Tadornini (shelducks), Aythyini (diving ducks), Mergini (eiders, scoters, mergansers), and Anatini (dabbling ducks), as well as other taxa of uncertain affinities, such as the Australasian *Malacorhynchus* and *Stictonetta*. Anatines diversified by the early Miocene at the latest, but the relationships of many fossils are uncertain owing to the poorly resolved phylogeny of extant Anatinae and the circumstance that many extinct taxa are only known from a few bones. Fossils that show similarities to the Tadornini were found in the late Oligocene or early Miocene of Australia (*Australotadorna*; Worthy 2009) and in the early Miocene of New Zealand (*Miotadorna*; Worthy and Lee 2008). However, many of the earliest anatines exhibit morphological features that have been interpreted as diving adaptations. That is true for fossils from the late Oligocene or early Miocene of Australia (*Pinpanetta*) and the early-middle Miocene of New Zealand (*Dunstanetta, Manuherikia*; Worthy and Lee 2008; Worthy 2009). Diving adaptations were also reported for two medium-sized anatines from the middle Miocene of Mongolia (*Sharganetta, Nogusunna*; Zelenkov 2011a). The affinities of these latter fossils are uncertain, but the New Zealand *Dunstanetta* and *Manuherikia* were considered to be most closely related to the Oxyurini, a globally distributed group of highly aquatic diving ducks. The position of these fossils is, however, dependent on whether presumed diving adaptations are included in the analyses, and in the resulting phylogenies (Worthy and Lee 2008) extant taxa with diving capabilities group together, which are not found to be closely related in molecular analyses (e.g., Donné-Goussé et al. 2002). It is possible that oxyurine anatines and other diving ducks were already diversified by the early Miocene and perhaps originated in the Australasian region, from where they dispersed northwards via Asia. Alternatively, however, some of the features shared by these early anatines and extant diving ducks may be plesiomorphic for Anatidae, or they may have evolved several times independently. This is especially true for one of the most distinctive characters, the absence of pneumatic openings in the proximal end of the humerus. Because the humerus is non-pneumatized in some early anseriforms outside Anatidae, such as *Presbyornis*, *Romainvillia*, and *Mionetta*, the absence of pneumatic openings may be plesiomorphic for Anseriformes, but the humerus is pneumatized in Anhimidae, Anseranatidae, and some basal Anatidae.

The interrelationships of the major groups of Anatinae are poorly resolved, but derived features of the humerus suggest a clade including Aythyini, Anatini, Mergini, and a few other taxa (Mayr and Pavia 2014). Fossil anatines with this humerus morphology first occur in middle Miocene localities. These include *Mioquerquedula* from the middle Miocene of Mongolia and, possibly, France (Zelenkov and Kurochkin 2012), as well as *Protomelanitta* from the middle Miocene of

Mongolia, which was tentatively assigned to Mergini (Zelenkov 2011a). A record of the Mergini also exists from the middle Miocene of Virginia (Olson 1985a). The diversification of dabbling ducks (Anatini), the most species-rich extant anseriform clade, appears to have started comparatively late, and reliably identified records of the widespread taxon *Anas* are unknown before the latest Miocene.

The biogeographic history of the Anatidae is likewise incompletely understood. Although duck-like anseriforms are quite abundant in some early Miocene European localities, their taxonomic diversity is low. This contrasts with the much more diverse waterfowl faunas from the early Miocene St Bathans fossil site in New Zealand (Worthy et al. 2007, 2008; Worthy and Lee 2008) and the middle Miocene Sharga locality in Mongolia (Zelenkov 2011a, 2012a, b; Zelenkov and Kurochkin 2012). Without a robust phylogenetic framework for these fossils, however, little can be said about their biogeographic significance, and the oldest substantial remains of modern-type Anatidae still stem from European fossil sites.

There are also a few fragmentary remains of putative anatids from the late Oligocene of Patagonia (Argentina), but once more these are too incompletely preserved to establish their relationships (Mayr 2009a). The earliest well-represented South American anatid-like anseriform is *Cayaoa* from the early Miocene of Patagonia (Argentina), a flightless, foot-propelled diving taxon of uncertain affinities, which has greatly reduced wings (Noriega et al. 2008). From the middle Miocene of South America, a putatively basal anatid taxon, *Ankonetta*, has been reported that shows similarities to the Dendrocygninae (Cenizo and Agnolín 2010).

Gastornithids: Giant Herbivorous Birds in the Early Paleogene of the Northern Hemisphere

The flightless gastornithids (Figure 7.4a) were forest-dwelling birds with robust legs and are among the most widely popularized birds from the early Paleogene of the Northern Hemisphere. The largest species reached a standing height of nearly 2 meters and had an estimated weight of about 175 kilograms (Andors 1992). Apart from their huge size, the most characteristic attribute of gastornithids is their oversized, deep beak, which has a strongly curved dorsal ridge but lacks a hooked tip. The greatly reduced wings and the absence of a sternal keel indicate a long evolutionary history of flightlessness for these birds.

In the mid-19th century the first gastornithid fossils had already been reported from the late Paleocene of France. Subsequently, numerous further remains were found in the Paleocene to middle Eocene of Europe and the early Eocene of North America and China (Mayr 2009a; Buffetaut 2013). The earliest specimens stem from the Paleocene of Germany (Mayr 2007), but gastornithid bones from the late Paleocene of France are only slightly younger (Martin 1992; Bourdon et al. 2016).

All gastornithids are now assigned to the taxon *Gastornis*, which includes the North American "*Diatryma*" and the Asian "*Zhongyuanus*" (Mayr 2009a; Buffetaut 2013). Various species have been described, but the distinctness of some of these is doubtful, and a revision of the fossils is overdue (Mayr 2009a). Some of these oldest gastornithids are distinctly smaller than the Eocene species and differ in presumably plesiomorphic features, such as the absence of a scapulocoracoid, which indicates that the earliest gastornithids are also the

Figure 7.4 Skeletons of giant flightless Cenozoic galloanserines. (a) The gastornithid *Gastornis* (Gastornithidae) from the Paleocene and early Eocene of the Northern Hemisphere (redrawn after Matthew and Granger 1917). (b) The dromornithid *Bullockornis* from the middle Miocene of Australia (redrawn after Murray and Vickers-Rich 2004).

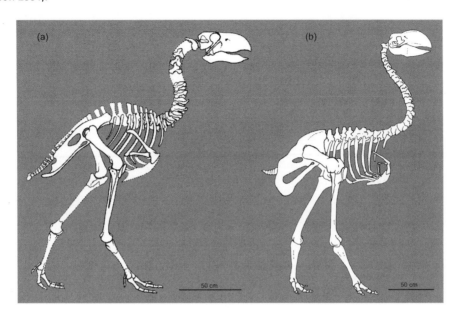

phylogenetically most basal representatives (Mayr 2007, 2009a).

The evolutionary origin of gastornithids is unknown, but they are likely to have lost their flight capabilities in an environment with a reduced predation pressure, which existed in insular Europe during the earliest Cenozoic (Mayr 2009a). The fact that the geologically oldest and phylogenetically most basal species occurred in Europe also suggests that gastornithids originated in the Old World and dispersed into North America in the earliest Eocene. That they took a route across a corridor in high northern latitudes is suggested by a record of these birds from the early Eocene of Ellesmere Island in the Canadian Arctic (Eberle and Greenwood 2012). The occurrence of gastornithid fossils in North America is, however, currently restricted to the early Eocene of the Rocky

Mountain region, and two species are recognized, which mainly differ in toe proportions (Andors 1992).

A *Gastornis* fossil also exists from the early Eocene of China (Buffetaut 2013). Because Europe and Asia were separated by the Turgai Strait in the early Eocene, it is most likely that the flightless gastornithids dispersed into Asia from North America rather than Europe. However, more data on the stratigraphic distribution of these birds in Asia is needed to establish this hypothesis firmly (Mayr 2009a; see also Buffetaut 2013).

The fossil record of gastornithids includes more than just skeletal remains. In southern France very large eggs occur in early Eocene sites, which are likely to stem from *Gastornis* (Mayr 2009a; Angst et al. 2015). Furthermore assigned to the taxon are footprints from the early Eocene of Washington State (Mustoe et al. 2012). Other putative *Gastornis* footprints from the middle Eocene of North America and the late Eocene of Europe are, however, younger than skeletal records of gastornithids in these areas and are of questionable authenticity (Mayr 2009a).

Gastornithids are sometimes portrayed as top predators in Eocene Northern Hemispheric ecosystems, but both the anatomical evidence and analyses of carbon isotopes of the bone apatite indicate a herbivorous diet (Andors 1992; Mayr 2009a; Angst et al. 2014). Sessile basipterygoid processes, features of the quadrate, and long retroarticular processes of the lower jaws have been listed in support of the galloanserine affinities of gastornithids (Andors 1992). Their exact relationships are, however, far from being well understood, and whether they are indeed the sister taxon of the Anseriformes (Andors 1992) needs to be reassessed; at least in details of the basipterygoid articulation, gastornithids are distinguished from all extant Galloanseres.

Dromornithids (Mihirungs or Thunderbirds): *Gastornis*-Like Birds from Australia

Like on other isolated land masses devoid of larger carnivorous mammals, giant flightless birds also evolved in the Cenozoic of Australia. The most remarkable of these are the Dromornithidae, stoutly built and heavy birds with greatly reduced wings, which at least superficially resembled the Northern Hemispheric gastornithids in their skeletal morphology (Figure 7.4b).

The fossil record of dromornithids spans a much longer geological period than that of gastornithids and it is also more comprehensive, with over 2,500 bones or fragments thereof having been found (Murray and Vickers-Rich 2004). The oldest remains assigned to these birds are tentatively identified impressions of pedal phalanges from the early Eocene of Queensland and putative

footprints from the late Oligocene of Tasmania (Murray and Vickers-Rich 2004). The earliest definitive dromornithid represented by skeletal remains is *Barawertornis* from the late Oligocene or early Miocene of the Riversleigh Formation. *Barawertornis* was a rainforest-dwelling but cursorial bird and, with a size similar to that of the Southern Cassowary (*Casuarius casuarius*), it is the smallest dromornithid (Nguyen et al. 2010).

Six further dromornithid species are known from Neogene deposits, and these belong to the taxa *Bullockornis* (middle Miocene), *Ilbandornis* (middle to late Miocene/early Pliocene), *Dromornis* (middle Miocene to early Pliocene), and *Genyornis*. The latter is the youngest and most completely known taxon, which occurred until the Pleistocene. *Genyornis* was possibly even depicted in Aboriginal rock art (Gunn et al. 2011), and its extinction may have coincided with the disappearance of the Australian **megafauna** (see, however, Grellet-Tinner et al. 2016, who question the identification of late Pleistocene eggshell remains).

Whereas the species of *Barawertornis* only weighed about 70 kilograms, the largest dromornithid species, *Dromornis stirtoni*, reached a standing height of 2.7 meters and had an estimated weight of up to 500 kilograms (Murray and Vickers-Rich 2004). Like gastornithids, dromornithids are considered to have been herbivorous, an assumption that is in concordance with the preservation of gastroliths in some dromornithid specimens (Murray and Vickers-Rich 2004). A herbivorous diet was also inferred from analyses of stable isotopes of putative *Genyornis* eggshells, but these fossils are now considered to stem from another large flightless bird, the megapode *Progura* (Grellet-Tinner et al. 2016; see the earlier discussion in Chapter 6 concerning alternative candidate eggs for *Genyornis*).

Like gastornithids, dromornithids converged on flightless palaeognathous birds in several features related to the loss of flight ability, such as the reduction of a sternal keel and the fusion of scapula and coracoid into a scapulocoracoid. Well-preserved skulls are known from *Dromornis* and *Bullockornis*, which have huge beaks with strongly arched dorsal ridges. Sessile basipterygoid processes and lower jaws with long retroarticular processes suggest galloanserine affinities of dromornithids, which are currently considered to be most closely related to the Anseriformes (Murray and Vickers-Rich 2004).

Despite their great overall similarity to gastornithids, dromornithids are clearly distinguished in a number of skeletal traits, including the shape of the palatine bones, the absence of a hind toe, and the loss of one phalanx of the fourth toe, which consists of four instead of five phalanges. Not least for biogeographic reasons, the similarities between gastornithids and dromornithids must be the result of convergent evolution, because there were no early Cenozoic land connections between Australia and the Northern Hemisphere that would have permitted the dispersal of large flightless birds.

Pelagornithids: Bony-Toothed Birds

Gastornithids and dromornithids were giant flightless birds. The present section introduces another group of putative galloanserines that includes fundamentally different birds, which were among the most advanced aerial specialists and feature some highly unusual specializations.

In 1873, the British anatomist Richard Owen announced the skull of a "dentigerous" (tooth-bearing) bird from the early Eocene of the British Isle of Sheppey, the description of which followed the discovery of toothed Mesozoic birds only a few years before. Instead of true archosaurian teeth,

however, the beak of this British fossil shows mere spikey excrescences, which emerge from the bony jaws and are neither situated in alveoles nor covered with enamel (Figure 7.5a, Plate 11b). During the 19th century, various other bones of these "bony-toothed birds" were described from the Isle of Sheppey and elsewhere, often without recognition of their true affinities. The taxonomic history of some of the various fossils is therefore very convoluted. Many of the names in the older literature are now considered synonyms of earlier, incorrectly identified taxa, and even the species that Owen (1873) originally assigned to *Odontopteryx* is now classified in the taxon *Dasornis* (Bourdon et al. 2010).

Not until the 1960s did it become evident that the first remains of these peculiar seabirds were in fact already discovered two decades before Owen's report, and *Pelagornis*, the namesake of the taxon Pelagornithidae to which bony-toothed birds are now assigned, is based on very large humeri from the early and middle Miocene of France, which were described in 1857. Until the discovery of a partial skeleton from the Miocene of California (Howard 1957), the postcranial skeletal morphology of bony-toothed birds nevertheless remained very poorly known. Even this latter fossil did not allow close examination of many skeletal details, and the anatomy of pelagornithids has become better understood only in the past few decades. Pelagornithids are now recognized as some of the most unusual Cenozoic birds. This is not only due to the peculiar and unique pseudoteeth, but to the fact that bony-toothed birds were also among the most efficient soaring birds and include some of, if not the, largest volant avian species that ever lived.

The remains of these pelagic birds are known from all continents and were found in late Paleocene to late Pliocene strata. The oldest definitive pelagornithid fossils stem from late Paleocene deposits of the Atlantic

Figure 7.5 Main skeletal elements of the bony-toothed bird *Pelagornis chilensis* (Pelagornithidae) from the late Miocene of Chile (a: skull, b: furcula, c: coracoid, d: carpometacarpus, e, f: humeri, g: radius and ulna, h: femur, i: tibiotarsus, j, k: tarsometatarsus). Note the extremely elongated humerus and very slender car-pometacarpus that characterize these highly specialized soaring birds.

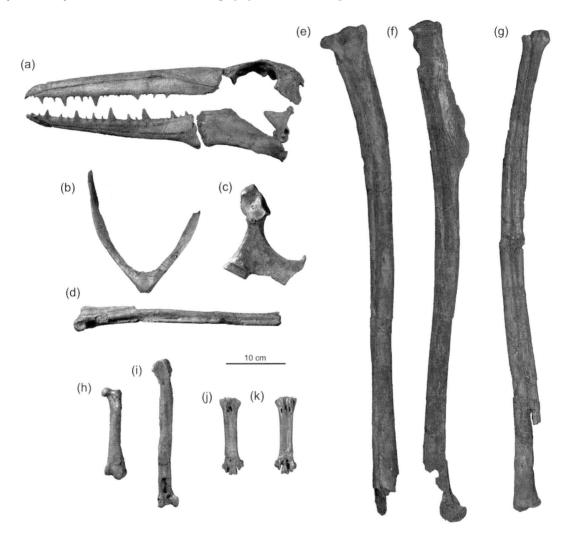

Ocean and the Tethys Sea, an epicontinental ocean covering large parts of Eurasia during the early Cenozoic. Remains of the late Pale-ocene/early Eocene *Dasornis* are abundant in phosphate deposits of Morocco (Bourdon et al. 2010). In these localities and in the British London Clay, *Dasornis* species of very different size occur, the largest of which reached a wingspan of more than 4 meters (Plate 11c; Mayr 2009a; Bourdon et al. 2010).

The wing skeleton of *Dasornis* was less specialized for sustained soaring than that of later pelagornithids, with some characteristics of the foot skeleton suggesting that the taxon was also more aquatic (Bourdon et al. 2010; Mayr and Zvonok 2012). Already by the middle Eocene, however, pelagornithids featured a more advanced morphology with an unusually narrow proximal end of the humerus, which restricted rotation of this bone (Olson 1985a; Bourdon et al. 2010; Mayr and Zvonok 2012). Towards the late Eocene they also achieved a near-global distribution, including Antarctica, the Pacific, and West Africa, where remains of the large *Gigantornis* were found (Mayr 2009a; Bourdon and Cappetta 2012; Cenizo 2012). Again, most records consist of fragmentary fossils, but *Lutetodontopteryx* from the middle Eocene of Ukraine is represented by all major bones and has particularly long and narrow pseudoteeth (Plate 11d; Mayr and Zvonok 2011, 2012). This taxon is about the size of the smaller *Dasornis* species and coexisted with another, much larger species, which may be closely related to *Gigantornis* (Mayr and Zvonok 2012). In other sites, too, pelagornithid species of very different size occurred together (Bourdon et al. 2010; Mayr and Smith 2010). By the Oligocene pelagornithids had disappeared from the Eurasian epicontinental seas, and it is unknown whether they still occurred in Antarctica after glaciation of this continent commenced towards the late Eocene.

The nomenclature and taxonomy of late Paleogene and Neogene pelagornithids are poorly resolved, because many species are based on insufficient material of sometimes uncertain stratigraphic origin. To simplify pelagornithid taxonomy it was suggested that all Neogene species be assigned to the taxon *Pelagornis*, which has nomenclatural priority (Mayr et al. 2013a; see also Bourdon et al. 2010), and this classification is followed here.

The most completely known Neogene species is *Pelagornis chilensis* from the late Miocene of Chile, which is based on a skeleton of a single individual and had a minimum wingspan of 5.2 meters (Figure 7.5, Plate 11a, b; Mayr and Rubilar-Rogers 2010). Even larger is *P. sandersi* from the late Oligocene

of South Carolina, for which a wingspan of at least 6.4 meters was estimated (Ksepka 2014). By the late Miocene and Pliocene, pelagornithids still had a wide distribution (e.g., Olson 1985a; Olson and Rasmussen 2001; Fitzgerald et al. 2012; Mayr et al. 2013a), but they seem to have disappeared from near-equatorial regions, which may have been due to the emergence of the Intertropical Convergence Zone with its unpredictable wind conditions. The latest records stem from the middle/late Pliocene (3.4–2.5 mya) of California and the late Pliocene (2.5 mya) of Morocco (Mourer-Chauviré and Geraads 2008; Boessenecker and Smith 2011).

Pelagornithids were highly specialized soaring birds with greatly elongated wings and short legs. All species reached at least the size of a small albatross. Some were truly gigantic, however, and very large species already coexisted with smaller ones from the early Eocene onwards. Despite their large size, however, pelagornithids were comparatively lightweight birds and the weight of *P. chilensis* was estimated at only 16–29 kilograms (Mayr and Rubilar-Rogers 2010).

As noted earlier, the derived shape of the proximal end of the pelagornithid humerus prevented rotation of this bone. For this reason, pelagornithids were probably not capable of sustained flapping flight and must have spread their wings against strong headwinds for take-off (Olson 1985a). Because the carpometacarpus is extremely narrow, the primary feathers are likely to have been very short (Mayr and Rubilar-Rogers 2010). Modeled flight properties indicate that very large pelagornithids were at the upper limit for lift to drag and glide ratios and probably exploited large soaring ranges (Ksepka 2014). Orientation of the bony labyrinth of the inner ear, which plays a leading role in the sense of balance, and the very steep articulation facets of the cranialmost cervical vertebrae indicate that the skull was held in a near-vertical position in flight (Milner and

Walsh 2009; Mayr and Rubilar-Rogers 2010). Pelagornithid prey, which is assumed to have consisted of fish or squid (Olson 1985a; Zusi and Warheit 1992), may therefore have been captured by skimming the sea surface in flight.

Pelagornithids exhibit some highly peculiar characteristics that are unparalleled by other birds, of which the pseudoteeth are the most outstanding. These projections of the cutting edges of the upper and lower beak are hollow outgrowths of the jaws. Differences exist in the formation and orientation of the pseudoteeth of pelagornithids, but they are fairly regularly arranged in all species and very large pseudoteeth are separated by smaller second- and third-order ones. Vascular furrows on their surfaces indicate that these projections were covered by the rhamphotheca, and a basal plate separates their hollow lumen from that of the jaws. It was therefore hypothesized that either pseudoteeth formation or hardening of the rhamphotheca occurred comparatively late in ontogeny, after growth of the jaw bones was complete (Louchart et al. 2013). Pseudoteeth are readily distinguished from true avian teeth in that they are not covered with enamel, do not rest in alveoles, and their histology is typical of bone (Louchart et al. 2013). If pseudoteeth go back on incompletely expressed tooth-specific developmental programs, they may nevertheless be homologous to true archosaur teeth on a molecular level, because early ontogenetic stages of alligator teeth are also mere outgrowths of the epithelial surface layers, and the pathways for tooth development can still be induced in chicken embryos (Mayr and Rubilar-Rogers 2010; Mayr and Zvonok 2012).

The pseudoteeth are, however, not the only unusual feature of pelagornithids. At least in the Neogene species, the tip of the rostrum exhibits a marked transverse furrow, which is unknown from other birds (Bourdon et al. 2010; Mayr and Rubilar-Rogers 2010; this transverse furrow is absent in *Dasornis*). In all species, furthermore, there is a marked longitudinal furrow along the sides of the upper beak, which indicates a composite rhamphotheca. The upper beak is very long and its ventral surface bears large pits, which encompassed the mandibular pseudoteeth. The pseudoteeth of the upper beak are positioned on the lateral surfaces of the mandible in the closed beak. The mandible of pelagornithids is reminiscent of that of some toothed Mesozoic birds in that it lacks a symphysis and exhibits an intraramal joint (Zusi and Warheit 1992). A hind toe is absent and the pedal phalanges are unusually wide, but it is not known whether the toes were connected by a web as in most extant aquatic birds (Mayr et al. 2013a).

The phylogenetic affinities of pelagornithids are not well understood. Some skull features support their classification in Galloanseres, with which they share sessile basipterygoid processes and a bicondylar mandibular process of the quadrate. The hypothesis that pelagornithids are the sister taxon of Anseriformes (Bourdon 2005) cannot be upheld, however, as they lack derived characteristics of crown group Galloanseres, such as long and blade-like retroarticular processes of the lower jaws. Some characteristics of pelagornithids, especially the absence of a ventral crest on the palatine, indicate a position outside Neognathae (Mayr 2011c) and raise the question of whether the features shared with galloanserines are possibly plesiomorphic for neognathous birds.

8

The "Difficult-to-Place Groups": Biogeographic Surprises and Aerial Specialists

In spite of the progress made in past years in unraveling the interrelationships of neornithine birds, a number of taxa defy a robust phylogenetic placement, and in particular the early divergences within Neoaves remain controversial. As noted in Chapter 5, analyses of nuclear gene sequences initially suggested a basally branching clade termed "Metaves" that encompassed various morphologically disparate groups. This clade is now considered an artifact of one particular gene locus, but most of its component taxa are still placed at the base of Neoaves (Figure 5.1). Of these, the South American Hoatzin, the sole living representative of the Opisthocomiformes, comes close to the epitomization of a phylogenetic enigma. Other "metavian" groups of uncertain higher-level affinities are doves and allies (Columbiformes), as well as Strisores, the clade including nightjars, swifts, hummingbirds, and their relatives.

There exist various other groups, whose exact positions in the avian tree of life are still controversial. Although recent genome-scale analyses, for example, supported a clade including turacos (Musophagiformes), cuckoos (Cuculiformes), and bustards (Otidiformes), this clade resulted in very disparate positions in the different phylogenies (Hackett et al. 2008; Jarvis et al. 2014; Prum et al. 2015).

The taxa included in the present chapter highlight the current limitations in our understanding of avian interrelationships and constitute a criterion

Avian Evolution: The Fossil Record of Birds and its Paleobiological Significance, First Edition. Gerald Mayr.
© 2017 John Wiley & Sons, Ltd. Published 2017 by John Wiley & Sons, Ltd.

for an assessment of the strength of new phylogenetic analyses. Irrespective of their insufficiently resolved affinities, however, some of these groups have a comprehensive fossil record. In several cases this provides insights into the biogeographic history that would not have been gained from a study of the extant representatives alone, and concerning flamingos and hummingbirds it also contributes to an understanding of major evolutionary transitions.

The Columbiform Birds: Doves, Sandgrouse, … and Mesites?

Columbiformes as traditionally recognized include the Columbidae (doves and pigeons) and Pteroclidae (sandgrouse), two groups that differ distinctly in their species diversity, habitat preferences, and way of life. The 16 extant species of the Pteroclidae only occur in Africa and Eurasia and are predominantly terrestrial denizens of open, arid landscapes. Columbidae, by contrast, comprise more than 300 species, have a worldwide distribution, and inhabit a large variety of habitats, from semideserts to tropical rainforests.

From a morphological point of view, a sister group relationship between Columbidae and Pteroclidae is well founded (e.g., Mayr and Clarke 2003; Livezey and Zusi 2007). A clade including both taxa also resulted from some earlier molecular analyses (e.g., Ericson et al. 2006), but more recent analyses of nuclear gene sequences placed the enigmatic Madagascan Mesitornithiformes (mesites) within Columbiformes, as the sister taxon of either Columbidae (Hackett et al. 2008) or Pteroclidae (Jarvis et al. 2014; Prum et al. 2015). Close affinities between mesites and columbiform birds certainly need to be scrutinized, as these birds share a number of derived characteristics, such as long schizorhinal nostrils, fused thoracic vertebrae (**notarium**), and similarities in the wing skeleton. Currently, however, there exists no morphological evidence for a

position of mesites within Columbiformes, and they lack most of the derived features shared by columbiform birds, such as a large crop and well-developed basipterygoid processes. Regardless of the question of columbiform monophyly, the higher-level relationships of these birds remain elusive. Analyses of nuclear gene sequences supported a sister group relationship to the clade including cuckoos, turacos, and bustards (Prum et al. 2015), or a basal position within Neoaves, as sister taxon of a clade including flamingos and grebes (Jarvis et al. 2014), whereas mitochondrial data suggested affinities to charadriiform birds (Pacheco et al. 2011).

The earliest well-dated stem group representative of the Pteroclidae is *Leptoganga* from the late Oligocene and the early Miocene of France (Mourer-Chauviré 1993). However, the occurrence of stem group Pteroclidae in Europe possibly goes even farther back in time, as three species of *Archaeoganga* from the late Eocene or Oligocene of France lack exact stratigraphic data. Like their extant relatives, the species of *Archaeoganga* occurred in arid habitats, although they were larger than extant Pteroclidae.

The oldest Columbidae are of comparable age and stem from the Australasian region. *Primophaps* from the late Oligocene or early Miocene of Australia was likened to the extant Australian taxon *Phaps* (Worthy 2012a). Another Australasian columbid,

Rupephaps from the early Miocene of New Zealand, was assigned to fruit doves (Ptilinopini) and considered to be most similar to the extant *Hemiphaga*, the only **endemic** extant columbiform taxon of New Zealand (Worthy et al. 2009). The exact affinities of *Gerandia* from the early Miocene of France still have to be determined. The earliest New World representative of the Columbidae is *Arenicolumba* from the early Miocene of Florida, a small species that is represented by hundreds of bones and was regarded as closely related to the African taxa *Oena* and *Turtur* (Steadman 2008).

The interrelationships of crown group Columbidae (Shapiro et al. 2002) do not provide a clear signal for the geographic origin of the group, but the distribution of the closely related sandgrouse (and mesites, for that matter) suggests that it was in the Old World. Already in the early Miocene, however, columbids had a wide distribution across the globe, and if the phylogenetic assignment of the above taxa is correct, the crown group must have undergone notable diversification by that time.

The Hoatzin: A South American Relict Species

The Hoatzin (*Opisthocomus hoazin*) is arguably among the most outstanding extant birds of tropical South America. It mainly occurs in the riparian lowland vegetation of the Amazonas and Orinoco basins and is superficially reminiscent of some galliform birds in external appearance. Hoatzins are obligate folivores, which process plant matter with a ruminant-like, microbe-assisted foregut fermentation in an unusually large crop. The mere space requirements of this muscular pouch led to a reduction of the cranial portion of the sternal keel and to a caudal shift of the pectoral muscles. Not least due to these anatomical peculiarities, hoatzins are notably poor long-distance fliers.

The relationships of the Hoatzin have been much debated and no consensus exists. Although the African turacos (Musophagiformes) show some skeletal similarities, a sister group relationship between Opisthocomiformes and Musophagiformes did not result from analyses of large molecular data sets, some of which placed hoatzins close to cranes and allies (Ericson et al. 2006; Hackett et al. 2008; Jarvis et al. 2014). Most recently, a sister group relationship to Telluraves, the landbird clade, was proposed based on analyses of genome sequence data (Prum et al. 2015). As yet, no morphological evidence has been put forward in support of any of these new molecular phylogenies.

The morphological and physiological distinctness of hoatzins suggests a long evolutionary history of their folivorous feeding specializations. Until a few years ago, however, the fossil record of this curious group of birds was limited to a fragmentary neurocranium from the middle Miocene of Colombia (*Hoazinoides*; Miller 1953). If correctly identified, this fossil documents the occurrence of hoatzins west of the Andes, but otherwise it does not tell us much about hoatzin evolution.

Meanwhile, further hoatzin fossils were identified, some of which revealed unexpected insights into the evolutionary history of these birds. Among the new fossils is the earliest South American record of Opisthocomiformes: *Hoazinavis* from the Oligo-Miocene of southeastern Brazil, which was found in an area where hoatzins likewise do not occur today (Figure 8.1d; Mayr et al. 2011b). *Hoazinavis* is smaller than the extant Hoatzin, but the known bones are otherwise very similar and indicate that South American hoatzins underwent few morphological changes in the past 22–24 million years.

Figure 8.1 Bones of fossil and extant hoatzins (Opisthocomiformes). (a) Coracoid of *Protoazin* from the late Eocene of France. (b) Coracoid and humerus of *Namibiavis* from the early Miocene of Namibia. (c) Tarsometatarsus of *?Namibiavis* from the middle Miocene of Kenya. (d) Fragmentary coracoid and humerus of *Hoazinavis* from the Oligo-Miocene of Brazil. (e) Coracoids of the extant *Opisthocomus* (left: juvenile; right: adult, in which furcula, coracoid, and sternum are co-ossified). (f) Humerus and (g) tarsometatarsus of *Opisthocomus*. The large pneumatic opening in the sternal end of the coracoid is a characteristic feature of the Opisthocomiformes.

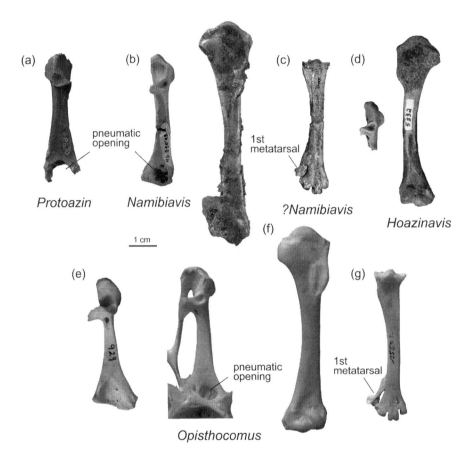

That hoatzins also had a long evolutionary history outside the New World is shown by fossils of *Protoazin* from the late Eocene of France (Mayr and De Pietri 2014). Even though only two fragmentary bones of this taxon have been found, these include the coracoid, which exhibits a characteristic derived morphology and allows a well-founded identification (Figure 8.1a).

Another fossil hoatzin, and one with an equally unexpected geographic distribution, is *Namibiavis* from the early Miocene of Namibia, which is represented by humeri and coracoids of several individuals (Figure 8.1b). These bones are similar to those of the roughly coeval South American *Hoazinavis*, but they differ in some plesiomorphic features, which suggest that *Namibiavis* is the sister taxon of a clade including *Hoazinavis* and the extant Hoatzin (Mayr et al. 2011b). A tarsometatarsus, which closely resembles the distinctive tarsometatarsus of the extant Hoatzin, has also been found in middle Miocene river sediments in Kenya (Figure 8.1c; Mayr 2014c). Unfortunately, the tarsometatarsus is unknown from the slightly older *Namibiavis*. Meaningful comparison between this taxon and the Kenyan opisthocomiform are therefore not possible, although close affinities are likely. The Kenyan tarsometatarsus exhibits a peculiar morphology in that the first metatarsal is fused to the shaft of the bone. This foot morphology fulfills the demands of a grasping foot, but would not be functional in a terrestrial bird. Hence, it can be assumed that these early African Opisthocomiformes were already strictly arboreal birds (Mayr 2014c).

The humeri of *Hoazinavis* and *Namibiavis* closely resemble that of the extant Hoatzin, especially with regard to the shape of the low deltopectoral crest. This indicates that these early hoatzins had limited flight capabilities, like their extant relative. Because the humerus morphology of the extant Hoatzin is functionally correlated with its large crop, stem group Opisthocomiformes likewise may have already featured a specialized alimentary tract and showed some degree of **folivory** (Mayr et al. 2011b).

Recognition of hoatzin fossils on both sides of the South Atlantic is of particular biogeographic interest, because separation of South America and Africa was already completed in the mid-Cretaceous, about 100 mya (Smith et al. 1994), and the distribution of stem group Opisthocomiformes is therefore more likely the result of dispersal than of vicariance. Birds have far better dispersal capabilities than flightless animals, but even small water bodies often pose insurmountable distribution barriers for arboreal taxa. At more than 1,000 kilometers in a straight line, the minimum distance between South America and Africa in the earliest Cenozoic was too large for active dispersal on the wing of a stem group opisthocomiform with flight capabilities similar to the extant species. Hence, it has been hypothesized that the transatlantic dispersal of hoatzins may have been due to rafting on floating vegetation islands, which was also assumed to have been the dispersal mode of the ancestors of Neotropic rodents and primates (Mayr et al. 2011b). Suitable flotsam is washed into oceans from the mouth of large rivers and can reach big dimensions, and as riparian, poorly flighted, folivorous birds, hoatzins are among the prime avian candidates for dispersal on floating vegetation. Because a westward journey on a floating raft was favored by the direction of Cenozoic currents, dispersal of stem-Opisthocomiformes from Africa to South America has been considered most likely (Mayr et al. 2011b).

Opisthocomiformes were long considered endemic for the Neotropic avifauna. However, the occurrence of stem group representatives of hoatzins in the late Eocene of Europe and in the Miocene of Africa shows their extant distribution in the Neotropic region to be relictual. Although the early evolutionary history of these birds remains poorly known, their origin was probably outside South America, and the classification of some early Cenozoic fossils from North American and European fossil sites may need to be reconsidered in the light of possible opisthocomiform affinities.

Turacos and Cuckoos

Africa houses comparatively few endemic extant avian higher-level taxa. One of these, the turacos (Musophagiformes), however, includes some of the most distinctive birds of the continent. These arboreal frugivores mainly occur in sub-Saharan regions. They are characterized by a deep beak and unique green and red feather pigments, which are unknown from other birds. The 23 extant species have semizygodactyl feet and were traditionally considered to be most closely related to Opisthocomiformes and cuckoos (Cuculiformes).

A comprehensive earlier analysis of nuclear gene sequences resulted in unexpected sister group relationships between Musophagiformes and Aequornithes – the "waterbird clade" – on the one hand, and between Cuculiformes and the core-Gruiformes (cranes, rails, and allies) on the other (Hackett et al. 2008). Whereas this analysis therefore did not support close affinities between turacos and cuckoos, more recent analyses of complete nuclear genomes resulted in a clade including Cuculiformes, Musophagiformes, and Otidiformes (bustards; Jarvis et al. 2014; Prum et al. 2015).

The earliest fossil taxon that shows some similarities to turacos is *Foro* (Foratidae) from the early Eocene of North America. This long-legged bird is the size of an average phasianid and is known from a well-preserved skeleton (Olson 1992b). It was considered to be most similar to Musophagiformes and Opisthocomiformes in the original description. Similarities to the Hoatzin are restricted to the skull, whereas the postcranial skeleton of *Foro* exhibits some derived characteristics that are also found in turacos (Olson 1992b; Mayr 2009a). However, judging from its long legs, *Foro* was much more terrestrial than extant turacos, and if affinities to the Musophagiformes can be established, this documents a major change in the ecological attributes of this group.

Regardless of the exact affinities of the peculiar early Eocene *Foro*, the fossil record suggests that crown group Musophagiformes evolved in Africa. The oldest modern-type turaco fossils, from the early Oligocene of Egypt, already resemble the phylogenetically basal extant taxon *Crinifer* in presumably plesiomorphic features and belong to a bird that was larger than all but the largest extant turaco species (Rasmussen et al. 1987). Neogene African turaco remains are known from the early and middle Miocene of Kenya (*Veflintornis*; Mayr 2014b). Turacos were also reported from the middle Miocene of southeastern France (Mlíkovský 2002). However, because there are no Paleogene musophagiform fossils in Europe, their occurrence in the Cenozoic of France is likely to be due to dispersal from Africa during the middle Miocene Climatic Optimum.

The globally distributed cuckoos comprise some 140 extant species of zygodactyl birds, which are widely different in external appearance and skeletal morphology. Cuculinae, the most species-rich cuculiform subclade, includes most of the brood parasitic species, the majority of which occur in the Old World. The other cuculiform clades include species that mainly forage on the ground, such as the Neomorphinae (New World ground cuckoos), as well as several taxa with restricted distributions, such as the Madagascan couas (*Coua* spp.).

Close affinities between cuckoos and the African Musophagiformes would be in line with the fact that the oldest putative cuculiform stems from the early or middle Eocene of Tunisia (*Chambicuculus*; Mourer-Chauviré et al. 2013a). This bird is, however, only known from the distal ends of tarsometatarsi, and even though these are very cuckoo-like, further bones are desirable to establish its affinities firmly. The same is true for a fossil from the late Eocene of Canada, which is solely based on a distal humerus (*Neococcyx*), and the affinities of *Eocuculus* from the late Eocene of Colorado and the early Oligocene of France, which was initially considered to be a cuckoo, also have yet to be determined (Mayr 2009a).

The earliest undisputed modern-type Cuculiformes stem from the early Miocene of North America – that is, Colorado (*Cursoricoccyx*) and Florida (*Thomasococcyx*) – and were considered to be most similar to the New World Neomorphinae (Olson 1985a; Steadman 2008). Again, these records consist only of a few fragmentary bones, and because even the late Cenozoic fossil record of cuckoos is poor, the evolutionary history of these birds remains largely unknown.

Bustards

As already detailed, recent molecular analyses indicated close affinities between cuculiform birds and bustards (Otidiformes; Hackett et al. 2008; Jarvis et al. 2014; Prum et al. 2015). These crane-like, cursorial birds were traditionally classified in the "Gruiformes," together with cranes and allies, and include nearly 30 extant species that are found in dry, open landscapes of all Old World continents.

There are no pre-Miocene records of bustards. The poorly preserved skeleton of "*Otis affinis*" from the middle Miocene of Germany, which is often considered one of the earliest bustards (Lambrecht 1933; Mlíkovský 2002), was misidentified and probably represents an ibis (unlike modern bustards, this fossil exhibits a well-developed hind toe, which is absent in extant bustards).

Apart from a few fragmentary late Miocene remains of doubtful affinities, the earliest definite bustards are from the Pliocene of Eurasia and North Africa, from where several fossils exist (Mlíkovský 2002; Mourer-Chauviré and Geraads 2010; Boev et al. 2013). This comparatively extensive late Neogene record contrasts with the virtual absence of fossils from earlier deposits, and either Otidiformes evolved their characteristics late, so that

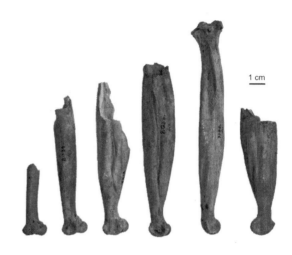

Figure 8.2 Tibiotarsi of the bustard *Gryzaja* (Otidiformes) from the early Pliocene of the Ukrainian Black Sea coast. The bone on the left approaches the normal proportions of an avian tibiotarsus, whereas the other bones show various degrees of the peculiar widening of the shaft that characterizes *Gryzaja*. Photographs by Leonid Gorobets.

1 cm

earlier fossils are not recognized as bustards, or they originated in a geographic area with a poor early and mid-Cenozoic fossil record.

Despite its limited general significance, the fossil record of otidiform birds includes a taxon that is highly interesting from an evolutionary point of view; that is, *Gryzaja* from the early Pliocene of the Black Sea coast of Ukraine and the adjacent Moldavia (Olson 1985a; Mlíkovský 2002). This bustard is characterized by a grotesquely widened and almost blade-like tibiotarsus shaft (Figure 8.2). Some individual variation in the degree of widening may suggest a pathologic origin, but the large number of bones with these deformations and their fairly regular shape speak against that assumption (Olson 1985a). The functional significance of this morphology, if any, remains to be determined.

The "Wonderful" Mirandornithes, or How Different Can Sister Taxa Be?

Many of the strongly supported clades that resulted from the new molecular analyses have already been proposed by earlier authors based on morphological studies. One of the most notable exceptions, however, is a sister group relationship between the long-legged, filter-feeding flamingos and the grebes, which are foot-propelled diving birds. A clade including these two taxa was obtained in nearly all recent analyses of various kinds of gene sequence data and was termed Mirandornithes, the "wonderful birds" (Sangster 2005). Flamingos and grebes have very disparate external appearances, but they share several derived anatomical similarities, including an unusually high number of primary feathers of the wing, nail-like pedal claws, and a chalky layer on the eggshell (Mayr 2004b, 2014d).

The closest extant relatives of Mirandornithes remain elusive, and analyses of molecular data either place flamingos and grebes at the neoavian base (Hackett et al. 2008; Jarvis et al. 2014) or suggest a sister group relationship to charadriiform birds (Prum et al. 2015; see also Mayr 2014d for morphological evidence of charadriiform affinities).

A fossil taxon that is of potential significance concerning the origin of Mirandornithes is *Juncitarsus* (Juncitarsidae), the two described species of which were found in early and middle Eocene localities in Wyoming and Germany (Plate 13b; Olson and Feduccia 1980b; Mayr 2014d). *Juncitarsus* measured about two-thirds the size of the smallest extant flamingo species. The taxon features greatly elongated hindlimbs, which are proportionally even longer than those of extant flamingos, but otherwise it is distinguished from both Phoenicopteriformes and Podicipediformes in some plesiomorphic features. These differences include the absence of extensive fusion of the thoracic vertebrae, a proportionally longer hind toe, and the absence of nail-like pedal claws (Mayr 2004b, 2014d). Unlike in grebes and flamingos, the long and straight beak of *Juncitarsus* furthermore has very long nostrils. Like flamingos, *Juncitarsus* probably occupied an ecological niche as a wading bird, but the different bill morphology and the preservation of gastroliths in one specimen document that it was otherwise distinguished from flamingos in its life habits (Mayr 2014d).

Palaelodids: The aquatic sister taxon of flamingos

Extant grebes and flamingos exhibit very disparate morphologies and ways of life. Some of these differences are bridged by the Palaelodidae, which were specialized swimming birds and had not yet evolved the highly specialized filter-feeding apparatus of crown group flamingos. The morphology of these peculiar stem group Phoenicopteriformes indicates that the immediate ancestor of flamingos and grebes was a swimming bird.

The earliest records of palaelodids are tentatively referred fragmentary fossils from the early Oligocene of Egypt (Rasmussen et al. 1987) and a large, but poorly represented taxon from the early Oligocene of Belgium (*Adelalopus*; Mayr 2009a). At least by the late Oligocene or early Miocene, palaelodids had achieved a near-global distribution, and they are very abundant in early Miocene localities in Central Europe. This is particularly true for fossil sites in the Saint-Gérand-le-Puy area in France, where thousands of bones of several differently sized species of the taxon *Palaelodus* were found (Mlíkovský 2002). *Palaelodus* remains are also known from the late Oligocene or early Miocene of Brazil and Australia, the early Miocene of New Zealand, and the middle Miocene of Mongolia (Alvarenga 1990; Baird and Vickers-Rich

1998; Worthy et al. 2010a; Zelenkov 2013). The large *Megapaloelodus*, which is distinguished from *Palaelodus* in some features, such as the lower count of thoracic vertebrae and the shape of the tarsometatarsal trochleae (Mayr 2014d), occurred in the early Miocene of Namibia, the late Miocene of Argentina, and the early Miocene to Pliocene of North America (Mourer-Chauviré 2008; Noriega and Agnolín 2008; Agnolín 2009a).

Although very large palaelodids from the Miocene of Europe are sometimes referred to *Megapaloelodus*, too, this referral has not been well established. The latest European records of palaelodids are from the middle Miocene of Germany (Mlíkovský 2002), but in Australia *Palaelodus* persisted into the Pleistocene (Baird and Vickers-Rich 1998).

Overall, the skeletal morphology of *Palaelodus*, the best-known taxon of the Palaelodidae, is quite similar to that of the Phoenicopteridae, but the legs are not as greatly elongated and the skull is very different from that of flamingos (Figure 8.3f, g). The tarsometatarsus of extant flamingos greatly exceeds the humerus in length, whereas it is much shorter in *Palaelodus*. The compressed shaft of this bone prompted earlier authors to

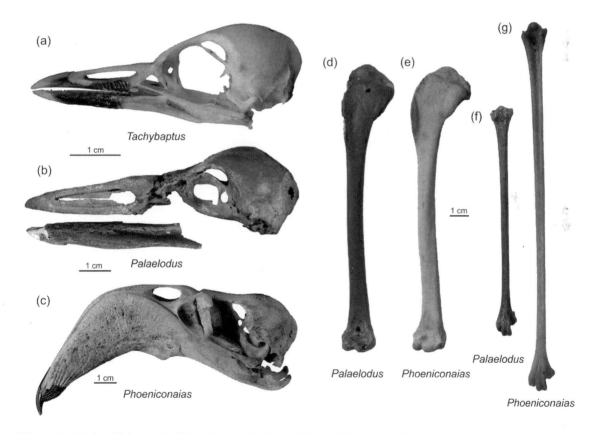

Figure 8.3 Skulls of (a) a grebe (*Tachybaptus*, Podicipedidae), (b) the early Miocene stem group phoenicopteriform *Palaelodus*, and (c) an extant flamingo (*Phoeniconaias*, Phoenicopteridae). (d, e) Humerus and (f, g) tarsometatarsus of *Palaelodus* (Palaelodidae) and *Phoeniconaias*. Note the intermediate bill morphology and much shorter legs of *Palaelodus* compared to extant flamingos. The skull and mandible of *Palaelodus* are not from the same individual; photograph of skull by Chris Torres.

assume more aquatic habits of palaelodids, and that these birds used their hindlimbs for aquatic locomotion is also suggested by the mediolaterally flattened pedal phalanges, which resemble those of albatrosses (Mayr 2015c).

Extant flamingos have a highly derived bill morphology and the distinct bend of their beak allows it to be opened with an equal width across its length, so that only small food particles are ingested. The deep lower jaws delimit a small space in which the thick tongue is moved back and forth, therefore acting like a piston pump. The expelled water is filtered by horny lamellae, which retain small algae and invertebrates. The beak of *Palaelodus*, by contrast, is short and straight, with a rounded tip and long nostrils, thereby resembling the beak of a crane, for which it was initially mistaken (Figure 8.3a–c; Cheneval and Escuillié 1992). The lower jaws of *Palaelodus* are very deep, but they are widely spaced and have narrow cutting edges, which would not have allowed the presence of filter-feeding lamellae in a way similar to extant flamingos (Mayr 2015c). The skull of palaelodids exhibits depressions for salt glands, and many of the sediments where palaelodid remains were found were deposited in saline or brackish paleoenvironments. Already in the 19th century it was assumed that *Palaelodus* fed on snails and caddisfly larvae, remains of both of which are abundantly represented in the lake deposits of the Saint-Gérand-le-Puy area in France, where palaelodid bones are very common.

Long-legged flamingos

Flamingos today occur on all continents except Antarctica and Australia, and the six extant species are assigned to the widely distributed taxon *Phoenicopterus*, the African *Phoeniconaias*, and the New World *Phoenicoparrus*. Flamingos have swimming capabilities, but usually they forage while standing in shallow areas of hypersaline lakes. Their extremely elongated legs, which are among the proportionally longest of all extant birds, therefore possibly represent an adaptation for the protection of the body against alkaline water.

The early Oligocene occurrence of palaelodids indicates that phoenicopterids must have also diverged by that time. Putative flamingo remains were indeed described from the late Eocene and early Oligocene of Europe and Asia (*Agnopterus*, *Elornis*; Mayr 2009a; Zelenkov 2013). All of these fossils are, however, based on very limited fossil material and contribute little to an understanding of flamingo evolution (Mayr 2009a). The same is true for fragmentary limb bones from the late Oligocene or early Miocene of Brazil, which were assigned to *Agnopterus* (Alvarenga 1990).

Miocene flamingos are much better represented and appear to have been diverse, although some of the described taxa are in need of a revision. *Harrisonavis* from the late Oligocene and early Miocene of France closely resembles extant flamingos in skeletal anatomy, but has a less curved beak, which identifies it as a stem group representative of Phoenicopteridae (Torres et al. 2015). *Leakeyornis* from of the early and middle Miocene of Kenya is likewise abundantly represented by cranial and postcranial remains (Rich and Walker 1983; Dyke and Walker 2008; Mayr 2014b). In light of the fact that *Harrisonavis* and *Leakeyornis* are clearly distinguished from crown group Phoenicopteridae, the referral of flamingo remains from the early Miocene of Thailand to the extant taxon *Phoeniconaias* (Cheneval et al. 1991) needs to be scrutinized. In any case, however, these latter fossils are of biogeographic interest because the extant distribution of flamingos does not include Southeast Asia.

Flamingos also have an extensive fossil record in Australia, where they no longer occur today. The earliest fossils from this continent stem from the late Oligocene or early Miocene and were assigned to the extinct taxon *Phoeniconotius* and the extant *Phoenicopterus* (Miller 1963). At least *Phoeniconotius* is very unlike extant Phoenicopteridae in details of the morphology of the distal tarsometatarsus and more closely resembles the coeval *Megapaloelodus*, whose exact affinities within Mirandornithes still need to be established (see previous section). A small species that was classified into the taxon *Phoeniconaias* also occurred in the Plio/Pleistocene of Australia, together with the extant *Phoenicopterus ruber*, and the disappearance of flamingos from Australia was ascribed to the increasing aridification of the continent towards the late Pleistocene (Miller 1963).

Extant flamingos lay only a single egg and their nests usually constitute mounds formed of mud. It is therefore of interest that a fossil nest with five eggs from the early Miocene of Spain was interpreted as a floating nest of a stem group representative of the Phoenicopteridae (Grellet-Tinner et al. 2012). That this nest was referred to the Phoenicopteridae rather than to the more aquatic Palaelodidae is mainly because the same strata also yielded flamingo bones. If correctly assigned to flamingos, the fossil shows that the breeding habits of early Miocene flamingos were unlike those of their extant relatives and resembled those of grebes, which also nest on floating vegetation. The nest of grebes is, however, built from reed, whereas the fossil nest consisted of twigs and may have possibly been washed in the water from a nearby shore.

Foot-propelled diving grebes

Like flamingos, the globally distributed grebes (Podicipediformes) are a species-poor taxon, which includes only some 22 extant species. Owing to their specialized foraging behavior – that is, foot-propelled pursuit dives in search of insects or fish – grebes have a distinctive skeletal morphology. However, although grebe bones are therefore easily recognized, fossils are comparatively rare.

Specimens from the late Oligocene of Kazakhstan and the late Oligocene or early Miocene of Australia have not yet been described (Mayr 2009a), and the earliest published taxon is *Miobaptus* from the early Miocene of the Czech Republic, which is known from a few bones (Švec 1982). The tarsometatarsus of *Miobaptus* exhibits an additional hypotarsal canal for a flexor muscle of the second toe, which among extant grebes only occurs in the taxa *Tachybaptus*, *Podilymbus*, and *Rollandia* (Ksepka et al. 2013a). However, being primitive for grebes, this feature does not necessarily support closer affinities between *Miobaptus* and any of the three aforementioned extant taxa.

Better documented by several partial skeletons is *Thiornis*, a medium-sized grebe from the middle Miocene of Spain, which overall resembles *Tachybaptus* in skeletal morphology (Olson 1995; Storer 2000); the exact interrelationships of *Thiornis* and *Miobaptus* have yet to be assessed. Partial grebe skeletons were also reported from the late Miocene of Nevada (Ksepka et al. 2013a). These fossils were not assigned to a particular species, but differ from extant grebes in proportionally shorter cnemial crests of the tibiotarsus.

As evidenced by the fossil record of its sister taxon, the Phoenicopteriformes, the evolutionary history of grebes must also go back to at least the late Eocene. The absence of fossils of these characteristic birds in the early Paleogene of Europe, together with the fact that the earliest known fossil grebes already exhibit a modern-type skeletal morphology, may indicate that the earliest stages of the evolution of Podicipediformes took place outside the Northern Hemisphere. In this regard, it may be worthwhile reconsidering the affinities of the putative gaviiform *Neogaeornis* from the Late Cretaceous of Chile (see Chapter 5), which, if indeed a neornithine bird, actually shows a greater resemblance to grebes than to loons.

Strisores: The Early Diversification of Nocturnal Avian Insectivores

It is now generally acknowledged that five groups of crepuscular or nocturnal birds, the oilbirds (Steatornithiformes), frogmouths (Podargiformes), potoos (Nyctibiiformes), nightjars (Caprimulgiformes), and owlet-nightjars (Aegotheliformes), form a clade with Apodiformes (swifts and hummingbirds), for which the name Strisores was introduced (Mayr 2002, 2010a). There is morphological evidence either for a sister group relationship between the Neotropic Steatornithiformes and all remaining Strisores, with Podargiformes branching next (Mayr 2010a), or for a sister group relationship between a clade formed by Steatornithiformes and Podargiformes and all other Strisores (Nesbitt et al. 2011). Molecular analyses, by contrast, supported a sister group relationship between Steatornithiformes and Nyctibiiformes (Hackett et al. 2008; Prum et al. 2015).

The basal divergences within Strisores are therefore controversial, but analyses of both morphological and molecular data congruently recovered a sister group relationship between the Australasian Aegotheliformes and apodiform birds, and a sequential divergence of other crepuscular or nocturnal Strisores (Figure 5.1; Mayr 2002; Ericson et al. 2006; Prum et al. 2015). Because all species of the Apodiformes are diurnal, this phylogenetic pattern raises interesting questions concerning the origin of dark activity (that is, a crepuscular or nocturnal way of life) in Strisores (Mayr 2010a). The most parsimonious explanation is a single origin of

dark activity in the stem lineage of Strisores and a reversal to a diurnal way of life in the stem lineage of Apodiformes. Possible support for a nocturnal stem species of Apodiformes does indeed come from the presence of very large orbital openings in the early Eocene stem group apodiform *Eocypselus* (Mayr 2010b). Otherwise, the morphology of apodiform birds does not indicate a nocturnal stem species, and the assumption of a multiple origin of dark activity may be a more plausible hypothesis (Mayr 2010a). A causal correlation has been suggested between the radiation of Strisores and the evolution of nocturnal lepidopterans (Mayr 2009a), but this does not explain dark activity in the frugivorous Steatornithiformes.

Strisores have a fairly comprehensive fossil record, which dates back into the earliest Cenozoic and documents that these birds were already diversified by that time. Many extant representatives of the clade are furthermore relict groups, which had a much wider distributions in the past.

Oilbirds and frogmouths

The Oilbird, *Steatornis caripensis*, occurs in the northern part of South America and is the sole extant representative of the Steatornithiformes. Unlike all non-apodiform Strisores, this highly specialized frugivore breeds in caves and has echolocation capabilities. The fossil record of Steatornithiformes is very limited, but it indicates a long evolutionary history of these birds, which is to be expected from their high degree of ecological specialization. *Prefica* from the early Eocene of North America documents the former occurrence of oilbirds outside South America and, like the extant Oilbird, has an extremely short tarsometatarsus. Although *Prefica* differs from the Oilbird in various plesiomorphic features, including a smaller size, the very similar shape of the lower jaws (the upper beak is unknown) suggests that it was already frugivorous like extant Steatornithiformes (Olson 1987).

It was recently proposed that *Euronyctibius* from the late Eocene of the Quercy region in France is another representative of the Steatornithiformes (Mourer-Chauviré 2013a). This taxon was previously assigned to the Nyctibiiformes and is mainly known from humeri, which aggravates a well-founded assessment of its phylogenetic relationships. Future finds are also needed to assess the affinities between *Euronyctibius* and the similar-sized putative podargiform *Quercypodargus*, which is only known from hindlimb bones and also stems from the late Eocene of the Quercy region, albeit from a different locality than the *Euronyctibius* remains.

Frogmouths (Podargiformes) are characterized by large and very wide beaks, and their distribution is today confined to Australasia and southern Asia. Even if the relationships of *Quercypodargus* are considered uncertain, Podargiformes have an early Cenozoic fossil record in Europe, and the stem group podargiform *Masillapodargus* from the early Eocene of Messel agrees well with extant Podargiformes in the characteristic bill shape and postcranial anatomy (Figure 8.4; Mayr 2015d).

Another Northern Hemispheric taxon for which podargiform affinities have been assumed is *Fluvioviridavis*, which occurs in the early Eocene of Europe and North America (Nesbitt et al. 2011). *Fluvioviridavis* exhibits a distinctive skeletal morphology, which originally led to its classification in its own higher-level taxon, Fluvioviridavidae. Initially considered to be of uncertain affinities, it was subsequently tentatively identified as a basal representative of Strisores (Mayr 2009a). *Fluvioviridavis* has a wide and flattened beak, which is, however, narrower and more pointed than in extant frogmouths, from which the fossil taxon distinctly differs in its postcranial

anatomy (Mayr 2015d). An assignment of *Fluvioviridavis* to the stem group of Podargiformes (Nesbitt et al. 2011) is not strongly based, and there were other – equally enigmatic – taxa with similar morphologies, such as the ineptly named *Palaeopsittacus* from the early Eocene of Europe (Mayr 2009a). Unfortunately, the known fossil material of these birds allows only limited comparisons, and future finds are required to establish their affinities firmly. Another taxon, *Eurofluvioviridavis* from the early Eocene of Germany, was erroneously likened to *Fluvioviridavis*, and is now considered to be a representative of Psittacopasseres, the clade including parrots and passerines (Nesbitt et al. 2011; Mayr 2015e).

"Flying insect nets": Potoos and nightjars

Morphological data support a clade including Caprimulgiformes, Nyctibiiformes, Aegotheliformes, and Apodiformes, which was termed Cypselomorphae (Mayr 2002; Nesbitt et al. 2011). With the exception of Nyctibiiformes, which, as already noted, group with Steatornithiformes, this clade is also obtained in some analyses of molecular data (Ericson et al. 2006; Hackett et al. 2008; see, however, Prum et al. 2015, who found an early divergence of the Caprimulgidae).

Possibly one of the most basal representatives of the Cypselomorphae is *Protocypselomorphus* from the early Eocene of Messel, whose swift-like beak, long wings, and short legs indicate an aerial insectivore (Mayr 2005a). Many of the extant cypselomorph representatives are likewise chasing flying insects and do so either on the wing or from a perch. Especially the Nyctibiidae and Caprimulgidae are characterized by short and very wide beaks, which function like an insect net.

Potoos (Nyctibiiformes) comprise seven extant species, which live in tropical forests of Central and South America. The fossil record shows them to be among the extant

Figure 8.4 (a) Partial skeleton of the frogmouth *Masillapodargus* from the early Eocene of Messel in Germany. Skulls of (b) *Masillapodargus*, (c) the extant podargiform *Batrachostomus*, and (d) the early Eocene North American *Fluvioviridavis*. (e–h) Humeri and (i–l) coracoids of *Masillapodargus*, the extant *Podargus* (Podargiformes) and *Steatornis* (Steatornithiformes), and the early Eocene *Fluvioviridavis* (specimens of the latter are from the London Clay). In the bones shown, *Fluvioviridavis* differs distinctly from *Masillapodargus* and more closely resembles steatornithiform than podargiform birds. Scale bars in (b–l) equal 1 centimeter.

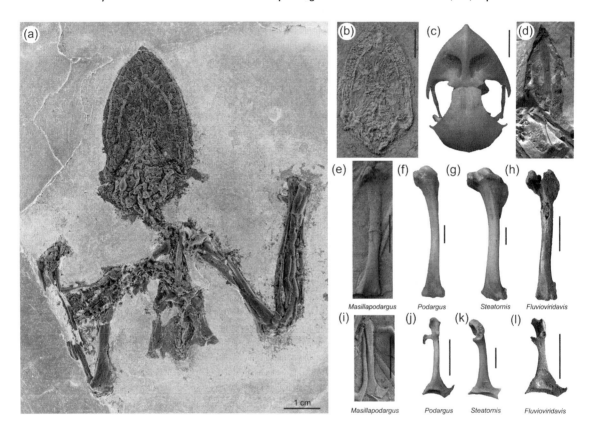

Neotropic birds with a relictual distribution, and stem group representatives occurred in the Paleogene of Europe. The evolution of Nyctibiiformes can be traced back to the early Eocene, and several complete skeletons of the stem group potoo *Paraprefica* were found in Messel (Figure 8.5, Plate 12c; Mayr 2005b). *Paraprefica* exhibits characteristic derived features of the Nyctibiiformes, such as a very short but extremely wide beak, greatly enlarged palatine bones, and a strongly shortened tarsometatarsus. After removal of the late Eocene *Euronyctibius* from Nyctibiiformes (see the preceding section), it is the only currently recognized fossil representative of potoos.

The occurrence of Nyctibiiformes in the early Eocene implies the presence by that time of their sister taxon, the nightjars (Caprimulgiformes). However, although nightjars today have a worldwide distribution and include almost 90 species, their fossil record is very sparse. There are tentative records from the early Eocene of North America and the late Eocene of Europe, but all are based on fragmentary bones, the identification of which has yet to be confirmed by more substantial fossils (Mayr 2009a). The

Figure 8.5 (a) Skeleton of the middle Eocene *Paraprefica* (Nyctibiiformes), and (b) skull, (c) mandible, and (d) tarsometatarsus of an extant potoo (*Nyctibius*). Characteristic derived features of the Nyctibiiformes are a very short beak, greatly enlarged palatines, and an extremely shortened tarsometatarsus.

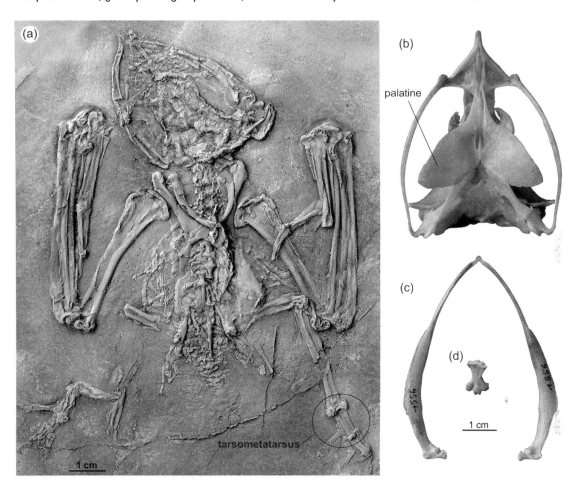

striking mismatch between the sparse fossil record of nightjars and the fact that they are the most species-rich and most widely distributed extant group of the Strisores may indicate a late diversification of these birds, or an origin in a geographic region with a poor fossil record.

Alternatively, stem group representatives of nightjars may exhibit a plesiomorphic morphology, which prevents their correct identification. Indeed, there is one group of Paleogene Cypselomorphae of which the exact phylogenetic affinities are still unresolved. These birds are the archaeotrogons (Archaeotrogonidae), small, short-legged birds that are abundant in some late Eocene and Oligocene fossil sites in France, where four species of *Archaeotrogon* have been recognized (Mourer-Chauviré 1980). As indicated by their name, the postcranial skeleton of archaeotrogons is somewhat similar to that of trogons (Trogoniformes), although their tarsometatarsus lacks specializations for a heterodactyl foot. *Archaeotrogon* is characterized by a carpometacarpus with a pointed, spur-like extensor process, which may have served for intraspecific combats. This spur on the carpometacarpus is also present in an undescribed archaeotrogon fossil from the early Eocene British London Clay, in which a nightjar-like beak is preserved, therefore supporting earlier hypotheses that archaeotrogons are in fact cypselomorph birds (see Mayr 2009a). A putative archaeotrogon from the early Eocene of Germany likewise has an owlet-nightjar–like beak, and some specimens show a distinct barring of the tail feathers, similar to that found in many extant Strisores (*Hassiavis*; Mayr 2004c; Plate 12b). It is not proposed here that archaeotrogons are particularly closely related to the Caprimulgidae, but such affinities certainly need to be taken into account once their exact relationships have been assessed in future studies.

The evolution of apodiform birds

The extant members of Apodiformes – that is, tree swifts (Hemiprocnidae), true swifts (Apodidae), and hummingbirds (Trochilidae) – are small to very small birds. These aerial specialists are characterized by an unusually long hand section of the wing and they often have feeble legs. As such, they differ distinctly from their extant sister taxon, the Australasian owlet-nightjars (Aegotheliformes), which possess rather short and rounded wings and long hindlimbs (as also illustrated by flamingos and grebes, very disparate morphologies of sister taxa are not all that uncommon, and may be the result of adaptive processes due to competition and **niche segregation**).

The fossil record of the Aegotheliformes is scarce. The oldest specimens stem from early Miocene New Zealand and consist of a few fragmentary bones (Worthy et al. 2007). The only other fossil aegotheliform, *Quipollornis* from the early or middle Miocene of Australia, is based on a partial skeleton and has proportionally shorter hindlimbs than extant Aegotheliformes (Rich and McEvey

1977). Apodiform birds, by contrast, have a surprisingly comprehensive fossil record that goes back to the early Eocene, by which time these birds were already quite diversified. Figure 8.6 gives an overview of the currently known taxa, their stratigraphic occurrence, and phylogenetic interrelationships.

The early Eocene *Eocypselus* (Eocypselidae) shows many plesiomorphic similarities to the Aegotheliformes and is one of the earliest apodiform birds known to date, as well as the most basally diverging apodiform taxon (Mayr 2003a, 2010b). *Eocypselus* includes two species, which occurred in Europe and North America, respectively, and are distinguished from crown group Apodiformes in a number of features (Figure 8.7; Mayr 2010b; Ksepka et al. 2013b). The humerus of *Eocypselus* is stouter than that of most non-apodiform birds, but it is less shortened than the humerus of crown group Apodiformes. The ulna is also more slender and proportionally longer, and several other differences in the wing and pectoral girdle skeleton indicate that *Eocypselus* was less aerial than extant Apodiformes. Furthermore, compared to extant swifts it has a relatively long beak and proportionally longer legs. *Eocypselus* may therefore have caught insect prey in a similar manner to owlet-nightjars, which conduct forays for flying insects from perches or snatch prey items from the ground (Mayr 2010b).

The Aegialornithidae constitute another taxon of presumably early diverging Apodiformes. These birds first occurred in the middle Eocene of Germany and are particular abundant in late Eocene fossil sites in France, where four species of *Aegialornis* have been distinguished (Mourer-Chauviré 1988b; Mayr 2009a). Some features of aegialornithids are more derived compared to those seen in eocypselids, but the humerus and ulna are less shortened than in crown group Apodiformes (Mayr 2003a, 2010b).

Figure 8.6 Phylogenetic interrelationships of extinct and extant apodiform birds (after Mayr 2015f). The gray bars indicate known temporal ranges, white bars denote uncertain ones.

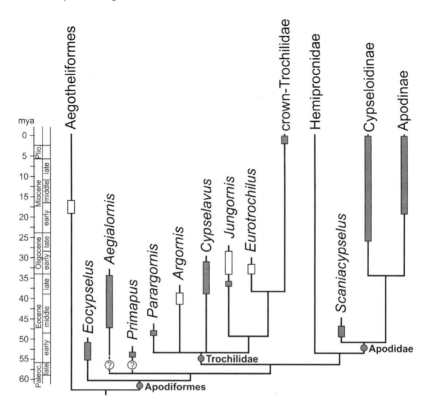

Hemiprocnidae (tree swifts) and Apodidae (true swifts)

Extant Swifts are classified in two taxa, the Southeast Asian tree swifts (Hemiprocnidae; four extant species) and the true swifts (Apodidae), which have a worldwide distribution and include almost 100 species. Most swifts use saliva in nest building. Whereas tree swifts glue their nests to tree branches, those of true swifts are usually attached to vertical surfaces, be it palm leaves, rocks, or the inside of hollow tree trunks. As an adaptation for this derived nesting behavior, true swifts have an unusual foot morphology with strongly shortened proximal phalanges of the toes, which results in a subequal length of the fore toes and, hence, increases the clinging capabilities of these birds.

Hemiprocnidae have no fossil record, and the earliest stem group representatives of the Apodidae are classified in the taxon *Scaniacypselus*. Fossils of this tiny, swiftlet-sized bird occurred in the early to late Eocene of Europe (Mayr 2009a). Specimens with preserved feather remains show that it was similar to modern swifts in external appearance, but unlike in many living swifts the tail is only slightly forked (Plate 12a). *Scaniacypselus* is a stem group representative of the Apodidae and differs from extant swifts in plesiomorphic features of the humerus, as well as in a proportionally longer ulna. This indicates that it was less aerial than its living relatives, some of which visit firm ground only during the breeding period and even sleep in flight

Figure 8.7 (a) Skeleton of the early Eocene apodiform bird *Eocypselus* from the Danish Fur Formation with interpretive drawing. (b–e) Humeri, (f–h) hand skeleton, and (i–k) sterna of *Eocypselus*, Aegotheliformes (owlet-nightjars), and extant apodiform birds. Note the more slender humerus of *Eocypselus* and the closer similarity of its hand skeleton and sternum to those of the Aegotheliformes.

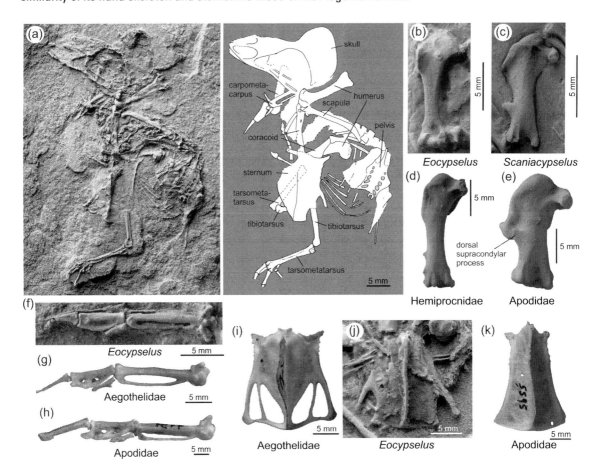

(Mayr 2015f). *Scaniacypselus* also has a proportionally shorter tarsometatarsus than many extant Apodidae and lacks shortened phalanges of the fore toes. In its length proportions, the tarsometatarsus of *Scaniacypselus* corresponds with that of extant Hemiprocnidae and Trochilidae, which suggests that a short tarsometatarsus is plesiomorphic for the Apodidae. Its elongation in the crown group is probably correlated with the clinging habits of true swifts, and *Scaniacypselus* may have built tree nests like extant Hemiprocnidae (Mayr 2015f). The

derived nesting behavior of the crown group representatives of true swifts is likely to have evolved as a strategy to reduce predation and may have contributed to their evolutionary success.

Crown group Apodidae are divided into two taxa, the species-poor New World Cypseloidini, which exhibit a less derived skeletal morphology than the much more species-rich and globally distributed Apodini. *Procypseloides* from the late Oligocene and early Miocene of France (Mourer-Chauviré et al. 2004) is known from a few bones. These show essentially modern-type morphologies and resemble the corresponding bones of the Cypseloidini, albeit this similarity may well be due to the retention of plesiomorphic characters in *Procypseloides* and the Cypseloidini. From the late Oligocene or early Miocene of Australia, an extinct swiftlet-like species was assigned to the extant taxon *Collocalia*, which was likewise identified in the early Miocene of New Zealand (Boles 2001; Worthy et al. 2007). Even though the few bones known of these birds probably do not allow an exact classification, they exhibit apodine-like morphologies and indicate the presence of crown group representatives of the Apodidae by the early Miocene.

The temporal occurrences of the various fossil swifts and swift-like birds correspond well with their position in the phylogenetic tree of Apodiformes (Figure 8.6), and apparently there was a strong selective pressure towards optimization of the wing apparatus for a highly aerial way of life, in order to expand both foraging times and ranges. The exact origin of the crown group is difficult to determine, but seems to have been towards the late Oligocene or early Miocene at the latest. One of the most interesting aspects of the evolution of Apodiformes however, concerns the transition to a nectarivorous (nectar-feeding) diet in the lineage leading to hummingbirds, and fossils described in the past two decades shed some light on the origin of this feeding specialization.

Hummingbirds

The predominantly nectarivorous – that is, nectar-feeding – hummingbirds include more than 300 extant species of small to tiny, long-beaked birds. Owing to their highly developed hovering capabilities, hummingbirds are able to stand still in front of flowers while flying. These locomotory characteristics are accompanied by complex wing movements and a highly specialized morphology of the flight apparatus.

Hummingbirds are iconic birds of the extant New World avifauna. The fossil record shows, however, that a major part of their evolution took place outside the Americas, and various stem group representatives are known from Paleogene fossil sites in Europe. The earliest such taxon is *Parargornis* from the early Eocene fossil site Messel in Germany (Figure 8.8; Mayr 2003b). *Parargornis* has a short and swift-like beak, which suggests an insectivorous diet. The humerus is greatly shortened as in other apodiform birds, whereas the feathering – short and broadly rounded wings and a fairly long tail – resembles that of owlet-nightjars. All extant birds with short and stocky humeri, by contrast, have very long primary feathers, and the combination of shortened humeri and short wings in *Parargornis* may reflect an early stage of the evolution of hovering flight. If so, hovering capabilities would have evolved earlier than nectarivorous feeding behavior in the stem lineage of Trochilidae, and may have enabled *Parargornis* to glean insects from the underside of leaves or around flowers (Mayr 2009a).

Identification of the unusual *Parargornis* as a stem group representative of the Trochilidae is bolstered by the occurrence of other basal hummingbird taxa in the late Eocene and early Oligocene of Eurasia. Of these, the middle Eocene *Argornis* from the Caucasus area and *Cypselavus* from the late Eocene and early Oligocene of France closely resemble *Parargornis* (Karhu 1999; Mayr 2009a). The tiny *Jungornis* from the late Eocene of France and the Oligocene of the Caucasus, by contrast, already shares characteristic derived features with modern hummingbirds, including a protrusion on the humerus head, which allows rotation of this bone in hovering flight (Figure 8.9; Karhu 1988; Mayr 2009a).

Figure 8.8 The earliest stem group hummingbird, *Parargornis* from the early Eocene of Messel in Germany (left: specimen coated with ammonium chloride to enhance contrast of the bones; right: actual fossil with feather preservation).

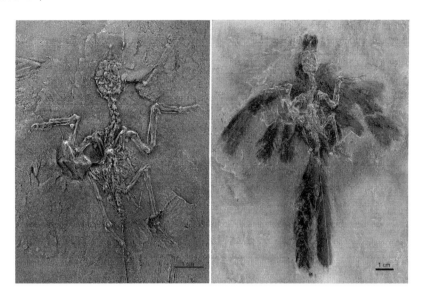

The skeletons of *Argornis*, *Cypselavus*, and *Jungornis* are too incompletely known to assess whether these birds may have resembled extant hummingbirds in external appearance, even more so as the skull of any of these taxa has not been found. Essentially modern-type stem group hummingbirds did, however, occur in the early Oligocene of Europe. These belong to *Eurotrochilus*, of which two species were described from fossil sites in Germany, Poland, and France (Mayr 2004d, 2009a; Mayr and Micklich 2010; Bocheński and Bocheński 2008; Louchart et al. 2008a). The skeleton of *Eurotrochilus* closely resembles that of extant Trochilidae (Figure 8.9), as does its hummingbird-like feathering (Plate 13a). The *Eurotrochilus* species are tiny birds with very long beaks, which, together with hovering adaptations of the wing skeleton, suggest a predominantly nectarivorous diet.

However, although the skeleton of *Eurotrochilus* exhibits most of the derived characteristics of modern Trochilidae, it differs in salient plesiomorphic features from its modern relatives, and this Oligocene hummingbird taxon undoubtedly is a stem group representative of the Trochilidae. The crown group of hummingbirds most likely originated in the mid-Cenozoic of South or Central America, from where these birds dispersed into North America (Bleiweiss 1998). So far, however, the only New World fossils of hummingbirds are a few bones of extant taxa from Quaternary cave deposits (Olson 1985a).

The origins of avian nectarivory

Birds are important flower pollinators in many extant tropical and subtropical ecosystems, and bird–flower interactions are among the textbook examples of coevolution. Nectarivory evolved multiple times independently in birds, and aside from hummingbirds it also occurs in parrots (lorikeets) and many passerines (sunbirds, sugarbirds, honeyeaters, flowerpeckers, and others).

Figure 8.9 (a) Skeleton of the stem group hummingbird *Eurotrochilus* from the early Oligocene of Germany with interpretive drawing. Lower row depicts (b, c) the coracoid, (d, e) humerus, and (f, g) carpometacarpus of *Eurotrochilus* and an extant hummingbird. The stocky humerus and large supracondylar process are apomorphies of the Apodiformes. Derived features of hummingbirds are the long beak, the distal protrusion of the humerus head, and the well-developed intermetacarpal process of the carpometacarpus.

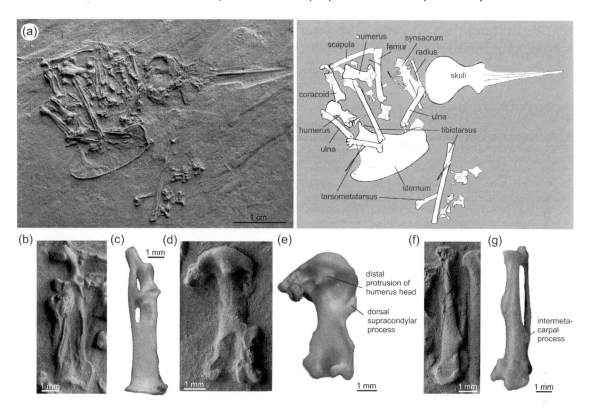

Nectarivorous birds are usually characterized by long and slender bills, but there are numerous exceptions. It is therefore very difficult to identify a nectarivorous bird just from skeletal remains, unless its phylogenetic affinities are well constrained. In a few instances, however, the fossil record provides evidence for a nectarivorous diet of fossil birds. The oldest hint on nectarivory in a fossil bird comes from a specimen of *Pumiliornis* from the early Eocene of Messel, in which large amounts of pollen are preserved as stomach contents (Plate 16d–f; Mayr and Wilde 2014). *Pumiliornis* was recently hypothesized to be a zygodactyl stem group representative of the Passeriformes (Mayr 2015e), and it is clearly not closely related to any of the extant nectarivorous taxa.

As detailed earlier, the origin of hummingbirds also dates back to at least the early Eocene. However, these early fossils have a broad, swift-like beak, and the early Oligocene *Eurotrochilus* is the oldest stem group hummingbird for which a nectarivorous diet can be assumed. Other extant nectarivorous avian groups – that is, passerines and parrots – likewise have no earlier fossil record. The occurrence of nectarivory in the much older, early Eocene *Pumiliornis* therefore suggests that the origin of bird-pollinated (ornithophilous) plants predates that of the extant groups of nectarivorous birds (Mayr and Wilde 2014).

In order to attract birds, plants developed various flower characteristics, such as a lack of scent (most birds have poorly developed olfactory senses), red or orange corollas (unlike most mammals, birds are capable of color vision), and the production of high amounts of nectar as a pollination reward. This flower morphology contrasts with that of typical insect-pollinated flowers, which are often blue or violet and provide less nectar. Hummingbird-pollinated flowers also differ from those of bat-attracting plants, which usually are white and emit strong olfactory signals. Because hummingbirds are capable of sustained hovering flight, many New World hummingbird-pollinated plants show a particularly characteristic flower morphology, which distinguishes them from bird-pollinated plants in the Old World. These traits include deep, pendulous flowers that do not provide perches on which a bird can sit. Intriguingly, there still are a few extant plants with such characteristics in the Old World today. It has been hypothesized that some of these flower morphologies may indeed go back to pollination by Old World hummingbirds in the early Cenozoic, and after the extinction of hummingbirds in the Old World, their pollination may have been taken over by insects or other birds with short-term hovering capabilities (Mayr 2004d, 2009a).

9 Shorebirds, Cranes, and Relatives

In the evolutionary history of birds, species foraging along shorelines, in wetlands, and in other aquatic or semi-aquatic habitats evolved multiple times in different lineages. Aside from taxa of the Aequornithes, which are discussed in the next chapter, many of these "wading birds" belong to the Charadriiformes (shorebirds and allies) and the polyphyletic "Gruiformes" (rails, cranes, and allies).

The traditional "Gruiformes" included various crane- or rail-like birds. Earlier authors already suspected that these birds constitute an unnatural assemblage, and close relationships of all of the traditional "gruiform" taxa were likewise not supported by any of the more recent analyses of molecular data. Sequence-based analyses and some morphology-based studies nevertheless congruently obtained a smaller clade of core Gruiformes, which includes cranes, rails, and their relatives. At least one recent genome-scale analysis furthermore resulted in a clade including these core Gruiformes and charadriiform birds (Jarvis et al. 2014; see, however, Hackett et al. 2008 and Prum et al. 2015, whose analyses did not detect such close affinities). Although an association of the taxa included in the present chapter is therefore not entirely far-fetched, it is not meant to reflect well-established phylogenetic affinities and is mainly for practical reasons.

The less specialized taxa of the Charadriiformes and core Gruiformes live in environments that already existed in the Mesozoic era. However, although fragmentary bones with shorebird- or rail-like morphologies are known from Cretaceous deposits, these are too incomplete for a reliable identification (see Chapter 5). As will be detailed in the following, the Cenozoic fossil record, by contrast, includes numerous well-preserved fossils that shed some light on the past diversity of these birds.

Avian Evolution: The Fossil Record of Birds and its Paleobiological Significance,
First Edition. Gerald Mayr.
© 2017 John Wiley & Sons, Ltd. Published 2017 by John Wiley & Sons, Ltd.

Charadriiformes: One of the Most Diverse Groups of Extant Birds

Few of the well-defined avian higher-level taxa are as diversified as the Charadriiformes; that is, the shorebirds and their allies. The spectrum of morphological and ecological specializations ranges from flightless, wing-propelled divers to long-legged waders and aerial specialists, and from highly aquatic forms to denizens of semideserts. The morphological disparity of these birds reflects the great diversity of aquatic and semi-aquatic habitats, but the diverse adaptations of charadriiform inhabitants of the intertidal zones are also likely to be due to **niche partitioning** in these spatially restricted areas.

Most charadriiform birds have long been recognized as closely related, owing to shared external and anatomical features. Their exact interrelationships were, however, only unraveled through analyses of gene sequence data, which provided a robust phylogenetic framework for the various groups. According to these studies, Charadriiformes can be divided into three major groups: the Lari (gulls, auks, and allies), Scolopaci (sandpipers and allies), and Charadrii (plovers and allies), with Charadrii being the sister taxon of a clade including Lari and Scolopaci (Figure 9.1; Paton and Baker 2006; Baker et al. 2007; Fain and Houde 2007).

Unfortunately, the early Cenozoic fossil record of charadriiform birds is surprisingly poor. One of the taxa that may be close to the charadriiform ancestry is *Scandiavis* from the early Eocene of Denmark, which is known from a well-preserved partial skeleton (Plate 9a; Bertelli et al. 2013). This taxon shares a few characteristic features with charadriiform birds, such as the lack of fusion between the ilia and the synsacrum of the pelvis, but there are also many differences. Charadriiform affinities may or may not be supported by future specimens preserving the unknown elements of the pectoral girdle and wing, which in most charadriiform birds exhibit a distinctive morphology.

Unpublished records of essentially modern-type Charadriiformes furthermore exist from the early Eocene of England, and fragmentary fossils were also described from the early Eocene of Germany and the middle Eocene of China (Mayr 2009a). Their exact affinities within Charadriiformes, however, have yet to be established. Almost the entire further fossil record of charadriiform birds postdates the Eocene, which is remarkable, because today many of these birds occur in aquatic habitats with a high fossilization potential.

The typical "shorebirds": Plovers, sandpipers, and allies

The two clades Charadrii and Scolopaci include most of the typical "shorebirds"; that is, species foraging along seashores and lakesides. The habitats of these birds are very diverse and several species even inhabit arid semideserts.

In addition to some species-poor groups without a fossil record, Charadrii encompasses plovers and allies (Charadriidae), oystercatchers (Haematopodidae), stilts and avocets (Recurvirostridae), as well as thick-knees (Burhinidae). Most of these taxa are characterized by the absence of a hind toe and by short fore toes, which represent adaptations to cursorial or wading habits. Concerning other derived skeletal features, however, it is difficult to characterize members of the Charadrii. There are no definitive pre-Oligocene fossils of these birds, and a possible record from the early Oligocene of France has not yet been studied (Mayr 2009a). The earliest published and phylogenetically well-constrained fossils of the Charadrii therefore are an extinct taxon of the Burhinidae (*Genucrassum*) and haematopodid-like fossils from the early Miocene of France (De Pietri et al. 2013a;

Figure 9.1 Interrelationships of major charadriiform groups, with temporal ranges of the crown group taxa (phylogeny based on Fain and Houde 2007 and De Pietri et al. 2011a).

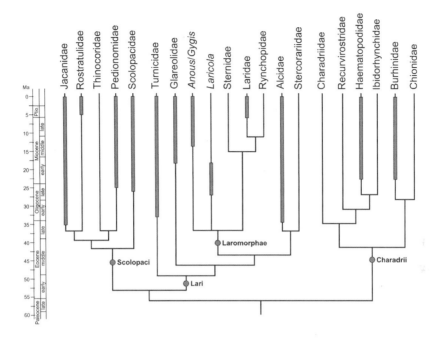

De Pietri and Scofield 2014). Early or middle Miocene fossils of putative Charadriidae have likewise been mentioned (Mlíkovský 2002), but these are in need of revision in light of the new phylogenetic hypotheses of charadriiform interrelationships.

Scolopaci includes the species-rich Scolopacidae (sandpipers, snipes, and allies), which are the sister taxon of a clade including the Jacanidae (jacanas), Rostratulidae (painted snipes), Pedionomidae (plains wanderer), and Thinocoridae (seed snipes), with each of the latter four taxa including only a few extant species (e.g., Paton and Baker 2006; Fain and Houde 2007; Mayr 2011b). The representatives of Scolopaci share a suite of characters that distinguish them from other Charadriiformes, including the absence of a foramen for the supracoracoideus nerve in the coracoid shaft and, possibly functionally correlated therewith, a derived morphology of the proximal end of the humerus (De Pietri and Mayr 2012). It may be the result

of a more straightforward identification that the published fossil record of these birds is more comprehensive than that of many other charadriiform groups.

The earliest fossils that can be assigned to one of these subclades of the Scolopaci belong to the Jacanidae. The eight extant species of this taxon have a pantropic distribution and are among the morphologically most aberrant Charadriiformes. Jacanas are characterized by extremely long toes, which enable them to forage on floating freshwater vegetation. Correlated with these habits, they share a distinctive tarsometatarsus morphology that allows a reliable identification of fossil remains. The earliest fossils belong to *Nupharanassa* and *Janipes* from the late Eocene and early Oligocene of Egypt (Rasmussen et al. 1987). These taxa are notable because some of their species are much larger than extant Jacanidae. Very large jacanas still existed in the middle Miocene of Africa, where a giant *Nupharanassa*-like species was

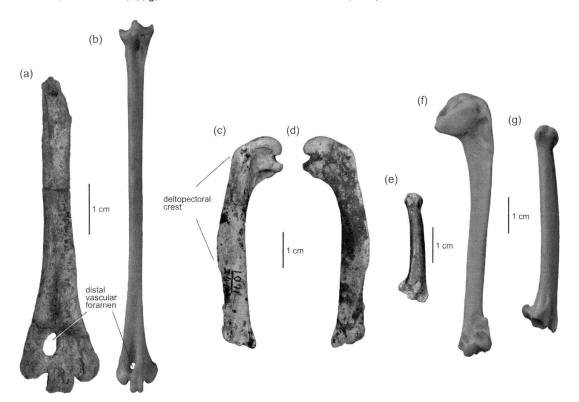

Figure 9.2 (a) Tarsometatarsus of a giant jacana (*Nupharanassa*; Jacanidae) from the middle Miocene of Kenya in comparison to (b) the extant African *Actophilornis*. The very large distal vascular foramen is one of the characteristics of jacanas. (c, d) Humerus of the mancalline auk *Mancalla* (Mancallinae). (e) Ulna of *Alcodes* (Mancallinae). (f, g) Humerus and Ulna of an extant auk (*Alca*).

identified in fossil material from Maboko Island in Kenya (Figure 9.2a; Mayr 2014b). This species, which is in fact the largest known jacana, measures twice the size of the extant African *Actophilornis* and coexisted with smaller species the size of extant Jacanidae. Why there are no similar-sized Jacanidae today remains to be assessed, and this is also true for the identity of the plants that could have borne the weight of these birds. Most extant jacanas live in the Old World, where they have a wide distribution across Africa, Asia, and Australia. The two extant species in the tropical Americas belong to the taxon *Jacana*. The earliest New World fossil record was likewise assigned to

this latter taxon and stems from the Pliocene of Florida (Olson 1985a), which may indicate a comparatively late dispersal of jacanas into the Americas.

There are only two extant species of Rostratulidae, which occur in the New and Old Worlds, respectively. Although rostratulid affinities of fragmentary remains from the early Miocene of the Czech Republic were considered (Mlíkovský 2002), the only reliably identifiable record is from the early Pliocene of South Africa (Olson and Eller 1989).

The earliest fossils of the likewise species-poor South American Thinocoridae are fragmentary remains from the late Miocene of Argentina (Agnolín et al. 2016). The Australian Pedionomidae include a single extant species, the endangered Plains-wanderer (*Pedionomus torquatus*). This species occurs in the grasslands of southeastern Australia, but an extinct pedionomid from the late Oligocene of Australia seems to have inhabited wooded habitats (De Pietri et al. 2015). Thinocoridae and Pedionomidae are geographically widely separated sister groups (Figure 9.1), and an extinct taxon that may belong to the same clade has been reported from the early Miocene of New Zealand (De Pietri et al. 2016).

The existence of Jacanidae in the late Eocene implies the occurrence of the earlier diverging Scolopacidae by that time. An uncontroversial identification of fragmentary remains of these birds is difficult, however, because the postcranial skeleton of scolopacids exhibits fewer specializations than that of the four other taxa of the Scolopaci. Some of the earliest records of putative Scolopacidae are too incomplete for a well-founded identification beyond the level of "indeterminate Scolopaci" (Mayr 2009a). Caution is particularly warranted, because some early Miocene Scolopaci still show unusual character combinations, which aggravate a well-founded phylogenetic placement. Scolopacidae are nevertheless likely to have diversified by the Oligocene at the latest, and fragmentary fossils from the late Oligocene of France were even assigned to Phalaropodinae (phalaropes); that is, a taxon of the crown group (Mayr 2009a).

Members of the Scolopaci were comparatively diversified in early Miocene deposits of the Saint-Gérand-le-Puy area in France. All fossils belong to extinct taxa, the exact affinities of which are difficult to determine (De Pietri and Mayr 2012). Some of these birds may occupy a basal position within Scolopaci, but there are also skulls that exhibit a very long beak, which is a characteristic of most Scolopacidae and serves for foraging in soft soil (De Pietri and Mayr 2012). From middle Miocene lacustrine strata in Europe, various essentially modern-type scolopacids have been described, such as *Mirolia*, a taxon assigned to the Calidrinae (sandpipers; Ballmann 2004). In the New World, scolopacid remains were reported from the late Miocene of California and the late Miocene or early Pliocene of Arizona (Bickart 1990; De Pietri and Mayr 2012). Because of the uncertain affinities of most of these taxa, clear biogeographic signals can, however, not be derived from these data.

A late radiation of marine birds: Gulls and allies

The third major charadriiform clade, the Lari, includes gulls (Laridae), terns (Sternidae), skuas (Stercorariidae), and auks (Alcidae), as well as the more aberrant pratincoles and coursers (Glareolidae) and buttonquails (Turnicidae). Molecular analyses have shown that the Turnicidae are the sister taxon of all other Lari (Paton and Baker 2006; Fain and Houde 2007). Buttonquails also constitute some of the most unusual Charadriiformes. These small, quail-like birds have a wide distribution over the warmer parts of the Old World, from southern Europe and Africa to Australia. Unlike most other Lari, buttonquails live in semiarid, open habitats, and in addition to various invertebrates their diet also includes seeds and other plant matter. The skeleton of extant buttonquails strongly departs from that of other charadriiform birds, which has long prevented recognition of their true phylogenetic affinities (traditionally, buttonquails were considered to be "gruiform" birds). The disparate morphologies of crown group Turnicidae and more typical charadriiform birds are, however, bridged by the earliest turnicid stem group representatives, which belong to the taxon *Turnipax* and occurred in the early Oligocene of France and Germany. Among others, these fossil species exhibit the peculiarly shaped

Figure 9.3 The stem group buttonquail *Turnipax* (Turnicidae) from the early Oligocene of Europe. (a) Skeleton with interpretive drawing (note the preservation of gastroliths next to the sternum). Coracoids of *Turnipax* fossils from (b) Germany and (c) France and of (d) an extant buttonquail (*Turnix*). (e) Foot of *Turnipax* (note the presence of a hind toe). (f–h) Humerus of *Turnipax* in comparison to an extant buttonquail and a typical charadriiform (*Vanellus*).

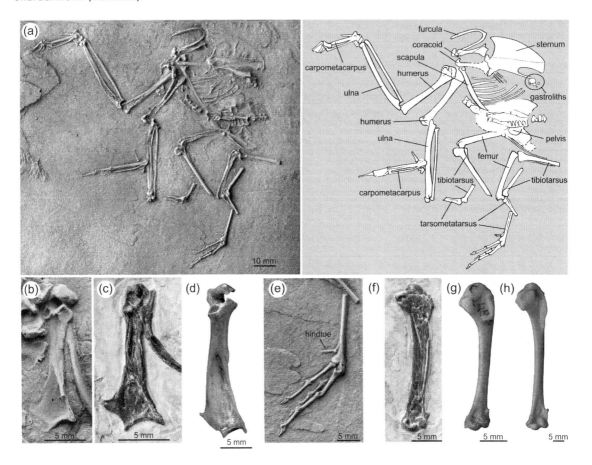

coracoid typical of buttonquails, but otherwise they have a less specialized and more typically charadriiform skeletal morphology than their extant relatives (Figure 9.3; Mayr and Knopf 2007). One skeleton of *Turnipax* is preserved with gastroliths that document an at least partially granivorous diet (Mayr and Knopf 2007). Unlike extant buttonquails, *Turnipax* still has a hind toe, which indicates differences to extant buttonquails in habitat preference and way of life. Crown group Turnicidae may not have diversified before the spread of open habitats towards the Neogene (Mayr and Knopf 2007), and the earliest modern-type buttonquails stem from the late Miocene of southeastern Europe, by which time these birds had a wider distribution than today (Zelenkov et al. 2016).

Glareolidae are another early diverging group of the Lari with an exclusively Old World distribution today. The earliest taxon that shows an overall similarity to pratincoles and coursers is *Boutersemia* from the earliest Oligocene of Belgium, which is known from various postcranial bones (Mayr and Smith 2001). Undisputed Glareolidae, which were assigned to the extinct taxon *Mioglareola* and the extant *Glareola*, occurred in the early and middle Miocene of Europe (Ballmann 1979; Mlíkovský 2002). A putative glareolid is also known from the early Miocene of Nebraska, and, if correctly identified, documents the former occurrence of glareolids in the New World (*Paractiornis*; Olson 1985a).

By far the most species-rich and most widely distributed clade of the Lari is termed Laromorphae and includes gulls, terns, and their relatives (Figure 9.1). Typical gulls and terns are characterized by a derived humerus morphology, the proximal end of which exhibits a double pneumotricipital fossa. A second fossa is, however, absent in the tern-like noddies, which were traditionally included in the Sternidae but in more recent analyses were found to be the sister taxon of a clade including all other extant laromorphs (Baker et al. 2007; De Pietri et al. 2011a).

Most extant Laromorphae inhabit near-shore marine habitats, although several species occur far inland in freshwater habitats. Definitive laromorphs first appear in the fossil record towards the late Oligocene and belong to the taxon *Laricola* (De Pietri et al. 2011a). The most abundant species, *L. elegans*, was originally identified as a gull and is known from thousands of bones from lacustrine deposits in Europe, mainly from the early Miocene of the Saint-Gérand-le-Puy area in France, where it seems to have formed large breeding colonies. The species of *Laricola* are small birds and have longer legs than extant gulls and terns, from which they are also distinguished by other plesiomorphic features, such as the absence of the second pneumotricipital fossa. In a recent revision it was concluded that *Laricola* is outside a clade including Sternidae and Laridae (De Pietri et al. 2011a). More closely related to

the latter clade is *Sternalara*, another laromorph taxon from the Saint-Gérand-le-Puy area, whose humerus agrees with that of extant Laromorphae in that it exhibits the double pneumotricipital fossa.

Feducciavis from the middle Miocene of Virginia (USA) is the earliest representative of the noddies, which – in addition to a single pneumotricipital fossa, and unlike most other charadriiform birds – exhibit a pneumatized humerus (Olson 2011). Remarkably, however, there are no definite records of Laridae or true Sternidae from the early and middle Miocene. Putative Laridae from the early-middle Miocene of New Zealand and the middle Miocene of North America (Olson and Rasmussen 2001; Worthy et al. 2007) are based on very fragmentary remains, and even an alleged gull from the late Miocene of Nebraska (*Gaviota*; Miller and Sibley 1941), which is only represented by a distal humerus, is clearly distinguished from gulls and all other extant Laromorphae. In some Pliocene localities, by contrast, marine gulls are abundantly represented (e.g., Olson and Rasmussen 2001). The fossil record therefore indicates a comparatively recent origin of these characteristic birds of extant coastal areas, which is also suggested by calibrated molecular data (Jetz et al. 2012).

Auks: Wing-propelled divers of the Northern Hemisphere

Of all extant charadriiform birds, auks (Alcidae) are the ones that are best adapted to a pelagic life off the coasts. These wing-propelled divers are only found in the Northern Hemisphere, and most of the 23 extant species occur in the North Pacific, where auks are likely to have originated (Olson 1985a; Smith and Clarke 2015). Molecular analyses show Alcidae to be nested within Lari, as the sister group of the Stercorariidae (Paton and Baker 2006; Fain and Houde 2007). Morphological characters in support of this hypothesis are few, but include short and strongly curved pedal claws (Mayr 2011b).

Auks are divided into the extinct flightless Mancallinae, or "Lucas auks," and their sister taxon Alcinae, which includes all extant species. Due to their use for underwater propulsion, the wings of auks are comparatively short, which results in a high wing loading and an energetically costly flight. Flightlessness is therefore positively selected for in predator-free environments and evolved at least twice independently within the group, in mancalline auks and in the more recently extinct Great Auk, *Pinguinus impennis*, which is a representative of the Alcinae.

The earliest fossils assigned to the Alcidae are from the late Eocene of North America: *Hydrotherikornis* from Oregon is based on a distal tibiotarsus and an unnamed species from Georgia is represented by a distal humerus (Chandler and Parmley 2003). The identification of *Hydrotherikornis* is controversial (Warheit 2002), but if future specimens corroborate its assignment to the Alcidae, auks would have already existed in both the Pacific and the Atlantic by the late Eocene. Whether these fossils are stem group representatives of the Alcidae or belong to the mancalline or alcine lineages cannot be determined.

A temporal gap exists between these specimens and the abundant Neogene fossil record of auks. *Petralca* from the late Oligocene of Austria, which is based on a poorly preserved partial skeleton, was misidentified and is a loon (Mayr 2009a; Wijnker and Olson 2009). Although auk fossils from the Oligocene of Japan have been mentioned in the literature (Smith 2011), these have not yet been described, and all fossil Alcidae discussed in the following are from Miocene or younger strata.

The flightless mancalline auks are only known from the North Pacific Basin and occurred from the middle Miocene to the late Pleistocene. Most fossils stem from California, where more than 4,000 bones of the taxon *Mancalla* were found, but there are also undescribed fossils from Miocene and Pleistocene localities in Japan (Smith 2011). The name *Mancalla* literally means "crippled wing," in reference to the greatly shortened forelimbs of these birds. Ulna and radius are extremely reduced compared with the hand and, most unusually, the ulna is even shorter than the carpometacarpus (Smith 2011). Furthermore, the deltopectoral crest of the humerus reaches a long way distally (Figure 9.2c, d), and this peculiar wing morphology is certainly related to a specialized use of the wings in underwater propulsion.

The taxonomy of mancalline auks was revised by Smith (2011), who altogether recognized six species in two taxa: *Miomancalla* ("*Praemancalla*"), with two late Miocene species, and *Mancalla*, with four species from Pliocene and Pleistocene localities. It remains to be seen whether this revision added to some further taxonomic confusion, because several species named by earlier authors were considered dubious and therefore not compared with newly described ones. The earliest mancalline records, from the middle Miocene of California, belong to *Alcodes*, which already exhibits the greatly foreshortened ulna characteristic of the Mancallinae (Figure 9.2e), but differs from later taxa in several presumably plesiomorphic features.

Because flightlessness released them from aerodynamic constraints, mancalline auks have a larger size than most crown group Alcidae and weight estimates range between 1 and 4 kilograms (Smith 2011). The largest species belongs to *Miomancalla*, a taxon characterized by unusually deep lower jaws. It was assumed that mancalline auks were specialized piscivores (Smith 2011). Extant auks show a diversity of bill morphologies and feeding specializations, and short-beaked species often feed on plankton, whereas the long-beaked species are piscivorous (Weir and Mursleen 2013).

According to recent analyses of morphological and molecular data, extant auks (Alcinae) fall into two clades: Fraterculini, to which auklets (*Aethia* and *Ptychoramphus*) and puffins (*Fratercula* and *Cerorhinca*) belong, and Alcini, which encompasses all other taxa, including the extinct flightless Great Auk (*Pinguinus*), as well as razorbills (*Alca*), guillemots (*Uria*), and little auks (*Alle*), among others (Smith and Clarke 2015). Whereas Mancallinae were restricted to the Pacific, the earliest fossils assigned to the Alcinae stem from the early Miocene of the western North Atlantic. Most of these belong to *Miocepphus*, which comprises four species from the early to late Miocene of the North American east coast. *Miocepphus* was regarded as ancestral to extant Alcinae (Wijnker and Olson 2009), but resulted as the sister taxon of the alcine taxon *Alle* in a recent phylogenetic analysis (Smith and Clarke 2015), in which case crown group Alcinae would have already diversified by the early Miocene. Another taxon, *Pseudocepphus* from the middle and late Miocene of Maryland, is distinguished from other Alcinae in humerus morphology and possibly represents an early diverging taxon of the Alcinae (Wijnker and Olson 2009). The exact relationships of the auklet-sized *Divisulcus* from the middle Miocene of Mexico, which is among the smallest known species of Alcidae, are unknown (Smith 2013).

The fossil record of putative Fraterculini dates back to the late Miocene, and specimens from southern California were assigned to extinct species of *Aethia* (Smith 2014). Because of the few known fossils, however, it remains elusive whether Fraterculini originated in the coastal areas of the Pacific.

In the Pliocene of the North Atlantic, alcine auks underwent considerable radiation and had a much greater diversity than today

(Olson and Rasmussen 2001). The single extant species of razorbills, *Alca torda*, is the survivor of a diversified group, and at least six extinct species of *Alca* occur in Pliocene deposits of the western Atlantic alone (Smith and Clarke 2011). Whether or not *Alca* is paraphyletic with respect to the flightless Great Auk (*Pinguinus*) has not yet been definitively resolved (Smith and Clarke 2011). If only extant auks are considered, the Great Auk evidently is most closely related to the Razorbill, but its affinities are less certain if fossil *Alca* species are also taken into consideration. It is possible, if not likely, that one of the large fossil species of *Alca* gave rise to the *Pinguinus* lineage, the earliest fossils of which stem from the early Pliocene of the North American Atlantic coast (Olson and Rasmussen 2001).

In terms of species diversity, the Neogene evolution of the Alcinae in the Atlantic paralleled that of the Mancallinae in the Pacific (Wijnker and Olson 2009; Smith and Clarke 2015). Both groups diversified into many species in the late Miocene and early Pliocene, but featured a significant decrease in diversity after the mid-Pliocene, which left only a single species of *Alca* in the Atlantic and one of *Mancalla* in the Pacific (Warheit 2002; Smith 2011; Smith and Clarke 2015).

Auks are ecologically linked to cold and productive oceanic upwellings. However, their greatest diversification appears to have been during periods of relative warmth, which may have been correlated with increased oceanic upwelling strengths (Smith and Clarke 2015). The geographic restriction of some alcid clades to different oceanic basins is notable, but agrees with the recent proposal that the Isthmus of Panama, which today separates the Atlantic from the Pacific, had already formed in the middle Miocene (Montes et al. 2015).

From Rail to Crane

Core Gruiformes include the sister taxa Ralloidea and Gruoidea, with Ralloidea comprising rails (Rallidae), flufftails (Sarothruridae), and finfoots (Heliornithidae), and Gruoidea being composed of trumpeters (Psophiidae), limpkins (Aramidae), and cranes (Gruidae). Only recently was it discovered that the very rail-like African flufftails (*Sarothrura* spp.), which have long been assigned to the rails, constitute the sister taxon of finfoots (Hackett et al. 2008; Mayr 2011a; García-R et al. 2014; Prum et al. 2015). Rallidae as traditionally recognized – that is, a clade including rails and flufftails – therefore exemplify one of those paraphyletic groupings of which highly specialized "outliers" (finfoots) have not been recognized as such.

The ecological attributes and habitat preferences of core gruiforms show considerable variation, but all are predominantly terrestrial and often associated with wetlands. Their last common ancestor probably was a medium-sized bird with a "rail-like" overall appearance.

Messel rails and other Paleogene rail-like birds

"Rail-like" birds already occur in Paleocene and early Eocene strata, and a particularly well-represented group are the Messel rails (Messelornithidae). In fact, *Messelornis* is by far the most abundant avian taxon in the early Eocene Messel fossil site in Germany, where hundreds of skeletons were found; *Messelornis* is also quite common in the early Eocene North American Green River Formation (Plate 14a, b; Hesse 1990; Weidig 2010; Grande 2013). Messel rails have furthermore been reported from the Paleocene of France (Mourer-Chauviré 1995), the early Eocene of Denmark (*Pellornis*; Bertelli et al. 2011), and the late Eocene and early Oligocene of France (*Itardiornis*; Mourer-Chauviré 1995).

Messel rails are considered to be the sister taxon of the Ralloidea and were terrestrial birds with long legs and short wings (Mayr 2009a; Bertelli et al. 2011). The well-known *Messelornis* is about the size of the Common Moorhen (*Gallinula chloropus*), and has a short, holorhinal beak and long tail feathers (Hesse 1990).

Much revisionary work has yet to be performed before the early evolutionary history of ralloids can be fully appreciated. The enigmatic rail-like taxon *Australlus* from the late Oligocene and middle Miocene of Australia, for example, is based on skull fragments and remains of most major postcranial bones. It is currently considered the sister taxon of swamphens (*Porphyrio*; Worthy and Boles 2010), but actually lacks a key apomorphy of Ralloidea; that is, a greatly reduced medial hypotarsal crest, with this crest being well developed in *Australlus*. The taxon is in need of a critical restudy and its bones show a resemblance to those of the messelornithid *Itardiornis* (Mayr 2013b). Initial assumptions that *Australlus* was flightless were based on comparisons with the limb proportion of putatively closely related rails, and may not be valid if rallid affinities are disproven.

The bones of Messel rails bear a resemblance to those of the Paleocene *Walbeckornis*, of which numerous well-preserved bones were found in a German fossil locality (Mayr 2007). The phylogenetic affinities of *Walbeckornis* are uncertain, although it is one of the candidate taxa for an early stem group representative of Ralloidea, or at least an early "rail-like" representative of core Gruiformes. Messelornithid-like birds are also known from the early Eocene of China, but these fossils have proportionally much longer toes, and the few known specimens do not allow a close examination of critical features for a definitive determination of their affinities (*Songzia*; M. Wang et al. 2012a; Plate 14c).

Rails, flufftails, and finfoots: The Ralloidea

Rails have a high dispersal potential and are a globally distributed and species-rich group that comprises ecologically diversified birds that live in various habitats. The nine species of flufftails, by contrast, are only found in forested environments of sub-Saharan Africa and Madagascar, and the three extant finfoot species are foot-propelled limnic birds that occur in the tropical regions of South America, Africa, and Asia.

Rail-like fossils from the early and middle Eocene of North America and Europe are too fragmentary for a reliable identification, and the exact stratigraphic occurrence of *Quercyrallus* from the middle Eocene to Oligocene of France is unknown (Olson 1977a; Mayr 2009a). The oldest stratigraphically well-constrained Rallidae are from the earliest Oligocene of Belgium (*Belgirallus*; Mayr and Smith 2001). The exact affinities of *Belgirallus* are unresolved, but the known bones closely resemble those of *Palaeoaramides* from the early Miocene of France, which is distinguished from crown group Ralloidea in presumably plesiomorphic humerus features (De Pietri and Mayr 2014b). Another early diverging rail-like taxon is *Rhenanorallus* from the late Oligocene or early Miocene of Germany, for which some similarities to the flufftail *Sarothrura* were noted (Mayr 2010c). A further *Sarothrura*-like taxon, *Paraortygometra*, occurs in the late Oligocene and early Miocene of France and, possibly, the early Miocene of Thailand (Cheneval et al. 1991; De Pietri and Mayr 2014b).

Outside Europe, the pre-Pliocene fossil record of rails is scant. In Africa, fragmentary remains of Rallidae were reported from the early Oligocene of Egypt (Rasmussen et al. 1987), and rails occurred in the early Miocene of New Zealand (Worthy et al. 2007). There are no published Paleogene or Miocene Rallidae from North America, and the earliest South American records are from the late Miocene/early Pliocene of Argentina (Tambussi and Degrange 2013).

The fossil record of finfoots consists of only two humerus fragments. One of these stems from the middle Miocene of North Carolina, USA, and is indistinguishable from the extant *Heliornis fulica*. Because finfoots do not cross oceanic barriers, the fossil suggests their dispersal into South America after formation of the Panamanian Isthmus (Olson 2003). Another finfoot humerus was described from the late Miocene of Chad (Louchart et al. 2005b). This fossil is more similar to the Asian taxon *Heliopais* than to the African *Podica*, but whether this resemblance is indeed indicative of closer affinities to *Heliopais*, as assumed by Louchart et al. (2005a), or whether it is due to the retention of primitive features needs to be scrutinized.

Gruoidea: The evolutionary history of cranes and allies

Psophiidae and Aramidae are species-poor Neotropic taxa with three and a single extant species, respectively, whereas the fifteen species of Gruidae have a much wider distribution on all northern continents as well as Africa and Australia. These three extant higher-level taxa of the Gruoidea show some diversity in their way of life that is reflected in differences in skeletal morphology and external appearance. Whereas the Psophiidae are rather short-beaked denizens of tropical forests, Aramidae and Gruidae inhabit wetlands and open grasslands. The representatives of all three gruoidean taxa are, however, comparatively long-legged birds and share derived features of the trunk skeleton, including a notarium and a long and narrow sternum.

Based merely on the interrelationships of extant Gruoidea, with Psophiidae being the sister taxon of a clade formed by Aramidae and Gruidae (e.g., Ericson et al. 2006; Hackett et al. 2008), it may be concluded that gruoideans originated in the Neotropic region. Disregarding the enigmatic and poorly known Late Cretaceous *Lamarqueavis* (Chapter 5), the early fossil record of the group is, however, an exclusively Northern Hemispheric one. Because this record includes putative gruoidean stem group representatives, Psophiidae and Aramidae possibly belong to the fair number of extant taxa with a relictual distribution in the Neotropic region. The habitat preferences of the forest-dwelling Psophiidae may well be primitive for Gruoidea, and the diversification of cranes is likely to be linked to the mid-Cenozoic spread of savannahs and open wetlands.

There are various early Paleogene taxa with presumed gruoidean affinities. Probably the phylogenetically most basal of these are the Parvigruidae, which include *Parvigrus* and *Rupelrallus* from the early Oligocene of Europe (Plate 13c; Mayr 2009a, 2013b). Parvigruids are medium-sized birds with rail-like limb bone proportions and relatively shorter legs than extant gruoideans. Derived features shared with extant Gruoidea include a greatly elongated and narrow sternum, as well as ossified tendons along the hindlimbs. Parvigruids exhibit long schizorhinal nostrils like extant Aramidae and Gruidae (the nostrils of the Psophiidae show the holorhinal condition), but some plesiomorphic characteristics indicate a position outside crown group Gruoidea (Mayr 2013b).

The evolutionary history of the Psophiidae and Aramidae is virtually unknown. A *Psophia*-like coracoid was reported from the late Eocene or Oligocene strata of the Quercy region in France, but the similarities of this fossil to extant trumpeters may be due to convergence (Mayr 2009a). A tarsometatarsus from the early Oligocene of South Dakota was assigned to the Aramidae (*Badistornis*;

Olson 1985a; Mayr 2009a), although comparisons of this fossil with the coeval European Parvigruidae still need to be performed.

The fossil record of cranes is more comprehensive, but many of the specimens are in need of a revision. The earliest fossils are classified in the taxon *Palaeogrus*, which was originally established for a distal tibiotarsus from the middle Eocene of Italy. There is also a late Eocene record of *Palaeogrus* from England (Mayr 2009a), but this is likewise too fragmentary for a firmly based identification. The earliest well-represented fossil cranes stem from the early and middle Miocene of France and Germany (Göhlich 2003a). These fossils were also attributed to *Palaeogrus*, but are likely to be more closely related to crown group Gruidae than to the Eocene fossils, in which case the taxon *Palaeogrus* as currently recognized would be paraphyletic. Another crane from the early Miocene of Germany, which is only known from a humerus, was assigned to *Balearica* (Mourer-Chauviré 1999). The similarities between this early Miocene crane and extant *Balearica* may well be plesiomorphic, and an ultimate assessment of its affinities depends on the discovery of more material.

Within crown group Gruidae, the African crowned cranes (*Balearica*) are the sister taxon of the remaining species, which, among others, share a hollow in the sternal keel that encompasses folds of the trachea. *Balearica*-like fossils occur in late Eocene and early Miocene strata of North America, although some of these have not yet been formally described and the restudy of others would be desirable (*Probalearica*, *Aramornis*; Olson 1985a). An extinct species of *Balearica* that is well represented by complete skeletons has been described from the late Miocene of Nebraska (Feduccia and Voorhies 1992). This species is smaller than all crown group Gruidae, and its morphology is very suggestive of that of crowned cranes in that the skull roof forms an inflated bulge.

Although it therefore seems likely that the extant distribution of crowned cranes in Africa is relictual, a reexamination of some of the North American *Balearica*-like fossils is nevertheless expedient. That crown group Gruinae (all extant Gruidae except *Balearica*) are of comparably recent origin is suggested by the circumstance that even fossils from the Pleistocene of the Mediterranean island of Menorca were identified as belonging to a basal representative of the clade (*Camusia*; Seguí 2002).

Outside Europe and North America, the fossil record of cranes is very sparse. *Eobalearica* from the middle Eocene of Uzbekistan, again only known from the distal end of a tibiotarsus, may be a pelagornithid (Mayr and Zvonok 2011), and a putative record of the Gruidae from the early Oligocene of Egypt is likewise too fragmentary for a definitive identification (Rasmussen et al. 1987).

Geranoididae and Eogruidae

An understanding of the evolutionary history of cranes and their allies may be promoted by a better knowledge of the skeletal morphology and phylogenetic affinities of two further taxa with putative gruoidean affinities, the Geranoididae and Eogruidae. The former was originally established for *Geranoides* from the early Eocene of Wyoming, which is only known from leg bones. The same is true for other putative geranoidids from the early and middle Eocene of Wyoming (Mayr 2009a), the best represented of which is the large and very long-legged *Palaeophasianus* (Figure 9.4a). Putative geranoidid fossils have also been described from the early Eocene of France (*Galligeranoides*; Bourdon et al. 2016), although these fossils show some similarity to the palaeognathous Palaeotididae (Mayr 2015b).

Geranoidids are medium-sized to large, long-legged birds. Their fossil record mainly consists of fragmentary hindlimb elements

Figure 9.4 Long-legged Eocene gruiform birds. (a) Femur, tibiotarsus, and tarsometatarsus of an unidentified species of the Geranoididae from the early Eocene of Wyoming (the fossil is in the collection of the American Museum of Natural History and consists of several fragments, which were assembled for the photo). (b) Tarsometatarsus of *Eogrus* (Eogruidae) from the middle Eocene of China.

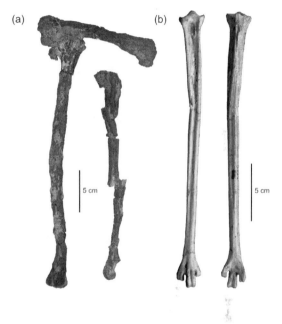

and it is therefore not certain that all taxa are closely related. Presumed affinities to the Gruoidea are mainly based on overall similarities and a resemblance of the bones of well-represented "geranoidids" to those of the Eogruidae, another taxon for which gruoidean affinities have been assumed.

The long-legged, cursorial eogruids first occur with *Eogrus* in the middle Eocene of Inner Mongolia (China). This taxon is known from hindlimb elements and a tentatively referred coracoid, and it was also identified in the late Eocene of Mongolia and Kazakhstan and the middle Miocene of China (Figure 9.4b; Olson 1985a; Clarke et al. 2005b; Mayr 2009a). Eogruids were comparatively abundant in the late Eocene and early Oligocene of Central Asia, where they persisted into the Pliocene. *Ergilornis* and *Sonogrus* occurred in the early Oligocene of Mongolia, whereas Neogene eogruids are represented by *Amphipelargus*, which was originally established for fossils from the late Miocene of Greece and Iran. Five other eogruid species from the early Miocene of Kazakhstan, the late Miocene of Iran, Moldova, and Ukraine, and the Pliocene of Kazakhstan and Mongolia are often assigned to *Amphipelargus*, but may be distinctive enough to merit classification in the separate taxon *Urmiornis* (Karhu 1997).

Like geranoidids, eogruids are mainly represented by leg bones. Skulls are not known for any species and only very few wing bones have been identified. The humerus morphology of *Ergilornis* suggests that this taxon was flightless (Kurochkin 1976), but flightlessness was not confirmed for the Eocene *Eogrus* (Clarke et al. 2005b).

Within eogruids, the tarsometatarsal trochlea for the second toe became progressively reduced, with this trochlea being very small in the early Oligocene *Ergilornis* and *Sonogrus* and almost completely lost in the Neogene *Amphipelargus/ Urmionis* (Olson 1985a). This characteristic foot morphology of Neogene eogruids allowed the assignment of didactyl foot tracks from the Pliocene of Iran to the Eogruidae (Lambrecht 1938).

A reduction of the tarsometatarsal trochlea for the second toe otherwise only occurs in the palaeognathous Struthioniformes, and it was assumed that eogruids are stem group representatives of ostriches (Olson 1985a), with which they coexisted in the late Miocene and Pliocene of Asia. This hypothesis is, however, not supported by the complex morphology of the tarsometatarsal hypotarsus of eogruids (in palaeognathous birds the hypotarsus is very simple), which were recovered as the sister group of Aramidae and Gruidae in the only phylogenetic analysis performed so far (Clarke et al. 2005b).

10 | Aequornithes: Aquatic and Semi-Aquatic Carnivores

Aequornithes includes specialized aquatic and semi-aquatic birds, which predominantly feed on fish and small vertebrates. The representatives of this clade – that is, loons (Gaviiformes), penguins (Sphenisciformes), tubenoses and allies (Procellariiformes), as well as most of the traditional "ciconiiform" and "pelecaniform" birds (storks, pelicans, and allies) – exhibit very disparate specializations, and their diversity spans the morphological spectrum from long-legged waders and highly aerial birds with feeble legs to pelagic, wing-propelled divers.

In most Aequornithes at least the bases of the three fore toes are webbed. Many representatives of this group furthermore have a compound rhamphotheca, which consists of several horny plates. The nostrils are often greatly reduced, but they are long and slit-like in the juveniles of most aequornithine taxa except loons. In the adult birds, elongated nostrils remain in the long-beaked ibises (Threskiornithidae), whereas in many Aequornithes with reduced nostrils the beak exhibits marked furrows that denote the former course of the narial openings.

Analyses of nuclear gene sequences supported loons, penguins, and procellariiform birds as early branching taxa (Hackett et al. 2008; Prum et al. 2015). The stem species of Aequornithes is therefore likely to have been a carnivorous bird, which lived in an at least semi-aquatic environment. The earliest fossil representatives of Aequornithes stem from strata around the Cretaceous/Paleogene boundary, which indicates a Mesozoic origin of the clade. These fossils show affinities to the aforementioned early diverging taxa, whereas there are no definitive records of "ciconiiform" and most "pelecaniform" birds before the late Eocene.

Avian Evolution: The Fossil Record of Birds and its Paleobiological Significance,
First Edition. Gerald Mayr.
© 2017 John Wiley & Sons, Ltd. Published 2017 by John Wiley & Sons, Ltd.

Loons: Foot-Propelled Divers of the Northern Hemisphere

Extant loons are foot-propelled diving birds. They predominantly feed on fish, but also take other vertebrate and invertebrate prey that is caught underwater. Loons visit freshwater lakes during the breeding season, but otherwise they are mainly found in marine habitats.

Except for differences in size and plumage coloration, the five extant species of loons are very similar, and all are classified in the taxon *Gavia*. As foot-propelled diving birds, loons are characterized by a derived hindlimb morphology with a very narrow pelvis, short femora, and greatly elongated cnemial crests on the proximal ends of the tibiotarsi, which enlarge the insertion sites of the hypertrophied leg muscles (Figure 10.1e). The tarsometatarsi are bilaterally compressed, so that drag is reduced in the underwater movement of the feet. Somewhat unexpectedly, loons also exhibit a derived morphology of the wing skeleton, with a flattened distal end of the ulna and a very low extensor process of the carpometacarpus. These features are usually found in birds that use their wings for underwater movement, and it was suggested that loons, too, exhibit such behavior "on a regular basis" (Olson 1985a: 214). Various footages of diving loons, however, suggest that this is not the case, although loons may have used their wings for subaquatic locomotion in an early phase of their evolution.

The extant distribution of loons is restricted to the Northern Hemisphere, but gaviiform affinities were suggested for some Late Cretaceous fossils from the Southern Hemisphere. None of these, however, can be confidently identified with the material at hand. *Neogaeornis* from the Late Cretaceous of Chile is only known from the holotype tarsometatarsus (Olson 1992a), which distinctly differs from that of early Cenozoic Gaviiformes. The known material of *Polarornis* from the Late Cretaceous of Antarctica (Chatterjee 2002) is even more fragmentary and does not allow a definitive identification, which remains true even after the recent report of further material of this taxon (Acosta Hospitaleche and Gelfo 2015).

The earliest unequivocal gaviiform stem group representative is *Colymbiculus* from middle Eocene marine strata in Ukraine (Figure 10.1; Mayr and Zvonok 2011, 2012; Mayr et al. 2013b). Like most other Paleogene stem group Gaviiformes, *Colymbiculus* was much smaller than extant loons. The larger deltopectoral crest of the humerus suggests that it made a different – aerial or subaquatic – use of its wings than extant loons, and the less specialized hindlimb morphology indicates that it was not adapted to foot-propelled diving to the same degree as later Gaviiformes.

Other Paleogene and early Neogene loons are more similar to the extant species. Most fossils were referred to the taxon *Colymboides*, which was first established for a species from early Miocene lacustrine deposits in France (Cheneval 1984). The earliest record of a *Colymboides*-like stem group gaviiform is a species from the late Eocene of England, which is based on only a few bones. Better known is another, tentatively referred, species from early Oligocene marine localities in Germany and Belgium (Mayr 2004a, 2009a; Mayr and Smith 2013). Fish bones in the stomach contents of an early Oligocene skeleton show Paleogene stem group representatives of the Gaviiformes to have already been piscivorous like their extant relatives (Mayr 2004a).

Apart from the early Miocene European fossils of *Colymboides* and possible records of this taxon from the middle Miocene of North America (Olson and Rasmussen 2001), all Neogene Gaviiformes were assigned to *Gavia*. In Europe, the earliest of these

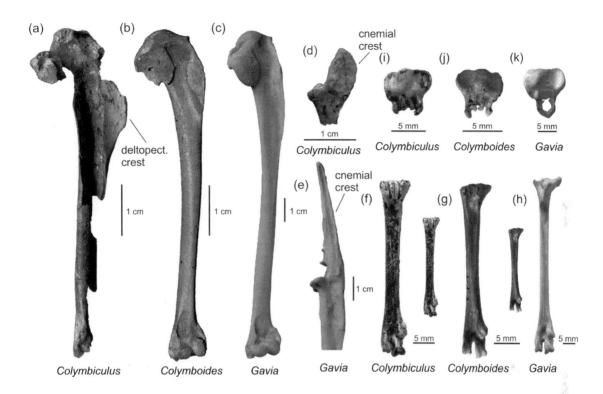

Figure 10.1 (a–c) Humeri, (d, e) proximal tibiotarsi, and (f–k) tarsometatarsi of an extant loon (*Gavia*), the middle Eocene *Colymbiculus*, and the early Miocene *Colymboides*. The right images in (f) and (g) show the actual size of the fossil tarsometatarsi relative to that of the smallest extant loon (h). The proximal ends of the tarsometatarsi in i–k illustrate the different hypotarsus morphologies. Note the extremely elongated cnemial crests of extant loons (e).

modern-type Gaviiformes are from the early and middle Miocene (Mlíkovský 2002), and the oldest definitive New World fossils of gaviiform birds likewise stem from early or middle Miocene deposits (Olson 1985a; Rasmussen 1998). That loons had greater past diversity is exemplified by the fact that three extinct species of *Gavia* occurred in the Pliocene Yorktown Formation of North Carolina alone (Olson and Rasmussen 2001).

Extant loons are mainly found in temperate or cold northern latitudes, but in the Paleogene and early Neogene stem group Gaviiformes inhabited subtropical lacustrine and marine environments. The different paleoenvironments of Paleogene Gaviiformes are illustrated by the remarkable discovery of a loon leg from the late Oligocene of Germany, which is preserved in association with a crocodilian tooth that stuck in the soft tissue (Plate 10c; Mayr and Poschmann 2009). This specimen documents the coexistence of loons and crocodilians in the late Paleogene of Europe, whereas these two taxa today occur in very different climatic zones and show little overlap in their distributions.

Some early Oligocene loons almost reached the size of the smallest extant species (Mayr 2009a). Most stem group Gaviiformes were, however, much smaller than their living relatives, and such small species existed at least from the middle Eocene to the early Miocene. Size increase in the stem lineage of Gaviiformes may be related to the retreat of loons to cold northern latitudes, where larger-bodied animals have a selective advantage owing to a more favorable – with regard to heat loss – body surface area to volume ratio (**Bergmann's Rule**).

Pelagic Tubenoses and Albatrosses

Few of the extant volant seabirds are as well adapted to a pelagic way of life as the Procellariiformes (albatrosses, tubenoses, and their allies). These exclusively marine birds mainly feed on fish and squid, although the diet of some species consists of planktonic invertebrates. Their most distinctive unifying feature is the presence of tubular nostrils, which are formed by the rhamphotheca above or lateral of the nostrils and probably play a role in olfaction or salt excretion. Another unique derived characteristic of procellariiform birds is a hind toe consisting of only the claw and the metatarsal, with the actual phalange in between having been lost. The humerus of many procellariiforms furthermore exhibits a large process on the dorsal side of the distal end, which is termed the "dorsal supracondylar process" and serves as the attachment site for tendons stabilizing the propatagium of these long-winged birds.

Procellariiformes include some 120 extant species, which are classified into Diomedeidae (albatrosses), Hydrobatidae (northern storm petrels), Oceanitidae (southern storm petrels), Pelecanoididae (diving petrels), and Procellariidae (fulmars, petrels, shearwaters). The interrelationships of these four taxa are still controversial. In several respects, albatrosses exhibit a more plesiomorphic morphology than the other crown group Procellariiformes. However, although albatrosses resulted as the sister taxon of all extant procellariiform birds in some analyses (Prum et al. 2015), other studies suggested that the Oceanitidae are the earliest diverging Procellariiformes (Hackett et al. 2008).

Procellariiformes have a worldwide distribution, but the greatest species diversity is found in regions of the Southern Hemisphere with productive marine upwellings, such as the Benguela and Humboldt currents along the coasts of southern Africa and western South America, respectively, as well as the seas around Antarctica and New Zealand. The earliest fossil representative is *Tytthostonyx* from the latest Cretaceous or earliest Paleocene of North America. This taxon is based on an incomplete humerus, which has a barely developed dorsal supracondylar process, but exhibits other derived characteristics of procellariiform birds (Olson and Parris 1987; Mayr 2015a). Although few fragmentary remains of putative procellariiform birds are furthermore known from the late Paleocene of Kazakhstan and the early Eocene of England (Mayr 2009a), the earliest well-represented fossils belong to the early Oligocene Diomedeoididae.

Diomedeoididae: A remarkable case of convergence in early procellariiform birds

In the early Oligocene, parts of Europe and the Middle East were covered by shallow epicontinental seaways. These marine environments were populated by a distinctive procellariiform taxon, the Diomedeoididae, of which numerous bones and skeletons have been found (Figure 10.2). Three species from the early Oligocene of Europe and Iran are currently recognized, with the stratigraphic origin of putatively late Oligocene and early Miocene diomedeoidid fossils being doubtful (Mayr et al. 2002b; Mayr 2009a; Mayr and Smith 2012a).

The taxonomic history of diomedeoidids is confusing. These birds are relatively abundant in the early Oligocene of Europe, and several names have been proposed for

Figure 10.2 (a) Skeleton of the procellariiform *Rupelornis* (Diomedeoididae) from the early Oligocene of Germany. (b, c) Foot, (d, e) coracoid, and (f, g) distal end of humerus of *Rupelornis* and extant Procellariiformes (b: *Nesofregetta*, Oceanitidae; e: *Fulmarus*, Procellariidae; g: *Lugensa*, Procellariidae). Note the widened pedal phalanges of *Rupelornis* and *Nesofregetta*.

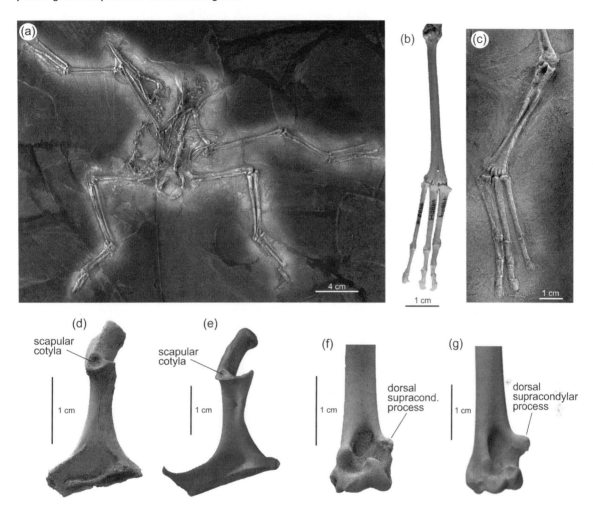

fragmentary remains whose true affinities were not recognized by earlier authors. All species are now classified in the taxon *Rupelornis*, which has nomenclatural priority over the previously used *Diomedeoides* (Mayr and Smith 2012a).

Diomedeoidids have an unusual foot morphology, which strikingly resembles that of the extant Polynesian storm petrel *Nesofregetta fuliginosa* (Oceanitidae) in the greatly widened pedal phalanges and the flattened and rounded claws (Figure 10.2b, c). To a somewhat lesser degree, such widened phalanges also occur in the extant oceanitid taxa *Fregetta* and *Pelagodroma*, but they are absent in other Oceanitidae. This indicates a convergent origin in diomedeoidids and southern storm petrels, because diomedeoidids are much larger than all extant Oceanitidae and lack the apomorphies of these birds (Mayr et al. 2002b; Mayr and Smith 2012a).

Diomedeoidids exhibit several plesiomorphic features, which suggest a basal phylogenetic position within Procellariiformes, either as the sister taxon of all crown group taxa or as that of the Diomedeidae (De Pietri et al. 2010; Mayr and Smith 2012a). Because the dorsal supracondylar process of the humerus is very poorly developed, they probably employed flap gliding like extant Oceanitidae rather than sustained gliding like most other extant procellariiforms (Mayr 2009a). Immersed in the water, the feet may have served as brakes or anchors to facilitate stationary positions.

Wide former distribution of albatrosses

The exact number of extant albatross species is controversial and ranges between 14 and 24. They are classified in four taxa, *Diomedea*, *Thalassarche*, *Phoebetria*, and *Phoebastria*, with only three species of the latter breeding in the North Pacific and none in the northern Atlantic. Today, albatrosses therefore have a predominantly Southern Hemispheric distribution. They usually occur in areas with productive marine upwellings, where they capture squid or fish from the sea surface. As highly pelagic birds that use a flight technique termed dynamic gliding, albatrosses depend on persisting wind systems, which accounts for their absence in the calm equatorial regions.

The identification of putative albatross remains from the middle Eocene of Uzbekistan (*Murunkus*, which reached only about one-third the size of the smallest extant albatross species) needs to be substantiated by additional specimens, which is also true for fossils from the late Eocene of Antarctica (Mayr 2009a). Certainly, however, albatrosses had a long evolutionary history in the North Atlantic, where they only occur as rare vagrants today. In fact, one of the earliest definite albatrosses is *Tydea* from the early

Oligocene of the North Sea Basin in Belgium (Mayr and Smith 2012b). This taxon is known from various postcranial bones and has the size of the extant black-browed albatross (*Diomedea melanophris*). Plesiomorphic features show *Tydea* to be outside crown group Diomedeidae. Although the close similarity of the wing bones suggests gliding capabilities, these were probably less sophisticated than in extant albatrosses (Mayr and Smith 2012b).

In the Neogene, albatrosses were diversified on both sides of the North Atlantic, where they no longer occur today. *Plotornis* from the early Miocene of southwestern France is a diomedeid stem group representative, which was smaller than the modern species, from which it also differs in a less specialized humerus morphology (Mayr and Smith 2012b). A record of *Plotornis* from the early Miocene of Italy documents the former occurrence of albatrosses in the Mediterranean (Mayr and Pavia 2014).

Albatrosses persisted in Europe into the late Pliocene, from which epoch an extinct species of *Phoebastria* was found in England (Olson and Rasmussen 2001; Dyke et al. 2007). Along the North American Atlantic coast, albatrosses occurred in the middle Miocene of Maryland (Olson 1985a). Two extinct and three extant species of *Phoebastria* were furthermore reported from the early Pliocene of North Carolina (Olson and Rasmussen 2001). The latest North Atlantic record is a Pleistocene breeding colony of *Phoebastria albatrus* on the Bermudas, which existed until 400,000 years ago and may have been extirpated by a mid-Pleistocene interglacial sea-level rise (Olson and Hearty 2003).

Various albatross fossils are also known from localities around the Pacific. Undescribed fossils from the late Oligocene of Washington State (USA) belong to a species that is smaller than any of the extant albatrosses (Mayr 2015a) and may be closely related to the European *Plotornis*. Large modern-type albatrosses first occurred in the early Miocene of Japan (Davis 2003), and fossils were furthermore found in the early Miocene of Argentina, the middle Miocene of California, and the late Miocene of Australia (Warheit 2002).

Other procellariiforms

Despite a similar external appearance, the two extant taxa of storm petrels – Oceanitidae (southern storm petrels; nine extant species) and Hydrobatidae (northern storm petrels; 16 extant species) – differ distinctly in their skeletal anatomy. Among others, the species of the Oceanitidae have much longer legs and shorter wings than those of the Hydrobatidae, and the traditionally assumed sister group between both taxa was not obtained in some analyses of nuclear gene sequences (Hackett et al. 2008; Prum et al. 2015). The fossil record of storm petrels is poor and the earliest specimens are Hydrobatidae from the late Miocene of California, which were assigned to the extant taxon *Oceanodroma* (Olson 1985a; Warheit 2002).

The wing-propelled diving petrels are among the most highly specialized crown group Procellariiformes and converged in their external appearance on some of the charadriiform auklets. The four very similar extant species are classified in the taxon *Pelecanoides* and only occur in the Southern Hemisphere. The oldest fossils of diving petrels stem from the early Miocene of New Zealand, and were assigned to an extinct species of *Pelecanoides* (Worthy et al. 2007). Another *Pelecanoides* species was described from the early Pliocene of South Africa, where diving petrels are no longer found today (Olson 1985a).

Procellariidae, the most widely distributed taxon of the Procellariiformes, includes some 80 extant species. The interrelationships of these birds are only incompletely understood, but five distinctive groups can be distinguished. The two most species rich of these are the Pterodromini, which include the gadfly petrels (*Pterodroma*), and the Puffini, which encompass the shearwaters (*Puffinus* and *Calonectris*). The other extant procellariid species are assigned to the Procellariini (petrels of the taxa *Procellaria* and *Bulweria*), the Fulmarini (fulmarine petrels), and the prions (*Pachyptila* and *Halobaena*).

Fragmentary remains of procellariid-like fossils are known from the late Eocene and early Oligocene of North and South America (Mayr 2009a), but only a partial skeleton of *Makahala* from the late Eocene or early Oligocene of Washington State (USA) allows meaningful comparisons with the extant taxa (Mayr 2015a). Despite an otherwise similar shape, the humerus of *Makahala* lacks the well-developed dorsal supracondylar process that characterizes extant Procellariidae, and the fossil taxon is therefore at best a procellariid stem group representative.

The earliest definite procellariid fossils are undescribed specimens from the late Oligocene of South Carolina (Olson 1985a; Ksepka 2014). The exact affinities of many Neogene fossils are in need of a revision, but it appears that crown group Procellariidae did not diversify very long before the Miocene. Most of the earliest fossils exhibit the wide and flattened humerus shaft that characterizes the Puffini. Extinct *Puffinus* species were described from the early and middle Miocene of North America, Japan, and Europe (Olson 1985a; Warheit 1992, 2002), and an extinct species of *Calonectris* was found in middle Miocene strata of eastern North America (Olson 2009). Fragmentary fossils of putative Fulmarini were reported from the middle and late Miocene of California (Olson 1985a), but whether their assignment to the extant taxon *Fulmarus* can be upheld remains to be seen once more material becomes available. A better-represented fulmarine taxon is *Pterodromoides* from the late Miocene of Menorca in the Mediterranean (Seguí et al.

2001), and the skull of a *Pachyptila*-like prion is known from the late Miocene of Chile (Sallaberry et al. 2007).

Penguins: More Than 60 Million Years of Flightlessness

The flightless, wing-propelled penguins (Sphenisciformes) are diving birds with a characteristic upright stance, and a standing height between 40 centimeters (Little Blue Penguin, *Eudyptula minor*) and slightly more than 1 meter (Emperor Penguin, *Aptenodytes forsteri*). Penguins occur on all Southern Hemispheric continents, but most of the 17 extant species breed in the Antarctic region. Although these birds are therefore usually associated with cold climates, one species, the Galapagos Penguin, occurs north of the Equator.

Penguins are among the most aberrant extant birds and their phylogenetic affinities are difficult to determine based on anatomical comparisons of the extant representatives. Most authors assumed close relationships to either Procellariiformes or Gaviiformes, but affinities to some "pelecaniform" birds were also proposed (Mayr 2005c; Ksepka and Ando 2011). Analyses of nuclear gene sequences supported a sister group relationship between Sphenisciformes and Procellariiformes (Hackett et al. 2008; Yuri et al. 2013; Jarvis et al. 2014; Prum et al. 2015), whereas smaller mitochondrial sets suggested a sister group relationship to the Ciconiidae (Pacheco et al. 2011).

Penguins are some of the most popular living birds, and a significant section of the avian fossils described in the past years also belongs to this group of birds. Many of these new finds are partial or complete skeletons, which provided novel insights into the evolutionary history of penguins (Jadwiszczak 2009; Ksepka and Ando 2011).

The fossil record

Penguins have an extensive fossil record, which is in part due to the fact that their limb bones are unusually robust, but probably also because these birds form large aggregations in marine environments with a high fossilization potential. Remarkably, every Southern Hemispheric continent has witnessed times during which more fossil penguin species coexisted than are found in each of these areas today (Ksepka and Ando 2011).

The oldest, albeit still undescribed, remains of stem group Sphenisciformes come from the latest Cretaceous of Chatham Island (see Mayr 2009a). The oldest published stem group sphenisciforms, which are also the phylogenetically earliest diverging ones, belong to the taxon *Waimanu* from the late Paleocene (58–61 mya) of New Zealand (Ksepka and Ando 2011). *Waimanu* includes two species of slightly different age and size, which were already flightless, wing-propelled divers with an upright stance and a standing height of about 80–100 centimeters (Slack et al. 2006). The skeleton of *Waimanu* exhibits many of the derived characteristics of extant penguins, from which, however, it still differs in numerous plesiomorphic features. In fact, *Waimanu* exhibits the most primitive morphology of the known stem group Sphenisciformes and, in addition to various other differences, its scapula is not greatly widened as in more derived sphenisciforms, the distal portions of ulna and radius are less flattened, and the tarsometatarsus is more elongated (Figure 10.3).

A sphenisciform taxon of comparable age to *Waimanu* is *Crossvallia* from the late Paleocene of Seymour Island (Antarctica), which is based on several bones of a single individual and even exceeds the *Waimanu* species in size (Jadwiszczak et al. 2013). The known bones of *Crossvallia* resemble those of *Waimanu*, but concerning a few features *Crossvallia* exhibits a somewhat more

Figure 10.3 Skeletal elements of various fossil and extant penguins (Sphenisciformes). Coracoids of (a) the Paleocene *Waimanu* and (b) the extant *Pygoscelis*. Humeri of (c) *Waimanu*, (d) the late Eocene *Icadyptes*, (e) the late Eocene *Pachydyptes*, (f) the early Oligocene *Kairuku*, and (g) the extant *Spheniscus*. Femora of (h) *Waimanu*, (i) the late Eocene *Archaeospheniscus*, (j) the late Eocene *Inkayacu*, (k) *Kairuku*, and (l) *Pygoscelis*. Tarsometatarsi of (m) *Waimanu*, (n) the late Eocene *Delphinornis*, (o) the late Eocene *Palaeeudyptes*, and (p) the extant *Eudyptes*. Note the stout humeri of *Icadyptes* and *Pachydyptes* and the stocky femora of *Inkayacu* and *Kairuku*. Not to scale.

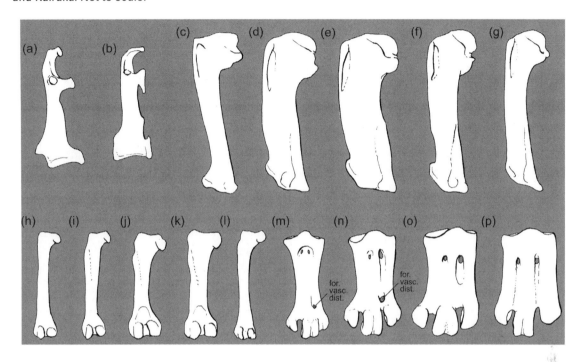

derived morphology (Acosta Hospitaleche et al. 2013a).

Early Eocene penguin fossils are comparatively rare. *Kaiika*, a stem group taxon from the early Eocene of New Zealand (Fordyce and Thomas 2011), was slightly larger than the extant Emperor Penguin and is known from a humerus, which again resembles that of *Waimanu*. Penguin remains were also reported from the early Eocene of Seymour Island and include a few medium-sized species (Jadwiszczak and Chapman 2011).

More substantial is the middle and late Eocene record, which shows that penguins were already very diversified by that time. Especially from Seymour Island, thousands of bones were collected in the vicinity of former rookeries. At least six taxa can be distinguished, which altogether encompass ten named species (*Anthropornis*, *Palaeeudyptes*, *Delphinornis*, *Mesetaornis*, *Marambiornis*, and *Archaeospheniscus*; Myrcha et al. 2002; Jadwiszczak 2006, 2013; Ksepka and Ando 2011). Most penguin fossils from Seymour Island consist of isolated bones, but of some species partial skeletons have been reported (Jadwiszczak 2012; Acosta Hospitaleche and Reguero 2014). The tarsometatarsi of *Delphinornis*, *Marambiornis*, and *Mesetaornis* exhibit a distal vascular foramen (Figure 10.3). This trait is lost in more derived Sphenisciformes and supports a position of these three taxa outside a clade including other Eocene and younger Sphenisciformes (Ksepka and Ando 2011). *Anthropornis*, *Palaeeudyptes*, and *Archaeospheniscus* include very large species with a standing height of more than 1.5 meters.

The earliest South American records of stem group Sphenisciformes are from middle Eocene strata (*Perudyptes*; Clarke et al. 2007; Ksepka and Clarke 2010a). The very large *Icadyptes* from the late Eocene of Peru is represented by a skull and wing bones, and has a minimum standing height of 1.5 meters (Clarke et al. 2007; Ksepka et al. 2008). *Inkayacu*, also from the late Eocene of Peru, is one of the best-represented Paleogene penguins and is known from an exceptionally complete specimen in which even parts of the wing feathers and the skin of the toes are preserved (Clarke et al. 2010). The skeleton of *Inkayacu* is most similar to that of *Palaeeudyptes*, with which its humerus shares a marked sulcus for the coracobrachialis nerve.

Penguins continued to be diversified in the Oligocene, but the fossils from this and later epochs are mainly from South America and New Zealand, and no post-Eocene penguin fossils were reported from Antarctica (Ksepka and Ando 2011). South America in particular has a very rich Neogene record of penguins, and the dramatic geological changes that accompanied penguin evolution on this continent are exemplified by the discovery of early Miocene *Palaeospheniscus* fossils in the Patagonian Cordillera, at an altitude of 1400 meters and some 500 kilometers away from today's Atlantic coast (Acosta Hospitaleche et al. 2013b).

The well-known taxon *Paraptenodytes* from the early Miocene of Argentina is still a stem group representative of Sphenisciformes (Bertelli et al. 2006). The oldest crown group penguin stems from the middle

Miocene (~11–13 mya) of the Pisco Formation in Peru and was assigned to the extant taxon *Spheniscus* (*S. muizoni*; Göhlich 2007). Another extinct *Spheniscus* species from the late Miocene of the Pisco Formation, *S. megaramphus*, has a much longer beak than any extant species of the taxon (Stucchi et al. 2003). *Madrynornis* from the late Miocene (~10 mya) of Argentina is considered to be the sister taxon of the extant *Eudyptes* (Acosta Hospitaleche et al. 2007), and a very large extinct species of *Pygoscelis* from the Pliocene of northern Chile has been hypothesized to be the sister taxon of the extant *Pygoscelis* species (Walsh and Suárez 2006). Molecular phylogenies calibrated with these and other fossils indicate a diversification of crown group Sphenisciformes some 20 million years ago (Subramanian et al. 2013).

Phylogenetic interrelationships and evolution
The phylogenetic interrelationships of stem group Sphenisciformes correspond well with the temporal occurrences of the various groups, with earlier taxa generally occupying a more basal phylogenetic position. Accordingly, the Paleocene *Waimanu* and *Crossvallia* are outside a clade including all other Sphenisciformes, and the taxa branching next are the Eocene *Delphinornis*, *Marambiornis*, and *Mesetaornis* (Figure 10.4; Ksepka and Ando 2011).

The interrelationships of other basal sphenisciform taxa are less well resolved and future studies will have to show whether the current taxonomy adequately reflects their phylogenetic affinities. *Inkayacu* from the late Eocene of Peru, *Kairuku* from the early Oligocene of New Zealand, and *Palaeeudyptes* from the late Eocene of Seymour Island, for example, share an unusually short and stout femur (Figure 10.3) and may be more closely related than is apparent from their current classification into separate genera. The same is true for *Icadyptes* from the late Eocene of Peru and *Palaeeudyptes*,

Figure 10.4 Phylogenetic interrelationships and temporal distribution of stem group Sphenisciformes (after Ksepka and Ando 2011 and Ksekpa et al. 2012). The geographic occurrences of the taxa are indicated in parentheses (ANT: Antarctica, NZ: New Zealand, SA: South America).

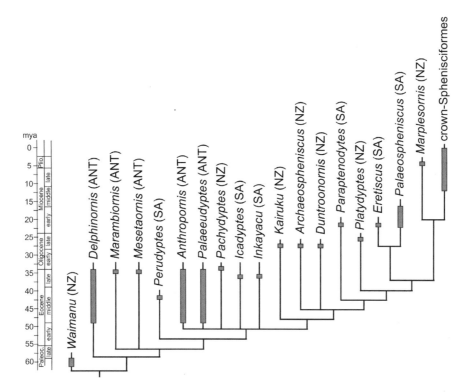

Pachydyptes, and *Anthropornis* from the late Eocene of Seymour Island, which exhibit a very stout and robust humerus (Figure 10.3).

Even the earliest fossil penguins were flightless and these birds must have lost their flight capabilities very early, around or even before the K/Pg boundary. However, although the oldest stem group Sphenisciformes already exhibit the basic skeletal characteristics of their living relatives, they differ in numerous plesiomorphic features from extant penguins, and only gradually were the characteristics of the crown group acquired. Unlike in extant penguins, the alular wing phalanges of the Paleocene *Waimanu* and the late Eocene *Inkayacu*, for example, are not fused with the carpometacarpus, which indicates the presence

of at least a rudimentary alula in these taxa (Ksepka et al. 2008; Ksepka and Ando 2011). In the early Oligocene *Kairuku*, the pygostyle has a more plesiomorphic shape than in crown group Sphenisciformes, which suggests that unlike in extant penguins the tail was not used to brace the standing bird (Ksepka et al. 2012; the pygostyle of earlier taxa is unknown).

Many Paleogene penguins are very large, and even the Paleocene *Waimanu* and *Crossvallia* already reached the size of the Emperor Penguin. From the early Eocene on, some species became truly gigantic, and *Anthropornis*, *Pachydyptes*, *Icadyptes*, *Palaeeudyptes*, and *Archaeospheniscus* reached standing heights of 1.5-1.7 meters and weights up to 80 kilograms (Jadwiszczak

2001; Ksepka and Ando 2011). Recently described remains of *Palaeeudyptes* suggest that this late Eocene taxon may even have had a standing height of 2 meters (Acosta Hospitaleche 2014). Such a very large size possibly evolved only once in the sphenisciform stem lineage and giant penguins existed until the late Oligocene (Clarke et al. 2007; Ksepka and Ando 2011). However, these giant forms coexisted with smaller ones and certainly the volant stem species of Sphenisciformes was a small to medium-sized bird (Ksepka and Ando 2011). Only after aerodynamic constraints ceased to exist, the giant size of stem group Sphenisciformes could have been positively selected for.

The size distribution of extant penguins, with the near-equatorial Galapagos Penguin being among the smallest species and some of the Antarctic species the largest, follows Bergmann's Rule, which postulates a larger body size of endothermic animals in colder climates. The marked size increase in stem group Sphenisciformes occurred, however, before the onset of global cooling. Fossils of the giant *Pachydyptes* were found in association with warm-water foraminiferans and the equally large *Icadyptes* lived in near-equatorial waters (Clarke et al. 2007). In fact, the average size of penguins appears to have decreased during the Cenozoic global cooling, and increased ocean productivity rather than changing climates may have led to a size increase in penguins (Clarke et al. 2007).

Extant penguins feed on fish as well as on planktonic prey, and there are distinct differences in bill morphology between the species at the extremes of these feeding spectra, with the beak of the planktonivorous *Eudyptes* being wide and with a very deep mandible, and that of the piscivorous *Aptenodytes* being long and pointed. Many early Cenozoic penguins have very long and dagger-like bills, which in *Icadyptes* and *Kairuku* reach nearly twice the length of

the neurocranium (Figure 10.5). Because a long beak is already present in *Waimanu*, it may be plesiomorphic for Sphenisciformes. Early stem group Sphenisciformes have well-developed temporal fossae for the jaw adductor muscles and may have used their beaks for spearing fish and other larger prey, although some diversity in feeding adaptations has been noted (Ksepka et al. 2008; Haidr and Acosta Hospitaleche 2012).

The fossil record sheds light on further aspects of the highly derived anatomy of penguins. The humerus of the extant species, for example, exhibits a complex vascular system, the humeral plexus, which was interpreted as a heat-retention structure. These vascular structures are, however, found in all sphenisciform species more closely related to the crown group than *Waimanu*. Therefore, they evolved at a time when penguins still lived in tropical climates. The humeral plexus may have enabled stem group Sphenisciformes to forage in cold offshore waters over an extended period of time, and it was one of the preadaptations that allowed the subsequent spread of these birds into areas with very cold climates (Thomas et al. 2011).

Examination of the feather remains preserved in the *Inkayacu* holotype revealed that this late Eocene taxon lacked the characteristic large and ellipsoidal melanosomes of extant penguins, and its plumage was reconstructed as grayish or brownish rather than black and white as in modern penguins (Clarke et al. 2010). The unique melanosome shape of modern penguins was hypothesized to be functionally correlated with the derived feather morphology of these birds, which itself was related to the hydrodynamic demands of underwater locomotion (Clarke et al. 2010). However, if the melanosome shape of penguins does indeed correspond with hydrodynamic constraints, one would expect the undersides of penguin flippers to be black, too, whereas they actually

Figure 10.5 (a) Skull of the long-beaked stem group penguin *Icadyptes* (Sphenisciformes) from the late Eocene of Peru (photograph by Daniel Ksepka). Skulls and mandibles of the extant (b) *Spheniscus* and (c) *Aptenodytes*. A greatly elongated beak is characteristic for many basal penguins and may be plesiomorphic for Sphenisciformes.

(a)

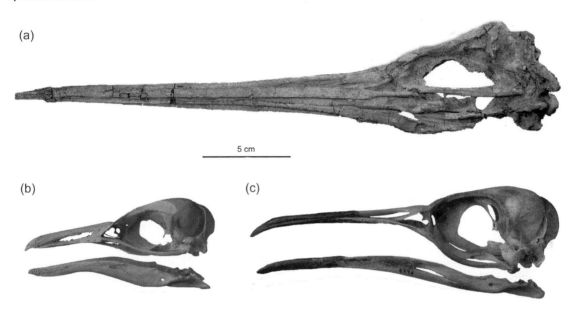

5 cm

(b)   (c)

are white in most species and therefore lack melanosomes altogether. Unlike the melanosomes, the *Inkayacu* feathers already have a modern shape, so that melanosome shape and feather morphology do not seem to be functionally correlated in penguins. This supports the traditional hypothesis that the marked contrast between the upper and lower sides of the penguin body is the result of **countershading**, which may have evolved in response to predation by pinnipeds (seals) and odontocete whales.

Paleobiogeography and paleodiversity

All penguin fossils stem from the Southern Hemisphere and were found within the geographic range of the extant species. The fossil evidence indicates that Sphenisciformes evolved in the New Zealand area in the latest Cretaceous or earliest Cenozoic, from where they rapidly spread into the Antarctic region

and South America (Ksepka and Thomas 2012). The question of why penguins did not disperse into the Northern Hemisphere is still unresolved. Oceanic circulation systems and the distribution of productive marine upwellings constitute geographic barriers today (Ksepka and Thomas 2012), but the origin of Sphenisciformes dates to geological periods with very different climatic regimes.

Penguins attained most of their characteristics before the glaciation of Antarctica started towards the late Eocene. More than by climatic events, their evolution may have been shaped by the absence of mammalian predators on some of the Southern Hemispheric continents, which facilitated the evolution of flightlessness in these birds. It is therefore probably no coincidence that the origin of flightless penguins appears to have been around the K/Pg boundary, and falls into a period in which most large terrestrial predators became globally extinct.

Paleogene penguin faunas were remarkably diversified and 10–14 sympatric species coexisted in the late Eocene of Seymour Island alone (Tambussi and Acosta Hospitaleche 2007; Ksepka and Ando 2011). The occurrence of so many species in a small geographic area may have been facilitated by the large size range spanned by these birds, from small species the size of the extant Macaroni Penguin, *Eudyptes chrysolophus*, to giant ones with a standing height of nearly 2 meters (Jadwiszczak 2001). Niche partitioning is also indicated by different bill morphologies and cranial specializations (Haidr and Acosta Hospitaleche 2012).

Penguins seem to have arrived very late in Africa, where the earliest fossils are four unidentified species from the late Miocene of South Africa, which greatly differ in size, covering the size range of the largest and smallest extant penguin species (Thomas and Ksepka 2013). At least three further species are known from the early Pliocene of South Africa and likewise belong to extinct taxa (*Nucleornis, Inguza, Dege*; Ksepka and Ando 2011). None of these species appears to be closely related to the African Penguin, *Spheniscus demersus*, the single extant penguin species breeding in Africa (Ksepka and Thomas 2012). The past diversity of penguins in Africa was explained by multiple dispersal events, which were favored by oceanic currents, and at least in the case of the ancestor of *S. demersus*, this dispersal was probably from South America (Ksepka and Thomas 2012). When exactly most African penguin species became extinct is unknown, but penguin diversity in Africa may have been affected by sea-level changes during the Pliocene, which led to the disappearance of secure breeding places on offshore islands (Ksepka and Thomas 2012).

The Polyphyletic "Pelecaniformes" and "Ciconiiformes"

Based on congruent and strongly supported phylogenies derived from molecular data, the traditional "Pelecaniformes" (pelicans and allies) and "Ciconiiformes" (storks, herons, and allies) are now widely recognized as polyphyletic assemblages. Although on the whole most representatives of both groups are recognized as closely related, convergent evolution led to similar morphologies in the taxa of different phylogenetic lineages.

Earlier authors classified into the "Pelecaniformes" six groups of marine or lacustrine birds with a throat pouch and a so-called totipalmate foot in which the hind toe directs medially (i.e., is not fully reversed) and all four toes are connected by a web. In the new molecular phylogenies, only frigatebirds (Fregatidae), gannets and boobies (Sulidae), cormorants (Phalacrocoracidae), and darters (Anhingidae) form a clade, which is termed Suliformes (Ericson et al. 2006; Hackett et al. 2008; Prum et al. 2015). Pelicans (Pelecanidae) are in a clade that also includes the Shoebill (Balaenicipitidae) and the Hamerkop (Scopidae), two distinctive African species that were traditionally assigned to the "Ciconiiformes."

Of the remaining "ciconiiform" taxa, herons (Ardeidae) and ibises (Threskiornithidae) were found to be the closest relatives of the clade including Balaenicipitidae, Scopidae, and Pelecanidae (Hackett et al. 2008; Prum et al. 2015). A derived morphological feature shared by these birds and the closely related Suliformes is a long hind toe, which could be functionally related to the circumstance that most species of these taxa build their nests in trees and therefore require some perching capabilities. Storks (Ciconiidae) were obtained as the sister group of a clade including all of these "ciconi-iform" and "pelecaniform" taxa (Hackett et al. 2008; Prum et al. 2015). Tropicbirds (Phaethontiformes), the last of the traditional "pelecaniform" groups, by contrast, appear to be only distantly related to the other "pelecaniforms" and "ciconiiforms."

Tropicbirds: A very old group of uncertain phylogenetic affinities

Tropicbirds are highly pelagic and predom-inantly piscivorous birds, which are today only found in tropical and subtropical seas. The three very similar extant species are classified in the taxon *Phaethon*. They are medium-sized birds with very short legs and wingspans of around 1 meter, which capture their prey with plunge-dives into the water. The tail forms a pair of long streamers that lend tropicbirds a characteristic appearance.

Tropicbirds are a very old group. The earliest possible record of these birds is *Nova-caesareala* from the Cretaceous/Paleocene boundary of North America, but this taxon is only known from a few fragmentary wing bones and phaethontiform affinities are not strongly based (Mayr and Scofield 2016). The earliest definite tropicbird fossils belong to the Prophaethontidae, which were found in late Paleocene to middle Eocene marine deposits of Eurasia, North Africa, and eastern North America and include the taxa *Prophaethon*, *Zhylgaia*, and *Lithoptila* (Bourdon et al. 2008a; Mayr 2009a, 2015g).

The phaethontiform affinities of pro-phaethontids are well established (Bour-don et al. 2005; Smith 2010). The skull of these birds is very similar to that of extant Phaethontiformes in the morphology of the neurocranium as well as in the reconstructed shape of the brain (Milner and Walsh 2009). Unlike in modern tropicbirds, however, the nostrils of prophaethontids are long and slit-like (Figure 10.6b, d), which presumably represents the plesiomorphic condition for Phaethontiformes (long and slit-like

nostrils also occur in the juveniles of extant tropicbirds; Olson 1977b).

The wing and pectoral girdle bones of prophaethontids resemble those of extant tropicbirds, but marked differences in hindlimb morphology suggest that they differed in their non-aerial locomotory spe-cializations from modern tropicbirds (Mayr 2015g). Unlike the short and wide pelvis of *Phaethon*, the pelvis of *Prophaethon* is long and narrow, the cnemial crests of the tibiotarsus are much larger, and the legs are not as greatly shortened as those of extant Phaethontiformes (Figure 10.6e–i; Harrison and Walker 1976; Bourdon et al. 2008a; Mayr 2015g). Overall, the pelvis and the leg bones of prophaethontids closely resemble those of albatrosses, indicating a similar hindlimb use. Like albatrosses, prophaethontids therefore probably seized their prey while swimming on the sea sur-face, a habit that may well be plesiomorphic for Phaethontiformes (Mayr 2015g).

With wingspans of about 1 meter, prophaethontids are of similar size to extant tropicbirds (Bourdon et al. 2008a). The concordances in wing and pectoral girdle morphology suggest that they were pelagic birds like their extant relatives (Bourdon et al. 2008a). The shape of the caudal verte-brae, however, indicates the absence of long tail streamers, and prophaethontids may have lacked the sophisticated aerial capabil-ities of their extant relatives (Mayr 2015g). The relative abundance of prophaethontid remains in the early Eocene London Clay (*Prophaethon*) and the late Paleocene and early Eocene of Morocco (*Lithoptila*) shows that these birds visited continental shores more often than their extant relatives do (Bourdon et al. 2008a; Mayr 2015g).

Together with the prophaethontid *Lithop-tila*, there occurred another, much rarer phaethontiform taxon in the early Eocene of Morocco, which is more similar to crown group Phaethontiformes (*Phaethusavis*;

Figure 10.6 (a, b) Skull, (e) partial pelvis (in matrix, with left femur in articulation), and (h) left foot of the early Eocene stem group tropicbird *Prophaethon* (Phaethontiformes) in comparison to the corresponding bones of (c, d, g, i) extant tropicbirds and (f) an albatross. Note the longer nostrils (denoted by arrows), narrower pelvis, and much smaller legs of *Prophaethon* (see Mayr 2015g for further details).

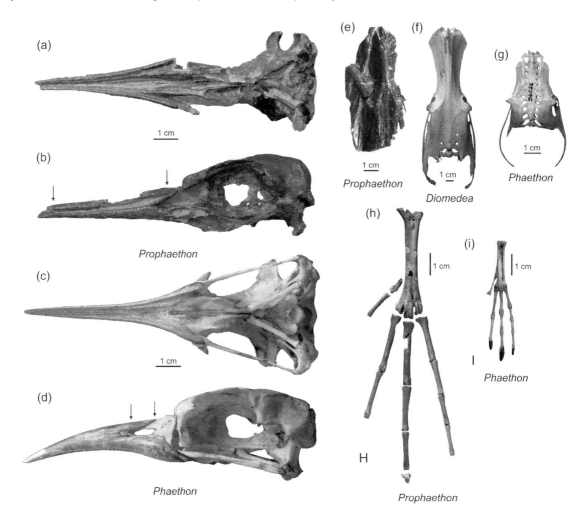

Bourdon et al. 2008b); an assessment of its exact affinities within Phaethontiformes, however, requires the discovery of additional bones. Very similar phaethontiform fossils were also found in the mid-Paleocene of New Zealand (Mayr and Scofield 2016).

Extant tropicbirds (Phaethontidae) breed on remote oceanic islands and are highly pelagic birds, which usually forage far offshore. These habits explain why fossil remains of phaethontids are very rare. Fairly modern-type tropicbirds were reported from the middle Miocene of Maryland (USA) and Belgium (*Heliadornis*; Olson 1985b; Olson and Walker 1997); a possible record also exists from the late Miocene of Austria (Mlíkovský 2002). These fossils document a wide mid-Cenozoic distribution of tropicbirds on the northern Hemisphere, which reached beyond the 50th northern latitude. Competition and predation at breeding sites may have led to the restriction of their ranges to the less productive seas of the tropical and subtropical regions, and their breeding on remote, predator-free islands (Mayr 2015g).

In tropicbirds, a throat pouch and totipalmate feet, the key "pelecaniform" characteristics, are only weakly developed. Although earlier molecular analyses did not support close affinities of tropicbirds to any of the other "pelecaniform" taxa, they yielded no congruent evidence for an alternative placement (e.g., Ericson et al. 2006; Hackett et al. 2008). The most recent analyses of large genomic data sets, however, came up with an entirely unexpected new hypothesis; that is, a sister group relationship between Phaethontiformes and the Southern Hemispheric Eurypygiformes (sunbittern and kagu; Jarvis et al. 2014; Prum et al. 2015). Eurypygiformes were long united with "gruiform" birds and include only two extant species, the rail-like South American Sunbittern (*Eurypyga*) and the more crane-like New Caledonian Kagu (*Rhynochetos*). A sister group relationship between Phaethontiformes and Eurypygiformes is not corroborated by morphological features and would imply an unprecedented degree of convergent evolution (Mayr 2015g). The clade formed by Phaethontiformes and Eurypygiformes was found to be the sister taxon of Aequornithes, and if these novel phylogenies are confirmed in future studies, the "seabird habitus" of tropicbirds probably evolved convergently to that of other taxa of Aequornithes. The fact that tropicbirds were even more aquatic in their early evolutionary history, as evidenced by the swimming adaptations of prophaethontids, would then be particularly remarkable, because their presumed sister taxon only includes highly terrestrial species.

Plotopterids: Penguin-like, flightless divers of the Northern Hemisphere

Among the most intriguing Cenozoic seabirds are the species of the Plotopteridae, which show a striking likeness to penguins in skeletal morphology. Remains of these distinctive flightless seabirds occurred in the late Eocene to middle Miocene of the North Pacific.

One of the earliest taxa, *Phocavis* from the late Eocene of Oregon, is only known from a single tarsometatarsus (Goedert 1988), but other plotopterids are represented by more substantial fossils. There was a particularly high diversity of these birds in the northeastern Pacific, along the western North American coast, during the late Eocene and Oligocene. The fossil record is particular rich in marine deposits of the Olympic Peninsula (Washington State, USA), from where two species of the taxon *Tonsala* have been reported, with the description of two further taxa pending (Olson 1980; Goedert and Cornish 2002; Dyke et al. 2011). The latest North American plotopterids belong to the comparatively small *Plotopterum* and the very similar *Stemec* from the late Oligocene of California and British Columbia, respectively (Howard 1969; Kaiser et al. 2015).

In Japan, plotopterids likewise had already appeared in the late Eocene (Sakurai et al. 2008), but the oldest described taxa, *Copepteryx* and *Hokkaidornis*, stem from the Oligocene (Figure 10.7; Olson and Hasegawa 1996; Sakurai et al. 2008; Kaiser et al. 2015). Most of these Oligocene Japanese plotopterids were very large birds, but a femur of the smaller *Plotopterum* was identified in middle Miocene rocks (Olson and Hasegawa 1985; Kaiser et al. 2015).

The wings of plotopterids are remarkably similar to the flippers of penguins, with which plotopterids share a remarkable flattening of the major wing bones and an unusually globose humerus head. The blade of the scapula is greatly expanded as in

Figure 10.7 (a) Reconstruction of the skeleton of the large plotopterid *Copepteryx* from the late Oligocene of Japan (Gunma Museum of Natural History, Japan; height of skeleton approximately 1 meter). (b, c) Partial skull (dorsal view) of the plotopterid *Tonsala* and an extant gannet (*Morus*, Sulidae). (d) Proximal humerus of an unnamed plotopterid species from the late Eocene or early Oligocene of Washington State, USA. (e) Carpometacarpus of a plotopterid from the late Eocene of Washington State. (f) Tarsometatarsus of the plotopterid *Phocavis* from the late Eocene of Oregon, USA. (g) Tarsometatarsus of the plotopterid *Hokkaidornis* from the late Oligocene of Japan. Photographs d–g by James Goedert.

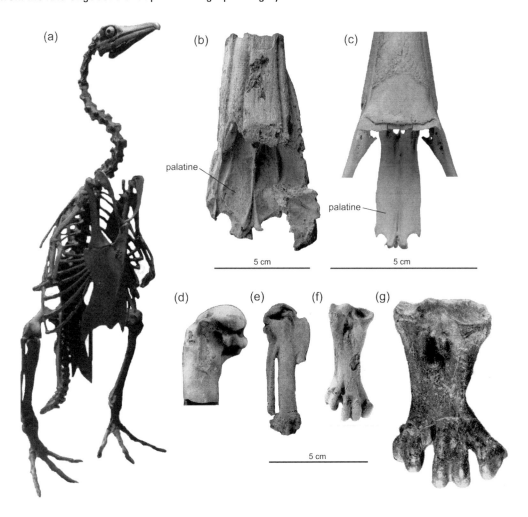

all sphenisciform birds except the basally diverging *Waimanu*, and, like in penguins, the tarsometatarsus is very short.

The skull of plotopterids has long remained undescribed, but meanwhile several well-preserved specimens are known (Kawabe et al. 2014; Mayr et al. 2015). In its proportions as well as in derived features, such as a wide nasal bar and a distinct nasofrontal hinge, the skull of plotopterids differs from that of penguins and resembles the skull of the suliform gannets and boobies (Mayr et al. 2015). Plotopterids are also distinguished from Sphenisciformes (including very early stem group representatives, such as *Waimanu*) in that the skull lacks supraorbital fossae for salt glands, the furcula exhibits a marked articulation facet for the acrocoracoid process of the coracoid, and the cranial end of the scapula is greatly elongated. The shape of their brain, by contrast, resembles that of penguins and differs from suliform birds (Kawabe et al. 2014).

Most authors attributed the similarities between Plotopteridae and Sphenisciformes to convergence and considered plotopterids to be closely related to the suliform Phalacrocoracidae and Anhingidae (Olson 1980; Olson and Hasegawa 1996; Smith 2010), but a sister group relationship between Plotopteridae and Sphenisciformes has also been proposed (Mayr 2005c; see also Mayr et al. 2015).

In some respects, plotopterids clearly differ from extant Suliformes. The nostrils, for example, are long and slit-like, whereas they are greatly reduced in all Suliformes, and the palatines do not exhibit the derived morphology found in Sulidae, Phalacrocoracidae, and Anhingidae, in which they are essentially flat, plate-like structures (Figure 10.7b, c). These plesiomorphic features of plotopterids support their position outside a clade including Sulidae, Phalacrocoracidae, and Anhingidae. Some of the derived characters shared with Phalacrocoracidae and Anhingidae, such as a greatly enlarged patella, are therefore likely to be the result of convergent evolution (Mayr et al. 2015).

A definitive classification of plotopterids does not only depend on a better understanding of the morphology of the late Eocene *Phocavis*, one of the earliest representatives of the group. Also critical for an assessment of the affinities of these birds would be the discovery of specimens in which the hind toe is preserved, which is expected to be very long, if plotopterids are indeed most closely related to the Suliformes.

The evolutionary origin of plotopterids remains poorly understood and so are the factors that caused their extinction, which broadly coincided with that of giant penguins. It was assumed that the demise of plotopterids was due to the evolution of pinnipeds, which may have preyed on plotopterids and probably also competed for breeding places on offshore islands (Goedert and Cornish 2002). At least regarding the North American coast, this hypothesis is in concordance with the fact that Cenozoic sea-level rising first led to the disappearance of offshore islands along the coasts of Washington and Oregon, whereas such islands persisted longer in California, from where some of the last North American plotopterids stem (Goedert and Cornish 2002). Whether a similar pattern can be observed in Japan has, however, not yet been assessed, and more data are needed for a full understanding of the reasons that led to the extinction of these unusual birds.

Frigatebirds and early suliforms

Frigatebirds are the sister taxon of the other Suliformes (e.g., Ericson et al. 2006; Hackett et al. 2008; Prum et al. 2015). The five extant species are highly aerial birds, which forage in tropical and subtropical oceans, and owing to their very short legs they can neither walk on land nor swim. Frigatebirds therefore usually catch their prey, mainly fish and squid, in flight, but also take seabird eggs and nestlings. As kleptoparasites they furthermore attack other marine birds and force them to regurgitate ingested food items.

The fossil record of frigatebirds dates back to the early Eocene, and skeletons of *Limnofregata* are among the more abundant avian remains in the North American Green River Formation (Olson 1977b; Olson and Matsuoka 2005; Smith 2010). The three species of this taxon (Stidham 2015) mainly differ in size and beak proportions, with the smallest being about the size of the males of the smallest extant frigatebird. Like its modern relatives, *Limnofregata* has an extremely short and stocky tarsometatarsus, a large, triangular deltopectoral crest of the humerus, and a short and wide pelvis. These features indicate that the taxon had already achieved a high degree of aerial specialization. However, *Limnofregata* differs from extant frigatebirds in other features, such as proportionately longer hindlimbs and a lack of fusion of the pectoral girdle bones, and these Paleogene stem group Fregatidae were probably less well adapted to soaring than their extant relatives (Olson 1977b).

Extant frigatebirds are exclusively pelagic birds, whereas the *Limnofregata* fossils were found in lacustrine deposits. These disparate habitat preferences may explain another difference between *Limnofregata* and its extant relatives; that is, the occurrence of long, slit-like nostrils in the Eocene frigatebirds. As in many other seabirds, the nostrils of extant Fregatidae are greatly reduced, but they are still slit-like in the juveniles (Olson 1977b), which is likely to represent the plesiomorphic condition for Suliformes (Mayr 2011a). A reduction of nostril size occurred repeatedly in the evolution of "pelecaniform" birds (see the discussion of Phaethontiformes), and at least in frigatebirds it may have been due to a transition from freshwater into marine habitats, in order to protect the nasal cavity against saltwater.

The *Limnofregata* fossils constrain the divergence between Fregatidae and the other suliform birds to at least the early Eocene. The fossil record of non-fregatid Suliformes from the earliest Cenozoic is, however, poor. One of the oldest possible representatives is *Protoplotus* from a lacustrine fossil site on Sumatra. The age of this taxon is debated, although Eocene or even Paleocene has been considered likely (Mayr 2009a). *Protoplotus* was originally assigned to the Anhingidae, but it is now classified in its own higher-level taxon Protoplotidae (van Tets et al. 1989). It is about the size of the smallest extant cormorants and resembles cormorants and darters in overall skeletal morphology. Unlike in extant Suliformes, however, the nostrils are also long and slit-like, which may indicate a basal position within the clade and conforms to the bird's lacustrine paleoenvironment. *Protoplotus* was assumed to have been an aquatic and possibly diving bird that captured fish and invertebrates (van Tets et al. 1989). The single known skeleton is preserved with numerous gastroliths and these were taken as evidence for an at least temporarily herbivorous diet (Lambrecht 1933; Zhou et al. 2004), which would be highly unusual for a suliform bird. However, other authors interpreted these gastroliths as ballast that assisted in diving (van Tets et al. 1989), and gastroliths are now also known in the Mesozoic *Yanornis*, for which a piscivorous diet is well documented by stomach contents (see Chapter 3).

Gannets and boobies

All extant representatives of the Sulidae are marine birds that perform plunge-dives to capture fish or squid. The ten globally distributed living species are classified in the taxa *Morus* (gannets) and *Sula* (boobies).

The earliest tentative records of sulids are from the early Eocene of Europe and consist of a skull from the lacustrine Messel oil shale (*Masillastega*) and a mandible from marine deposits in Romania (*Eostega*). The sulid affinities of these fossils are, however, only weakly based, and the same is true for remains from the Oligocene of Europe (Mayr 2009a). Unambiguous records are known from the Miocene onwards, and all stem from marine fossil sites. Of particular interest is the occurrence of very small species in the early and middle Miocene of Europe and North America, which are classified in the taxon *Microsula* (Olson 1985a; Göhlich 2003b). It is likely that a small size is plesiomorphic for Sulidae, but the exact affinities of *Microsula* have yet to be assessed.

Figure 10.8 Skulls of (a) the sulid *Ramphastosula* from the early Pliocene of Peru and (b) an extant booby (*Sula*). Photograph of *Ramphastosula* by Marcelo Stucchi.

Mainly from the Americas, many Miocene and Pliocene sulids have been described, and there was a high diversity of these birds in the early Pliocene Pisco Formation in Peru, where five species were distinguished (Warheit 2002; Stucchi 2003; Stucchi and Urbina 2004). Besides tentative records of modern species, the avifauna of the Pisco Formation includes a very small species of *Sula*, which was about 10–15% smaller than the smallest extant species (*S. sulita*; Stucchi 2003). A particularly remarkable taxon from the Pisco Formation is *Ramphastosula*, which is only known from skull remains and differs considerably from extant Sulidae in bill shape, with the upper beak being deeper and more curved (Figure 10.8). Because of this difference in beak morphology, *Ramphastosula* may not have been adapted to plunge-diving to the same degree as extant Sulidae (Stucchi and Urbina 2004).

The extant species of *Sula* are found in the tropical regions of the globe, whereas *Morus* occurs in the temperate zones of the Northern and Southern Hemispheres. Today there are only three *Morus* species, in South

Africa, Australasia, and on both sides of the North Atlantic. However, *Morus* appears to have been more diversified in the past and no fewer than three species have been described from the late Miocene and the Pliocene, respectively, of the North American east coast (Olson and Rasmussen 2001).

Cormorants

The Phalacrocoracidae are globally distributed and predominantly piscivorous birds, which live in freshwater and coastal marine habitats, and are particularly diverse in the northern and southern latitudes of the globe. In addition to cormorant-like beak fragments from the late Eocene of England and the early Oligocene of Egypt, there are complete skeletons of small cormorants from the early Oligocene of France in private collections, which still await description (Mayr 2009a). In the late Oligocene, cormorants were already diversified and widespread, and three distinct taxa occurred, *Nambashag* in New Zealand and *Borvocarbo* and *Oligocorax* in Europe.

Presumably one of the phylogenetically earliest diverging cormorant taxa is *Borvocarbo* from the late Oligocene/early Miocene of Europe (Mourer-Chauviré et al. 2004; Mayr 2010c). *Borvocarbo* was somewhat smaller than the smallest extant cormorants. Although its skeletal morphology is poorly known, the coracoid differs from that of extant Phalacrocoracidae and Anhingidae in the presence of a plesiomorphic, concave articulation facet for the scapula. Flattening of this facet is a recurrent theme in the evolution of neornithine birds, but the underlying functional reasons remain elusive in the case of cormorants.

Much better represented is a contemporaneous cormorant from the late Oligocene of Germany, of which several complete skeletons have been found (Plate 10b). These fossils were originally also classified in *Borvocarbo*, but they are now tentatively referred to the taxon *Oligocorax*, which is otherwise known from early Miocene localities in Europe (Mayr 2015h). *Oligocorax* slightly exceeds the extant Pygmy Cormorant (*Microcarbo pygmeus*) in size, and its phalacrocoracid affinities can be established with several derived characters, including a stout femur, a very large patella, and a long fourth toe (Smith 2010; Mayr 2015h). *Oligocorax* nevertheless exhibits a number of plesiomorphic features that support its position outside crown group Phalacrocoracidae (Mayr 2015g).

Another presumably basal phalacrocoracid is *Nectornis*, which occurs in the early Miocene of France and the middle Miocene of Turkey and Kenya (Cheneval 1984; Mayr 2014b). *Nectornis* includes comparatively small, lacustrine species. It differs from crown group Phalacrocoracidae in some skeletal features that indicate different locomotory characteristics, including a differently shaped proximal end of the humerus and a more slender femur (Mayr 2015h). Cormorants similar to *Nectornis* also occurred

in the late Oligocene or early Miocene of Australia (*Nambashag*; Worthy 2011).

The fossil record indicates that cormorants evolved in the Old World, from where all Paleogene fossils stem, and the earliest representatives were very small and probably lived in a limnic environment. From the mid-Miocene on, only modern-type cormorants were reported from various localities around the globe. In South America, Phalacrocoracidae first occurred in the late Miocene or early Pliocene of Chile (Walsh and Hume 2001).

Darters

Darters or snakebirds (Anhingidae) are predominantly piscivorous birds, which live in tropical and subtropical freshwater habitats of the Americas, Africa, Asia, and Australia. Depending on whether Old World darters are considered subspecies of a single species or treated as three distinct species, the extant diversity of the Anhingidae is confined to two or four very similar species, which are all classified in the taxon *Anhinga*. Darters are aquatic birds with dense and heavy limb bones. Due to the reduced buoyancy resulting from these adaptations, only the head and long neck of the swimming birds are visible, with the rest of the body being immersed in the water. Prey is generally caught underwater and stabbed with the pointed beak.

The earliest fossil darter is *Anhinga walterbolesi* from the late Oligocene of Australia. This species is known from a tarsometatarsus, which differs from the corresponding bone of extant darters in several features (Worthy 2012b). From the early Pliocene of Australia another extinct species of *Anhinga* was described, which was smaller than all extant darters (Mackness 1995).

Today, no more than one species of Anhingidae is found on the same continent, but darters were particularly diversified in the Miocene and Pliocene of South America, where several giant species evolved. These very large darters belong to the taxa *Meganhinga* (early Miocene of Chile) and *Giganhinga* (late Miocene of Argentina and Pliocene/Pleistocene of Uruguay), as well as to *Macranhinga*, which includes two or three species from the early to late Miocene of Peru, Argentina, and Brazil (Alvarenga 1995; Campbell 1996; Noriega and Alvarenga 2002; Alvarenga and Guilherme 2003; Areta et al. 2007; Cenizo and Agnolín 2010). One giant species was also reported from the late Miocene of southern North America (Becker 1987a). The latest giant taxon, *Giganhinga*, represents the largest darter that ever lived, for which a weight of 17–25 kilograms has been estimated (Rinderknecht and Noriega 2002; Areta et al. 2007). It is assumed that these large darters performed pursuit diving and did not spear their prey like the extant species (Noriega 2001). At least *Giganhinga* and *Meganhinga* were probably flightless (Alvarenga 1995).

The occurrence of giant darters in South America goes back to a mid-Cenozoic radiation restricted to that continent. Together with the large forms there also existed small species in the late Miocene of Brazil, one of which was smaller than all extant Anhingidae and was assigned to the taxon *Anhinga* (Alvarenga and Guilherme 2003). The factors that led to the evolution of gigantism in South American darters and to the loss of flight capabilities in some species are poorly understood, but they are likely to have been related to reduced predation pressure (see Chapter 13). Although flightlessness and giant size evolved in other aquatic birds, giant darters exemplify a rare case of these traits occurring in freshwater taxa. Giant darters have not been found on other continents and the pre-Pliocene absence of large mammalian carnivores in South America may have played a role in their evolution. Large mammalian carnivores

were, however, also absent in Australia, where darters remained comparatively small. Whether characteristics of the composition of South American crocodilian faunas also contributed to the gigantism of darters may be worthy of future study.

The earliest North American record of a darter is *Anhinga subvolans* from the early Miocene of Florida. Three species of Anhingidae were reported from the Pleistocene of North America, again exemplifying the higher diversity of these birds in the past (Becker 1986a, 1987a). Darters furthermore have a fossil record in Europe, where they do not occur today. These specimens stem from the late Miocene and belong to an extinct species, *Anhinga pannonica*, which was also tentatively identified in the middle and late Miocene of various African localities and in the late Miocene of Pakistan (Mlíkovský 2002; Dyke and Walker 2008; Louchart et al. 2008b). The occurrence of darters in Europe may go back to a Miocene dispersal from Africa, and may have temporally coincided with range extensions of other African bird groups in that period.

The phylogenetic interrelationships of fossil darters are poorly understood, owing to the fragmentary record of most taxa. Classification of all but the giant species in the taxon *Anhinga* is unlikely to reflect the correct interrelationships of these birds, which is particularly true for the earlier species. It is therefore to be expected that the taxonomy of Cenozoic Anhingidae will undergo substantial revisions once more fossils are known.

Hamerkops, shoebills, and pelicans

In sequence-based analyses, Scopidae, Balaenicipitidae, and Pelecanidae form a clade, and the most recent of these studies suggest that the latter two taxa – that is, the Shoebill and pelicans – are sister taxa (Prum et al. 2015). The fossil record is rather limited, but indicates that the initial divergences within the clade occurred before the early Oligocene.

The long-legged Scopidae and Balaenicipitidae include a single extant species each, which lives in sub-Saharan wetlands. The Hamerkop, *Scopus umbretta*, is a comparatively small bird that forages on small vertebrates and invertebrates. The Shoebill, *Balaeniceps rex*, by contrast, is tall and has a particularly massive beak, which is used to capture fish and other vertebrate prey.

The fossil record of Scopidae and Balaenicipitidae is very poor and consists of fragmentary bones that do not provide many insights into the evolutionary history of these birds. The only fossil hamerkop is an extinct species of *Scopus* from the early Pliocene of South Africa, which was slightly larger than the extant Hamerkop and may have been somewhat better adapted to swimming (Olson 1984). Balaenicipitidae have been reported from the early Oligocene of Egypt (*Goliathia*) and the late Miocene of Tunisia and – albeit on a rather speculative basis – Pakistan (*Paludiavis*; Olson 1985a; Rasmussen et al. 1987). However, these fossils merely show that shoebills were already large birds by that time and that they occurred outside their extant east African range.

The earliest definitive fossil record of pelicans is a skull from the early Oligocene of France, which, apart from its smaller size, closely resembles that of extant pelicans (Louchart et al. 2011). With the possible exception of the likewise smaller *Protopelicanus* from the late Eocene of France, which is only known from a femur (and may or may not belong to the same taxon as the aforementioned skull), there are no pre-Oligocene fossil records of the Pelecanidae (Mayr 2009a). Pelicans are today found on all continents except Antarctica, but their greatest diversity is in the Old World, with only two of the seven extant species occurring in the Americas. The abrupt appearance of modern-type pelicans in the early Oligocene of Europe supports an origin outside this continent. Because Scopidae and Balaenicipitidae are today found only in Africa, pelicans are likely to be of African origin, too, especially if Balaenicipitidae and Pelecanidae are sister taxa, as indicated by the new molecular phylogenies and some shared morphological features.

The Pelecanidae include limnic and marine extant species, but most early Cenozoic fossils were found in lacustrine sediments, and an origin in a freshwater habitat seems likely for this group. Some early Neogene pelicans are well represented, which is particularly true for *Miopelecanus* from the early Miocene of France and Germany, the two species of which were likewise smaller than the smallest extant Pelecanidae (Cheneval 1984). *Miopelecanus*-like pelicans also occurred in the middle Miocene of Kenya (Mayr 2014b), whereas other species from the Miocene of Europe were assigned to the extant taxon *Pelecanus* (Mlíkovský 2002). An extinct species of *Pelecanus* was furthermore described from the Miocene of Australia (Olson 1985a).

The earliest record of pelicans from the Americas is from the late Miocene of the Pisco Formation in Peru (Altamirano-Sierra 2013). An extinct species from the early Pliocene of North Carolina was larger than any extant New World pelican, and was considered to represent an extinct lineage rather than being ancestral to one of the two extant New World species (Olson 1999b). Given the long evolutionary history of modern-type Pelecanidae in the Old World, their absence in the early and mid-Cenozoic New World fossil record is notable and suggests a late dispersal of these birds into the Americas.

Herons

Ardeidae constitute one of the most species-rich extant groups of Aequornithes, and the more than 60 extant representatives have a global distribution. Most herons are long-legged and long-necked birds, although presumably basally diverging extant taxa, such as *Cochlearius*, *Nycticorax*, and *Tigrisoma* (Sheldon et al. 2000), are rather small in size and more stoutly built.

Most of the early fossil record of herons stems from the Old World, but these fossils closely resemble the crown group representatives and therefore do not elucidate the evolutionary origin of Ardeidae. The only aberrant taxon is *Xenerodiops* (Xenerodiopinae) from the early Oligocene of Egypt, which was likened to herons, but is based on a few fragmentary bones (Rasmussen et al. 1987). The same early Oligocene deposits of Egypt also yielded definitive heron remains, some of which show a – possibly plesiomorphic – resemblance to the extant *Nycticorax* (Rasmussen et al. 1987).

Heron fossils were also reported from the Eocene or Oligocene of France (*Proardea*) and the early Oligocene of Mongolia (Mayr 2009a). These specimens were poorly described and their position within Ardeidae is unresolved. The single Paleogene New World taxon for which affinities to herons were proposed is *Gnotornis* from the early Oligocene of South Dakota (Olson 1985a), but the fossil material, a distal humerus, is likewise too scant to allow unambiguous identification.

By the beginning of the Miocene, herons were already diversified and widely distributed, and the named taxa include *Proardeola* from the early Miocene of France as well as *Pikaihao* and *Matuku* from the early Miocene of New Zealand (Worthy et al. 2013b). A large-sized taxon from the early Miocene of Libya, *Zeltornis*, was considered similar to *Nycticorax*, even though it is solely known from a fragmentary coracoid (Mlíkovský 2003). Herons were also reported from the early Miocene of Thailand and the middle Miocene of Kenya and Mongolia (Cheneval et al. 1991; Zelenkov 2011b; Mayr

2014b). The Mongolian record was assigned to the extant taxon *Ardea*, but a well-founded assessment of the exact affinities of all of these fossils depends on the discovery of additional fossil material and a robust phylogeny of extant Ardeidae. Likewise in need of a critical restudy are some of the late Cenozoic fossils (Olson 1985a; Mlíkovský 2002).

The fossil record of herons is still very scant, but their early appearance and comparatively high diversity in the Cenozoic of Africa are notable, so much the more as this contrasts with the otherwise rather poor avian fossil record of the continent. Whether this indicates an African origin of herons needs to be considered in future studies. The fact that several of the earliest fossil representatives of Ardeidae show similarities to *Nycticorax* may furthermore suggest that an overall habitus similar to that of night herons is plesiomorphic for the Ardeidae.

Ibises

Most extant Threskiornithidae are fairly long-legged birds, which use their greatly elongated beaks to probe for food in soft soil or water. The fossil record shows, however, that ibises underwent significant morphological changes in their evolutionary history, and early Cenozoic stem group representatives were very differently built. The best-known taxon of these, *Rhynchaeites* from the early Eocene of Germany, is represented by more than a dozen skeletons (Peters 1983). Its beak is very similar to that of extant ibises in proportions, but the tip lacks openings for sensory nerves, which indicates that *Rhynchaeites* was a less tactile forager (Mayr 2009a). The legs are much shorter than those of extant ibises and the skeleton differs in several other aspects, such as a cup-like rather than shallow scapular articulation facet of the coracoid and a very different shape of the sternum. All of these features are likely to be plesiomorphic for ibises and set *Rhynchaeites* not only apart from extant ibises, but also from all extant close relatives of ibises, which underscores its potential significance for an understanding of the evolutionary history and phylogenetic affinities of these birds. There are records of various other early Cenozoic ibis-like birds from Europe, Asia, and North America (Mayr 2009a; Mayr and Bertelli 2011; M. Wang et al. 2012b; Smith et al. 2013), but most of these are very fragmentary or poorly preserved and do not allow a detailed analysis of their affinities.

The earliest well-represented modern-type ibis is the small *Gerandibis* from the early Miocene of France, of which numerous bones have been found (De Pietri 2013). A phylogenetic analysis suggested a sister group relationship between *Gerandibis* and a clade including the extant taxa *Plegadis* and *Eudocimus*, the bones of which bear some resemblance to those of *Gerandibis* (De Pietri 2013). Close affinities between these two extant taxa were, however, not supported by an analysis of molecular data (Ramirez et al. 2013). Because of the potential significance of *Gerandibis* for the calibration of molecular phylogenies, its phylogenetic relationships possibly need to be revisited once a robust phylogeny of the extant taxa exists. Otherwise, the fossil record of ibises is largely a latest Cenozoic and Quaternary one, which includes some bizarre island forms of evolutionary interest (see Chapter 13).

Storks
The extant members of the globally distributed Ciconiidae are medium-sized to large birds with long legs, necks, and beaks. The 19 living species are classified into three groups: wood storks and openbills (Mycteriini), true storks (Ciconiini), and marabous, saddlebills, and jabirus (Leptoptilini). All of these ciconiid subclades have extant representatives in the Old and the New Worlds. In each case, however, their diversity is higher in the Old World, from where the New World was colonized at least four times independently. The comparatively small Mycteriini are likely to be the sister taxon of all other Ciconiidae (De Pietri and Mayr 2014a). The fossil record of storks, however, mainly includes representatives of the Leptoptilini and Ciconiini, whereas the evolutionary history of the Mycteriini is poorly documented.

With a record of a putative stork from the middle Eocene of China being too fragmentary for a reliable identification (Mayr 2009a), the oldest definitive ciconiid fossils are from the late Eocene and early Oligocene of Egypt (Lambrecht 1933; Rasmussen et al. 1987). Among this material is a well-preserved skull of the early Oligocene *Palaeoephippiorhynchus*, which resembles the similar-sized skull of the extant Saddlebill (*Ephippiorhynchus senegalensis*), especially regarding the slightly upturned tip of the beak. *Palaeoephippiorhynchus* is generally assigned to the Leptoptilini, but its exact affinities within the taxon have yet to be determined.

Another leptoptiline stork is *Grallavis* from the early Miocene of France and Libya, which is very well represented, with all major skeletal elements being known. *Grallavis* is smaller than extant Leptoptilini, but a phylogenetic analysis supported a position within crown group Leptoptilini, as the sister taxon of marabous (De Pietri and Mayr 2014a). Whether coeval storks from the early Miocene of Florida (Olson 1985a) are possibly closely related to *Grallavis* has yet to be assessed.

Marabous are scavengers and the three extant species of *Leptoptilos* occur in sub-Saharan Africa and southern and southeastern Asia. However, they had a wider distribution in the past, and extinct *Leptoptilos* species have been described from the early to late Miocene of North Africa, the Pliocene of India and Ukraine, and the Pleistocene of northeastern China (Miller et al. 1997; Louchart et al. 2005c, 2008b; Zhang et al. 2012a). Some of these fossil species exceeded extant marabous in size, and the mid and late Cenozoic diversity and a wide distribution of marabous are likely to have been related to the wider past distribution of mammalian megafaunas (Zhang et al. 2012a). An extinct species of *Leptoptilos* has also been identified in the late Miocene of Argentina (Noriega and Cladera 2008). Extinct species of saddlebill storks (*Ephippiorhynchus*) have been reported from the late Miocene of Chad and Pakistan (Louchart et al. 2008b).

The earliest record of the Ciconiini is an extinct species of *Ciconia* from the late Oligocene or early Miocene of Australia, which is about the size of the extant White Stork, *C. ciconia* (Boles 2005). Species of *Ciconia* also occurred in the Neogene of Australia (Boles 2005), which is noteworthy because the single extant Australasian stork species belongs to *Ephippiorhynchus*, whereas there are no Ciconiini in Australasia today. An extinct small species of *Ciconia* has been described from the early Miocene of Kenya (Dyke and Walker 2008; Mayr 2014b), and several other extinct species of the Ciconiini are known from late Miocene to Pleistocene Old World sites (Louchart et al. 2005c).

In summary, the fossil record suggests an Old World origin of the Ciconiidae. Because storks had already attained their characteristic morphological attributes by the early Oligocene, they must have diverged from their sister taxon much earlier. If storks are indeed the sister group of all other Aequornithes and with penguins being known from deposits around the K/Pg boundary, this divergence had probably already occurred in the latest Mesozoic.

Late Cenozoic Turnovers in Marine Avifaunas

The composition of early Cenozoic marine avifaunas is clearly distinguished from that of present ones, where the most species-rich and most widely distributed groups of marine birds are ducks and geese, the procellariiform tubenoses, cormorants, and the charadriiform gulls, skuas, and auks. As detailed in the preceding chapters, some of the latter groups seem to have entered marine habitats comparatively late, and various species-poor seabird groups that are now largely restricted to the Southern Hemisphere had a wider distribution and greater diversity in the past, such as tropicbirds, frigatebirds, and albatrosses. Frigatebirds, which are today found exclusively in marine environments, furthermore occurred in freshwater sites in the early Eocene, and, unlike the highly pelagic extant tropicbirds, early Eocene phaethontiforms frequented near-shore habitats.

Several seabird taxa, such as Sphenisciformes, Procellariiformes, and Phaethontiformes, have a long evolutionary history in marine environments, whereas no records of large marine Laridae and Stercorariidae exist from deposits that predate the Pliocene. It is during this latter epoch that some of the most dramatic changes in global seabird communities appear to have taken place. In the Northern Hemisphere in particular, there was a significant loss of diversity in marine birds during the Pliocene (Warheit 1992, 2002; Olson and Rasmussen 2001; Olson and Hearty 2003; Smith 2011), and groups that have become locally or globally extinct include pelagornithids, albatrosses in the North Atlantic, and mancalline auks in the North Pacific.

Oceanic circulation systems were subjected to considerable changes during the Cenozoic, owing to major geographic events such as the opening of the Drake

Passage between South America and Antarctica, which led to the formation of the circum-Antarctic current in the late Eocene (Scher and Martin 2006). The Gulf Stream, one of the major oceanic currents of the Northern Hemisphere, appears to have formed in the Pliocene, and its origin was long associated with the emergence of the Panamanian Isthmus, which today separates the Atlantic from the Pacific (e.g., Haug and Tiedemann 1998). Most recently, however, a mid-Miocene origin of the Isthmus was proposed (Montes et al. 2015), which is in better agreement with the fact that the Miocene seabird communities of the Atlantic and Pacific already were very different. Neither plotopterids nor mancalline auks are found outside the North Pacific region, which indicates the presence of physical or ecological dispersal barriers, and in the case of plotopterids these must have existed even before a mid-Miocene origin of the Panamanian Isthmus.

The factors that shaped seabird evolution are not yet well understood and apart from changes in oceanic circulation systems and marine productivity, predation at the nesting sites may have impacted seabird communities. Even the most pelagic birds need to visit land to raise their chicks. Because of the long incubation and nesting periods and the low number of offspring, seabirds are particularly prone to predation at their nesting grounds. Many extant seabirds therefore breed in remote or inaccessible areas, such as oceanic islands, steep cliffs, or regions with unfavorable climatic conditions at high latitudes, where competition for breeding sites can be severe. Introduced mammalian predators, be it cats or rats, are known to have devastating impacts on extant marine avifaunas, and the late Cenozoic occurrence of skuas and gulls, which are important egg and nestling predators in extant seabird colonies, may have further tightened the competition for suitable breeding sites. Changing sea levels during the Cenozoic finally resulted in varying availabilities of offshore islands as breeding grounds. The disappearance of safe breeding grounds on offshore islands, for example, was considered to be a cause of the local extinction of plotopterids (Goedert and Cornish 2002), and Pliocene fluctuations in sea-level stands (Dwyer and Chandler 2009) may have likewise affected some of the ecologically less flexible seabird taxa, such as the flightless mancalline auks and the giant pelagornithids.

11 | Cariamiforms and Diurnal Birds of Prey

The diet of most birds includes at least some animal component, which often consists of insects or other invertebrates. Truly carnivorous birds that prey on land vertebrates evolved in various unrelated groups, and some of the most specialized species belong to the two taxa discussed in the present chapter, the Cariamiformes (seriemas and allies) and the presumably polyphyletic "Falconiformes" (diurnal birds of prey).

Cariamiformes include only two extant species, but the taxon gave rise to the "terror birds" (Phorusrhacidae), one of the predominant groups of avian carnivores in the Cenozoic of South America. The traditional "Falconiformes," on the other hand, encompass New World vultures (Cathartidae), secretarybirds (Sagittariidae), hawks, eagles, and allies (Accipitridae), as well as falcons (Falconidae). Although New World vultures were at times affiliated with storks, close relationships were not supported by virtually all analyses in the past few years, which established a clade including New World vultures, secretarybirds, and hawks and allies. On the other hand, however, analyses of nuclear gene sequences resulted in an unexpected sister group relationship between falcons and a clade including parrots and passerines. Seriemas, which were traditionally united with cranes and allies in the "Gruiformes," were found to be the sister taxon of the clade including these latter three groups (Ericson et al. 2006; Hackett et al. 2008; Jarvis et al. 2014; Prum et al. 2015).

In these novel sequence-based phylogenies, raptorial birds are therefore scattered across Telluraves, which was taken as evidence for a raptorial stem species of this clade (Jarvis et al. 2014). This hypothesis depends not only on the correctness of current phylogenies, but also on the living habits of the stem group representatives of these groups.

Avian Evolution: The Fossil Record of Birds and its Paleobiological Significance, First Edition. Gerald Mayr.
© 2017 John Wiley & Sons, Ltd. Published 2017 by John Wiley & Sons, Ltd.

Unfortunately, the early origins of both cariamiforms and diurnal birds of prey remain poorly known from a paleornithological point of view, although fossils do shed some light on the past diversity of these birds.

Seriemas and Allies: Two Species Now, Many More in the Past

South America is home to two species of long-legged cursorial birds, which are classified in the taxa *Chunga* and *Cariama* (Figure 11.1i-m), and which constitute the sole extant representatives of the seriemas (Cariamidae). Seriemas have a long evolutionary history in South America and their oldest representative, *Noriegavis* from the early-middle Miocene of Patagonia (Argentina), is already very similar to the extant species (Mayr and Noriega 2015). However, seriemas represent only one small segment of the past cariamiform diversity and the two extant species are mere relics of a once much more diversified avian group.

Other than in South America, stem group Cariamiformes have been reported from North America and Europe, where they existed until the early Miocene. As outlined in the following, the fossil record of these birds is quite comprehensive. However, the interrelationships of the various taxa are still poorly understood and the geographic origin of Cariamiformes is elusive.

Phorusrhacids: The "terror birds"

From a biogeographic point of view, the South American Phorusrhacidae, or "terror birds," are the prime candidates for the sister taxon of the Cariamidae. In South America, these distinctive flightless birds existed from the Paleocene to the Pliocene, and a few species may have even survived into the late Pleistocene. Phorusrhacids are well characterized by their massive, raptor-like beaks, the reduction of the wing bones, and

the mediolaterally compressed pelvis. Most species have long hindlimbs, which indicate a cursorial way of life. The majority of the fossils stem from Argentina, but there are also records from Brazil and Uruguay. Currently, about 20 species are distinguished, which are classified in the taxa Mesembriornithinae ("Hermosiornithinae" sensu Agnolín 2013), Psilopterinae, Patagornithinae, Phorusrhacinae, and Brontornithinae (Alvarenga and Höfling 2003; Alvarenga et al. 2011; Degrange et al. 2015a).

The earliest fossils are fragmentary hindlimb elements of *Paleopsilopterus* from the late Paleocene of Brazil (Alvarenga 1985a). This incompletely known taxon was assigned to the Psilopterinae, which include small and gracile species with slender leg bones, but its classification needs to be verified by additional specimens, particularly as even an assignment to the Phorusrhacidae has been questioned (Agnolín 2009b).

The next oldest phorusrhacid remains are from an unnamed, presumably psilopterine species from the late Eocene of Argentina (Acosta Hospitaleche and Tambussi 2005). Other psilopterines are younger. The taxon *Psilopterus* itself occurs in the late Oligocene to middle Miocene of Argentina and includes some of the smallest phorusrhacid species, which are the size of the extant *Cariama cristata* and exhibit a less reduced ulna and radius than other phorusrhacids (Figure 11.2).

The species of the Patagornithinae and Phorusrhacinae share a long and slender mandibular symphysis as well as elongated tarsometatarsi (Alvarenga and Höfling 2003; Alvarenga et al. 2011). Both taxa are mainly distinguished in the body sizes of

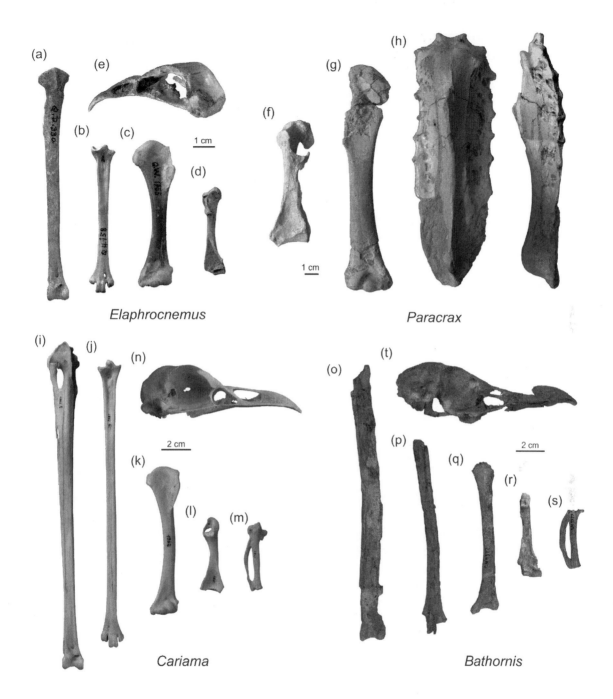

(a)

(e)

(b)

(c)

1 cm

(d)

(f)

(g)

(h)

1 cm

Elaphrocnemus

Paracrax

(i)

(j)

(n)

(g)

(o)

(t)

2 cm

(k)

(p)

(q)

2 cm

(l)

(m)

(r)

(s)

Cariama

Bathornis

Figure 11.1 Skeletal elements of stem group and extant Cariamiformes. (a–e) *Elaphrocnemus* from the late Eocene of France. (f–h) *Paracrax* from the early Oligocene of South Dakota. (i–n) The extant *Cariama*. (o–t) *Bathornis* from the middle Eocene of Wyoming. (e, n, t: skulls; d, f, l, r: coracoids; c, g, k, q: humeri; a, i, o: tibiotarsi; b, j, p: tarsometatarsi; h: sternum in ventral and lateral view; m, s: carpometacarpi). Note the greatly reduced sternal keel of *Paracrax*, the short acrocoracoid process of the coracoid of the flightless *Bathornis*, the short legs of *Elaphrocnemus*, and the very different humerus and coracoid morphologies of the North American *Bathornis* and *Paracrax*.

their representatives, with patagornithines being smaller than phorusrhacines, and their respective monophyly was not supported by some analyses (Agnolín 2009b, 2013; Alvarenga et al. 2011). The earliest of these "textbook phorusrhacids" is the patagornithine *Andrewsornis* from the late Oligocene of Argentina. *Patagornis* (Figure 11.2; Patagornithinae) and *Phorusrhacos* (Phorusrhacinae) are well-represented taxa from the early and middle Miocene of the Santa Cruz Formation of Patagonia, and *Andalgalornis* occurred in the late Miocene and early Pliocene of northwestern Argentina. Exceptionally large phorusrhacines are *Devincenzia* from the late Miocene and early Pliocene of Argentina and Uruguay, and its very similar sister taxon *Kelenken* from the middle Miocene of Argentina (Bertelli et al. 2007). After formation of the Panamanian Isthmus, at least one phorusrhacine taxon, *Titanis*, dispersed into North America, where it was found in Pliocene (5–1.8 mya; MacFadden et al. 2007) deposits in Texas and Florida. Like *Devincenzia* and *Kelenken*, *Titanis* reached a very large size. It also ranks among the youngest phorusrhacids, with the age or identification of even younger records from the Pleistocene of Uruguay (Tambussi et al. 1999; Alvarenga et al. 2010) having been contested (Agnolín 2013). The late Miocene and Pliocene Mesembriornithinae include *Mesembriornis* ("*Hermosiornis*"), *Llallawavis*, and, possibly, *Procariama* (the latter is sometimes assigned to the Psilopterinae, e.g., Vezzosi 2012).

Brontornis (Brontornithinae) from the early and middle Miocene of Argentina is particularly large and massively built, and reached a standing height of more than 2 meters (Alvarenga and Höfling 2003). The affinities of this poorly known taxon are contentious, and some authors have even questioned its classification in Phorusrhacidae and considered it to be a representative of the Galloanseres, which resembled the Australian Dromornithidae (Agnolín 2007; Degrange et al. 2012, 2015a). Others, however, reaffirmed its assignment to the Phorusrhacidae (Alvarenga et al. 2011). The controversial status of *Brontornis* is due to the fact that the fossil material is quite limited, and the taxon differs from typical phorusrhacids in some features, such as the bicondylar mandibular process of the quadrate and the lack of curved, raptor-like claws. Two further taxa referred to the Brontornithinae are *Physornis* from the late Oligocene of Argentina and *Paraphysornis* from the late Oligocene or early Miocene of Brazil (Alvarenga and Höfling 2003). *Physornis* is only based on fragmentary remains, but of the very large *Paraphysornis* a nearly complete skeleton was found, which exhibits the typical phorusrhacid morphology. Like *Brontornis*, *Paraphysornis* is more massively built than other phorusrhacids and has a shorter and stouter tarsometatarsus, which indicates less cursorial and possibly carrion-feeding habits (Alvarenga and Höfling 2003).

The interrelationships of the various phorusrhacid taxa are controversial, but Psilopterinae or Mesembriornithinae are considered to be among the earliest branching taxa (Alvarenga et al. 2011; Agnolín 2009b, 2013; Degrange et al. 2015a). An early divergence of psilopterines is suggested by the small size of *Psilopterus* and the presence of a well-developed furcula in this taxon (Degrange et al. 2015b). Unfortunately, the earliest known putative phorusrhacid, *Paleopsilopterus*, was not included in any of the phylogenetic analyses performed so far.

Phorusrhacids cover a wide size range, with the weight of small *Psilopterus* species having been estimated at about 8–9 kilograms (Degrange and Tambussi 2011) and that of the giant *Paraphysornis* at about 180 kilograms (Alvarenga and Höfling 2003). Some species may have been sexually dimorphic (Alvarenga and Höfling 2003). The wings of phorusrhacids are very short, but it was assumed that smaller psilopterines may have still possessed limited flight capabilities (Tambussi and Acosta Hospitaleche 2007).

Phorusrhacids were raptorial birds, with tall and mediolaterally compressed beaks with a sharply hooked tip (Figure 11.2), and they may have been among the top terrestrial predators in the Cenozoic of South America. Skull properties indicate that they lost cranial kinesis, and either foraged on prey that could be swallowed whole or killed prey items by multiple strikes with the huge beak (Degrange et al. 2010).

The extinction of phorusrhacids is likely to have been linked to the "Great American Interchange" after formation of the Isthmus of Panama, which resulted in the immigration of Northern Hemispheric taxa into South America (e.g., Marshall et al. 1982). Phorusrhacids could have been affected by direct predation from mammalian carnivores, by the disappearance of their prey (of which the endemic South American notoungulate mammals, which went extinct at a similar date, presumably constituted a high proportion), or by both. However, the fact that at least one phorusrhacid taxon, *Titanis*, immigrated into North America after the Panamanian Isthmus formed challenges the idea that these birds succumbed to predation by mammalian carnivores. Alternatively, it has been hypothesized that competition with condors played a role in the extinction of phorusrhacids (Tonni and Noriega 1998), which would be a viable hypothesis if these birds were scavengers, but also fails to explain the North American occurrence of *Titanis*.

Were there phorusrhacids outside the Americas?

Putative phorusrhacid remains were reported from the late Eocene of Seymour Island in Antarctica, but their identification is now disputed and all fossils have been attributed to other avian groups (Cenizo 2012). Because there was a land connection between South America and Antarctica until the late Eocene (e.g., Scher and Martin 2006), the occurrence of phorusrhacids in the early Paleogene of

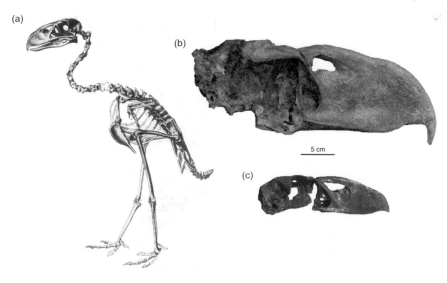

Figure 11.2 Phorusrhacids from the early and middle Miocene of Argentina. (a) Skeleton of *Psilopterus* (Psilopterinae; photograph from Lambrecht 1933). Skulls of (b) *Patagornis* (Patagornithinae) and (c) *Psilopterus*.

Antarctica would not be unexpected on biogeographic grounds.

More puzzling are records of putative phorusrhacids in the Old World. Their presence in the Cenozoic of Europe was first proposed based on a study of several bones from unknown, middle Eocene to Oligocene, stratigraphic horizons of the Quercy region in France (Mourer-Chauviré 1981). These fossils, various wing and pectoral girdle elements, bear a striking similarity to phorusrhacids, but they also exhibit distinct differences (Alvarenga and Höfling 2003). The humerus shows a close resemblance to that of another putative European phorusrhacid, from the early Eocene of Messel in Germany, which is now assigned to the taxon *Strigogyps* (Peters 1987; Mayr 2005d, 2009a).

The three known species of *Strigogyps* are rooster-sized birds with robust legs and reduced wings, which indicate flightlessness or at least very weak flight capabilities (Figure 11.3). A characteristic of these birds is the absence of a supratendinal bridge on the distal end of the tibiotarsus, which is typically reduced in birds with well-developed extensor muscles of the toes and may indicate that *Strigogyps* used its feet to manipulate food items. With regard to this and other features, such as the structure of the hypotarsus of the tarsometatarsus, *Strigogyps* is clearly distinguished from phorusrhacids (Mayr 2009a). Close affinities to the carnivorous Phorusrhacidae furthermore conflict with the fact that the stomach contents of one of the Messel specimens consist of plant matter (Mayr and Richter 2011).

Strigogyps may be closely related to another early Eocene taxon, *Salmila* (Salmilidae), which was described from the Messel fossil site (Figure 11.3; Mayr 2009a). *Salmila* is much smaller than *Strigogyps* and was fully flighted. Especially the leg bones show some similarities to *Strigogyps* in the morphology of the hypotarsus and the presumptive absence of an ossified bridge on the distal tibiotarsus, but as yet there is no conclusive evidence for close

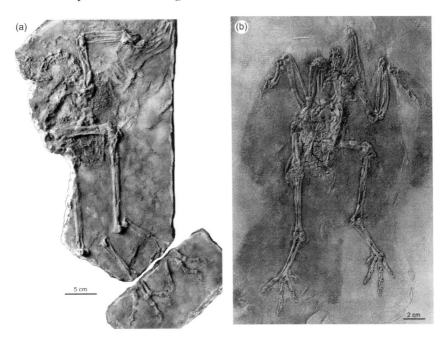

Figure 11.3 Skeletons of (a) *Strigogyps* and (b) *Salmila* from the early Eocene Messel fossil site in Germany.

affinities. *Strigogyps*-like birds have also been reported from the middle Paleocene of China (*Qianshanornis*, Qianshanornithidae; Mayr et al. 2013c; Plate 14d) and the early Oligocene of Egypt (Stidham and Smith 2015).

More recently, another European phorusrhacid candidate taxon was proposed – that is, *Eleutherornis* – which was previously considered to be a "ratite" (Angst et al. 2013). *Eleutherornis* was first reported from the middle Eocene of Switzerland, where rhea-sized pelvis fragments and pedal phalanges of this taxon were found. Later, it was suggested that a putative gastornithid from the middle Eocene of France, which is based on a fragmentary tarsometatarsus and pedal phalanges, may also belong to *Eleutherornis* (Mayr 2009a). Unfortunately, the fossil material of *Eleutherornis* is too fragmentary for a well-founded phylogenetic assignment. The few known bones, especially the curved claws, resemble those of *Strigogyps*, and from a biogeographic point of view, close affinities to the latter taxon are certainly more likely than a relationship to the South American phorusrhacids.

A putative phorusrhacid was also reported from the middle Eocene of Algeria (Mourer-Chauviré et al. 2011b). *Lavocatavis* is represented by a single femur and was a large and undoubtedly flightless bird, which was the size of one of the medium-sized phorusrhacids. Although its femur is indeed very similar to that of phorusrhacids, this bone is unknown for *Eremopezus*, another large flightless bird from the Eocene of northern Africa, which was assigned to the Palaeognathae (Chapter 6). Whatever are the true affinities of *Eremopezus*, there remains a possibility that *Lavocatavis* is more closely related to this African taxon than to phorusrhacids.

In summary, no unequivocal record of phorusrhacids exists from the Old World. No well-established and widely accepted overland dispersal routes are furthermore known that would have allowed the distribution of flightless cariamiforms between Europe and South America in the early Paleogene. Given its potential biogeographic implications, a possible occurrence of phorusrhacids outside the Americas therefore has to be established with stronger evidence than currently exists.

European and North American cariamiforms

The controversial presence of phorusrhacids notwithstanding, undisputed stem group cariamiforms occurred in Europe during the early and mid-Cenozoic. A particularly well-represented group of these are the Idiornithidae. In the Quercy region in France, where idiornithids were first recognized, they are the most abundant medium-sized birds and occur in middle Eocene to late Oligocene strata (Mourer-Chauviré 1983a; Mayr 2009a). Until recently, these seriema-like birds were classified into the taxon "*Idiornis*." However, it has now been recognized that humeri described as *Dynamopterus* in the late 19th century belong to the same species as other bones assigned to "*Idiornis*." Because *Dynamopterus* has nomenclatural priority, all currently recognized "*Idiornis*" species were transferred to this taxon (Mourer-Chauviré 2013b).

In the Quercy material alone, six species of *Dynamopterus* were distinguished and additional ones were described from the early and middle Eocene of Germany (Mourer-Chauviré 1983a, 2013b; Mayr 2009a). All correspond well with extant Cariamiformes in skeletal morphology, with one of the characteristic cariamiform features being a block-like hypotarsus of the tarsometatarsus that lacks canals for the flexor tendons of the toes.

Most idiornithids are rather small and the tarsometatarsus of the smallest species measures only one-third of that of the extant *Cariama cristata*. The legs of most species are long and gracile, but those of some early Eocene species are more stoutly built, which may reflect less cursorial habits in forested paleoenvironments (Mayr 2009a). At least these early Eocene species also differ from the extant Cariamidae in the possession of a well-developed hind toe.

The latest European Cariamiformes belong to *Ibidopodia* from the early Miocene of the Saint-Gérand-le-Puy area in France (De Pietri and Mayr 2014c). This taxon was originally compared with ibises and is only known from tarsometatarsi and a tentatively referred tibiotarsus and carpometacarpus.

Elaphrocnemus, another taxon from the Quercy region, was for a long time also assigned to the Idiornithidae and constitutes the most abundant medium-sized avian taxon in the 19th-century collections from these localities. Three late Eocene and Oligocene species are known, which have the size of small to medium-sized phasianids (Figure 11.1a–g; Mourer-Chauviré 1983a). The tarsometatarsus of *Elaphrocnemus* is proportionally shorter than that of *Dynamopterus* and extant Cariamidae, and unlike in the latter two taxa the humerus of *Elaphrocnemus* exhibits a prominent deltopectoral crest. Presumably, *Elaphrocnemus* was therefore less cursorial than *Dynamopterus* and other Cariamiformes and had better flight capabilities. A tentatively referred skull from the Quercy region shows that *Elaphrocnemus* also has a more slender beak than extant Cariamiformes (Mayr and Mourer-Chauviré 2006). The phylogenetic affinities of *Elaphrocnemus* have not yet been convincingly resolved. Although the taxon exhibits some derived similarities to Cariamiformes, it also shows close resemblances to opisthocomiform birds, which is especially true for the very hoatzin-like humerus (Mourer-Chauviré 1983a). In any case, *Elaphrocnemus* lacks the characteristic block-like hypotarsus shared by idiornithids, phorusrhacids, and cariamids, and is likely

to be outside a clade including these three taxa (Mayr and Mourer-Chauviré 2006; Mayr 2009a). A small *Elaphrocnemus*-like bird has been described from the late Paleocene of Brazil, with the exact relationships of this insufficiently known taxon likewise being uncertain (*Itaboravis*; Mayr et al. 2011a).

Cariamiform affinities have also been assumed for various fossils from middle Eocene to late Oligocene strata of the North American Great Plains (Cracraft 1968; Olson 1985a; Mayr 2009a). These birds were for a long time assigned to the Bathornithidae, which in its traditional composition almost certainly is a polyphyletic taxon. The best-represented species are *Bathornis* ("*Neocathartes*") *grallator* from the middle Eocene of Wyoming (Figure 11.1s–x) and *Paracrax wetmorei* from the early Oligocene of South Dakota (Figure 11.1h–l), of both of which partial skeletons have been found (Wetmore 1944; Cracraft 1968; Olson 1985a).

B. grallator appears to have been flightless, as evidenced by the shape of the wing bones and the reduced acrocoracoid process of the coracoid. The long legs also indicate predominantly terrestrial habits. One phylogenetic analysis suggested a sister group relationship between *Bathornis* and the Phorusrhacidae (Agnolín 2009b). Whether this phylogeny can be upheld will have to be seen, once the entire material of North American cariamiforms has been subjected to a detailed revision. In at least one feature, the presence of a well-developed hind toe, *Bathornis* is distinguished from phorusrhacids and seriemas, in which the hind toe is greatly reduced.

The species of *Paracrax* were large to very large birds, and the largest species, *Paracrax gigantea*, reached more than twice the size of the extant *Cariama cristata*. The skeletal morphology of *Paracrax* is so different from that of *Bathornis* that both taxa certainly do not belong to the same higher-level taxon. The wing bones of *Paracrax* are much more robust than those of *Bathornis* and the sternum exhibits a greatly reduced keel (Figure 11.1h), which shows a striking similarity to that of the extant Hoatzin. If this resemblance in sternal morphology is due to similar functional constraints, *Paracrax* may have also had a very large crop and a herbivorous diet (see Chapter 8). Although it was suggested that *Paracrax* forms a clade with the Idiornithidae and Cariamidae (Agnolín 2009b), its humerus shows a greater similarity to that of *Elaphrocnemus* and hoatzins. The hindlimbs of *Paracrax* are unknown and it is possible that some species currently assigned to *Bathornis*, which are only known from hindlimb elements and differ from *B. grallator* in the morphology of the tarsometatarsal hypotarsus, actually belong to *Paracrax* (Mayr 2009a).

Obviously, cariamiform birds lost their flight capabilities several times independently, in the South American phorusrhacids, some European taxa, and the North American *Bathornis*. A prerequisite for flight loss in these birds is the circumstance that they were cursorial and foraged on the ground, which favored the abandonment of a volant lifestyle in environments with reduced predation pressure (see also Chapter 13). Most likely the stem species of the clade including *Bathornis*, *Dynamopterus*, phorusrhacids, and extant seriemas was already a long-legged terrestrial bird with weak flight capabilities. A possible candidate taxon for such an early stem group representative is *Gradiornis* from the late Paleocene of Germany, which is, however, only known from a few bones (Mayr 2007).

Diurnal Birds of Prey: Multiple Cases of Convergence among Raptorial Birds

Despite a likeness in external appearance, diurnal birds of prey distinctly differ in anatomical features. The superficial resemblance of New World and Old World vultures has long been recognized as the result of convergent evolution, but the non-monophyly of Old World vultures was firmly established only through more recent analyses of molecular data.

Some earlier authors already doubted the monophyly of the traditional "Falconiformes" as a whole. Typically, however, the aberrant New World vultures (Cathartidae) and secretarybirds (Sagittariidae) were the critical taxa, whereas close affinities between falcons (Falconidae) and hawks and allies (Accipitridae) were rarely cast into question (see, however, Olson 1985a). The novel affiliation of falcons with seriemas, parrots, and passerines in analyses of nuclear gene sequences is therefore certainly among the major surprises of molecular avian systematics (Ericson et al. 2006; Hackett et al. 2008; Jarvis et al. 2014; Prum et al. 2015), although it should be noted that some analyses of mitochondrial sequences supported the traditionally assumed sister group relationship between Falconidae and Accipitridae (Pacheco et al. 2011).

Falcons

A candidate taxon for an early Paleogene falconid stem group representative is *Masillaraptor*, which was initially described from the early Eocene Messel oil shale in Germany (Figure 11.4; Mayr 2006c, 2009b), but of which a hitherto unidentified skull was also found in the early Eocene North American Green River Formation (Grande 2013: Figure 134A). *Masillaraptor* has a long beak with a straight dorsal ridge, which curves just before the tip. The legs are long and the pedal claws rather weak. As in extant Falconidae and Accipitridae, some phalanges of the toes are shortened (Figure 11.4d). Overall, *Masillaraptor* more closely resembles Falconidae than Accipitridae in certain features (Mayr 2009b), but the known specimens are not

Figure 11.4 The falconiform-like *Masillaraptor* from the early Eocene Messel fossil site in Germany. (a, b) Skeletons of two individuals. (c) Skull. (d) Detail of foot. Note the shortened central phalanges of the fourth toe (arrows).

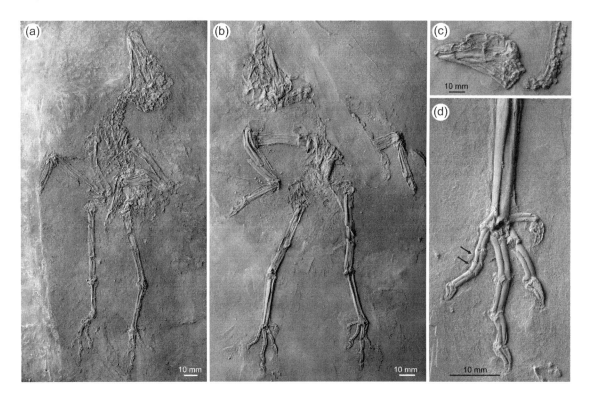

well enough preserved for a well-founded assessment of the phylogenetic affinities of this distinctive taxon, which also shows some resemblance to phorusrhacids in skull shape. Putative falconids were described from the early Eocene of England and Seymour Island (Antarctica). These records consist only of distal tarsometatarsi and their identification can therefore not be regarded as well supported (Mayr 2009a), although falconid affinities of the Antarctic fossil were recently reaffirmed (Cenizo et al. 2016).

Extant Falconidae are divided into three clades. The earliest diverging group are the South American Herpetotherinae (laughing falcons), which unlike other falcons lack a notarium. The Herpetotherinae are the sister taxon of a clade including the likewise South American Polyborinae (caracaras and

forest falcons) and the globally distributed Falconinae (true falcons). Based on the interrelationships of the extant taxa, a South American origin of crown group Falconidae is therefore likely, and this assumption is also supported by the provenance of the earliest definite falconid fossils.

One of these, *Thegornis* from the early Miocene of the Patagonian coast of Argentina, was hypothesized to be the sister taxon of *Herpetotheres*, one of the two extant members of Herpetotherinae (Noriega et al. 2011). If this placement is correct, *Thegornis* would not only be the oldest crown group representative of the Falconidae, it would also be a crown group representative of one of the three extant subclades of the taxon. Fossils of the Polyborinae first occurred in the late Miocene of Argentina and were assigned to the extant taxon *Milvago* (Cenizo et al. 2012). The earliest representative of the falconid clade Falconinae, which includes all of the Northern Hemispheric species, is *Pediohierax* from the middle Miocene of Nebraska (Becker 1987b). In the Old World, Falconinae first appeared in the late Miocene of Europe and Asia, and all of these species were assigned to the extant taxon *Falco* (Boev 2011; Li et al. 2014b).

The fossil record of falcons therefore indicates that crown group Falconidae originated in the late Paleogene or early Neogene of South America, from where falcons dispersed into Europe via North America in the mid to late Cenozoic. Falcons of the clade Herpetotherinae are more adapted to a life in forested regions than are the species of the Polyborinae and Falconinae, and the northward dispersal of falcons may be due to the mid-Cenozoic spread of open landscapes in the Northern Hemisphere (Fuchs et al. 2015).

New Word vultures and teratorns

The seven extant species of New World vultures occur in the Americas. Two new World vulture-like taxa were, however, reported from the Paleogene of France, where one taxon (*Diatropornis*) was found in middle and late Eocene sites, whereas the other (*Parasarcoramphus*) is of uncertain age, middle Eocene to Oligocene (Mourer-Chauviré 2002). Owing to the limited fossil material of both taxa (which are only represented by tarsometatarsi), their assignment to the Cathartidae is largely based on overall similarity, and there is a possibility that these birds are stem group representatives of a more inclusive clade that includes another group of New World vulture-like birds, the Teratornithidae (Mayr 2009a).

The extinct Teratornithidae are only known from the Americas and include the largest known raptorial birds. Their affinities have not yet been subjected to a formal phylogenetic analysis, but these birds most likely are the sister taxon of the Cathartidae, with which they share some derived skull features, such as a co-ossification of the lacrimal, ectethmoid, and frontal bones of the skull. Teratorns appear to have originated in South America and their earliest record is *Taubatornis* from the late Oligocene or early Miocene of Brazil, which is much smaller than later teratorns, but is only known from a few fragmentary bones (Olson and Alvarenga 2002).

Later teratorns, however, have a substantial fossil record and include some truly spectacular species. The most widely publicized of these is the giant *Argentavis magnificens* from the late Miocene of Argentina, whose size exceeded that of all other diurnal birds of prey. By comparison with extant species, the wingspan of *Argentavis* was estimated at 6–8 meters (Campbell and Tonni 1980, 1983), with the most realistic value being about 6 meters (Ksepka 2014). The estimated weight of *Argentavis* was 70 kilograms (Chatterjee et al. 2007), which is more than three times as much as the weight of the heaviest extant volant bird and also significantly more than the weight of the largest pelagornithids (see Chapter 7). Although aerodynamic models suggest flight capabilities of *Argentavis*, the take-off of this heavy bird would have required favorable conditions (Chatterjee et al. 2007). It was argued that such may have been present in the late Miocene of the Argentinean pampas region, because the Andean uplift had just commenced by that time, for which reason there were much stronger westerly winds east of this mountain chain than there are today (Campbell and Tonni 1983). However, the flight capabilities of *Argentavis* have mainly been inferred from the presence of feather attachment knobs on the ulna and no distal wing elements are known. Whether *Argentavis* was indeed capable of flight therefore remains somewhat conjectural, and it is completely possible that the taxon, which lived in a paleoenvironment without large mammalian predators, only had very limited flight capabilities.

North American teratorns are restricted to Pliocene and Quaternary fossil sites of the southern USA and belong to the taxa *Aiolornis*, *Cathartornis*, and *Teratornis*. Most fossils stem from the Quaternary Rancho la Brea tar pits in California, where numerous skeletal remains, especially of *Teratornis*, have been found (Figure 11.5). These North American teratorns are smaller than *Argentavis*, but they still were large to very large, with the wingspan of *Teratornis* having been estimated at 3–4 meters (Campbell and Tonni 1983). The stratigraphic occurrence of teratorns suggests that they dispersed into North America in the Pliocene after formation of the Panamanian Isthmus, although a record from the Pleistocene of Cuba (*Oscaravis*) shows that these birds were capable of overwater dispersal (Suárez and Olson 2009).

The earliest New World fossils assigned to the Cathartidae – that is, true New World vultures – are fragmentary hindlimb bones of small species from the late Eocene of Colorado (*Phasmagyps*; see, however, Olson 1985a for the uncertain identification of this taxon) and the late Oligocene or early Miocene of Brazil (*Brasilogyps*; Alvarenga 1985b). Various more substantial cathartid fossils are known from the Miocene on. The earliest representative of the condor lineage, which includes the two largest extant species of New World vultures, is *Hadrogyps* from the middle Miocene of California (Emslie 1988). *Perugyps* and *Kuntur*, the earliest South American taxa of condor-like cathartids, stem from the late Miocene of the Pisco Formation in Peru, and it has been assumed that condors originated in North America and dispersed into South America before formation of the Panamanian Isthmus (Stucchi and Emslie 2005; Stucchi et al. 2015; but see Montes et al. 2015 concerning an earlier, mid-Miocene formation of this land connection).

Being scavengers, the evolution of teratorns and cathartid vultures was certainly shaped by the availability of food items, and the disappearance of the American mammalian megafaunas towards the Pleistocene is likely to have impacted the diversity of these birds. Why, however, teratorns became extinct towards the late Pleistocene, whereas cathartids did not, is an unresolved question. On average, teratorns were larger than cathartids and there are no late Cenozoic fossils of teratorns from South America, with all Pliocene and Pleistocene records being from southern North America and Cuba. Whether these facts are causally related to their extinction still needs to be scrutinized. It is furthermore worth noting that, at least from the mid-Cenozoic on, neither cathartids nor teratorns dispersed into the Old World, where the ecological niches for large avian scavengers were occupied by aegypiine Old World vultures.

(a)

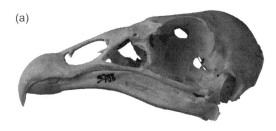

(b)

Figure 11.5 Skulls of (a) an extant New World vulture (*Cathartes*) and (b) the teratorn *Teratornis* (Teratornithidae) from the Pleistocene of the Rancho La Brea Tar Pits in California, USA. Not to scale.

Secretarybirds, ospreys, accipitrid vultures, hawks, and eagles

The African Secretarybird, *Sagittarius serpentarius*, is the only extant representative of the Sagittariidae. This long-legged bird mainly preys on terrestrial animals and is a characteristic inhabitant of arid landscapes south of the Sahara. Secretarybirds had a wider distribution in the past, and two species of the taxon *Pelargopappus* were reported from the Oligocene and early Miocene of France (Mourer-Chauviré and Cheneval 1983). Both have elongated hindlimbs like the extant secretarybird, but because the hind toe is much better developed in the fossils, their foraging strategies probably were different. The earliest and so far only African fossil record of the Sagittariidae is *Amanuensis* from the early Miocene of Namibia (Mourer-Chauviré 2003). Unfortunately, the exact interrelationships of *Pelargopappus*, *Amanuensis*, and extant Sagittariidae are unknown, and a well-founded assessment of the area of origin of secretarybirds is therefore not possible.

The majority of extant diurnal birds of prey belong to the Accipitridae, which include hawks, eagles, vultures, and allies. Molecular data provide a well-resolved phylogenetic framework (Lerner and Mindell 2005), and a consensus exists that the anatomically distinct Osprey (*Pandion*) is the sister group of all extant accipitrids. This species has a nearly global extant distribution and is often classified in its own higher-level taxon, Pandionidae. All Paleogene fossils of ospreys stem from the Old World. The earliest records, from the late Eocene of England and the early Oligocene of Germany, only consist of pedal claws (Mayr 2009a). These fossils are nevertheless diagnostic, because the pedal claws of ospreys have a characteristic shape owing to the specialized foraging mode of these birds, which capture fish with feet-first plunges into the water. A few fragmentary osprey bones were also reported from the early Oligocene of Egypt (Rasmussen et al. 1987). Neogene Pandionidae are somewhat better represented, even though the fossil record is nevertheless sparse, consisting of two extinct species of *Pandion* from the middle and late Miocene of California and Florida, respectively (Becker 1985).

The diverse and globally distributed Accipitridae include about 240 extant species. Among the earliest branching taxa are Elaninae (kites), Gypaetinae (gypaetine vultures), Perninae (honey buzzards), Milvinae (kites), Haliaeetinae (sea eagles), and Aegypiinae (aegypiine vultures), which are outside a clade including most of the well-known "typical" accipitrids, such as Accipitrinae (hawks), Buteoninae (buzzards), and Aquilinae (true eagles; Lerner and Mindell 2005). With the fossil record of ospreys going back to the late Eocene, Accipitridae must have also existed before the Oligocene. One of the earliest taxa with possible accipitrid affinities is *Horusornis* (Horusornithidae), an enigmatic kestrel-sized bird from the late Eocene of the Quercy region in France (Mourer-Chauviré 1991). *Horusornis* was a comparatively long-legged bird with a specialized foot morphology suggesting a particularly flexible intertarsal joint. Unlike in all other diurnal birds of prey there is no ossified supratendinal bridge on the distal tibiotarsus, which indicates a strong development of the digital tendons. Not least due to its unusual morphology, the interrelationships of *Horusornis* are difficult to determine. Although the taxon is comparatively well represented by many bones, more data on critical skeletal features are needed for a well-founded classification. A very flexible intertarsal joint is otherwise only known from the accipitrid taxon *Pengana* from the late Oligocene or early Miocene of Australia (Boles 1993a) and, among extant accipitrids, the South American *Geranospiza* and the African *Polyboroides*. These two extant taxa are unrelated to each other and to *Horusornis* and use their feet to extract prey (mainly nestlings of other birds) from tree cavities.

Fragmentary accipitrid-like tarsometatarsi were described from the middle Eocene of England (*Milvoides*) and the early Oligocene of Belgium (Mayr 2009a), but these are too incomplete for a well-founded determination of their exact affinities. Resemblances between the tarsometatarsus of *Aquilavus* from the late Eocene or Oligocene of the Quercy region and that of extant kites (*Milvus* spp.) may well be plesiomorphic (Mayr 2009a). Likewise doubtfully indicative of close relationships are the similarities between the tarsometatarsus of *Palaeohierax* from the late Oligocene of France and that of the extant Palm-nut Vulture *Gypohierax* (Zhang et al. 2012b), and those between a partial tarsometatarsus of a large accipitrid from the late Eocene of Egypt and extant Haliaeetinae (Rasmussen et al. 1987). Putative Accipitridae were also reported from the Oligocene of Mongolia and Kazakhstan (Kurochkin 1976), but these are either undescribed or based on very fragmentary specimens of little diagnostic value. However, even though these early accipitrid-like fossils cannot be phylogenetically constrained, it is notable that most bear similarities to early diverging accipitrid taxa, such as Milvinae and Haliaeetinae. In contrast, fossils showing morphological features typical of more advanced accipitrids – that is, taxa of the clade including hawks and eagles – have not yet been reported from Paleogene deposits.

Molecular analyses indicate that Old World vultures are paraphyletic (Lerner and Mindell 2005) and a vulture-like habitus evolved twice independently within Accipitridae, in the Gypaetinae (Bearded Vulture and allies) and in the Aegypiinae (Cinereous Vulture and other "typical" vultures). Most vultures are scavengers and the evolution of large vultures is therefore likely to have been intimately connected with the mid-Cenozoic spread of open habitats and the evolution of mammalian megafaunas. Both Gypaetinae and Aegypiinae today only occur in the Old World, and the occurrence of "Old World" vultures in the Cenozoic of North America was long considered one of the biogeographic surprises of the avian fossil record. However, all of these New World fossils show affinities to the Gypaetinae. With these being the sister taxon of the Perninae, which include several New World taxa (Lerner and Mindell 2005), a New World occurrence of early gypaetines is therefore less unexpected than it may seem at first glance, especially if the fossil taxa were shown to be basal gypaetine representatives.

Gypaetinae include only three extant species: the Bearded Vulture (*Gypaetus barbatus*), the Palm-nut Vulture (*Gypohierax angolensis*), and the Egyptian Vulture (*Neophron percnopterus*). One North American accipitrid taxon for which affinities to gypaetines were suggested is *Neophrontops*, which has been reported from middle Miocene to late Pleistocene localities (Rich 1980). Of the five described species of this taxon, in particular the well-represented *Neophrontops americanus* from the late Pleistocene of the Californian Rancho La Brea tar pits is very similar to the extant *N. percnopterus*. Another North American vulture-like accipitrid is *Palaeoborus*, of which three species were described from early and middle Miocene strata of the Great Plains. *Anchigyps* from the late Miocene of Nebraska is based on a partial skeleton and was considered most similar to *Gypohierax* (Zhang et al. 2012b). It is very possible that these North American vultures are early stem group representatives of the Gypaetinae, with the similarities to *Neophron* and *Gypohierax* being plesiomorphic. In any case, however, the absence of pre-Pliocene gypaetine-like fossils outside North America indicates that gypaetine vultures originated in the New World, from where they dispersed into the Old World. With regard to possible competition with the widespread and more species-rich aegypiine vultures, it may therefore be no coincidence that *Gypohierax* and *Gypaetus* exhibit specialized feeding ecologies, with their diet mainly consisting of fruits of the oil palm and bone marrow, respectively, whereas *Neophron* is a rather opportunistic feeder.

Aegypiine vultures, by contrast, are the sister taxon of the Old World Circaetinae (Lerner and Mindell 2005) and have not been reported from the New World. The Old World evolutionary history of the Aegypiinae is, however, not yet well understood. A beak fragment from the middle Miocene of Kenya was compared with the extant *Aegypius monachus* (Walker and Dyke 2006), from which it is clearly distinguished (Manegold et al. 2014). Several putative Old World vultures were also reported from the Miocene of China. Of these, the early Miocene *Qiluornis*, which is based on a partial skeleton, and the well-represented late Miocene *Gansugyps*, of which complete skeletons were found (Z. Zhang et al. 2010), are likely to be stem group representatives of the Aegypiinae, whereas the affinities of the large middle Miocene *Mioaegypius*, which is only known from a tarsometatarsus, are uncertain (Manegold et al. 2014). The earliest unequivocal record of a crown-group aegypiine vulture is an extinct species of *Aegypius* from the early Pliocene of South Africa (Manegold et al. 2014).

The oldest fossils of reasonably well-constrained representatives of the clade including hawks, buzzards, and eagles stem from the early Miocene. Some earlier Accipitridae from the Oligocene of North America were assigned to the extant taxon *Buteo*, but the known remains are too fragmentary for a well-founded assessment of their affinities (Mayr 2009a). Other accipitrid remains from the Oligocene of North America are

likewise in need of a revision, with the partial holotype skeleton of the Osprey-sized *Palaeoplancus* (Palaeoplancinae) being the most substantial of these records (Mayr 2009a). The oldest remains of large eagle-like accipitrids similar to the extant taxon *Aquila* come from the middle Miocene of France and Australia (Gaff and Boles 2010). *Apatosagittarius*, an unusual, long-legged accipitrid from the late Miocene of Nebraska, appears to have converged on secretarybirds in its foraging behavior (Feduccia and Voorhies 1989). The oldest definite accipitrid fossils from the Southern Hemisphere are fragmentary remains from the late Oligocene or early Miocene of Australia (Boles 1993a), the early Miocene of New Zealand (Worthy et al. 2007), and the early Miocene of Namibia and Kenya (Mourer-Chauviré 2008; Walker and Dyke 2006).

In summary, there is a fair record of accipitrids from the Oligocene of Europe and North America, with some specimens possibly dating back into the middle Eocene. Unfortunately, the phylogenetic affinities of many of these fossils are poorly constrained, owing to their fragmentary preservation and the fact that molecular analyses have only recently provided a framework for the interrelationships of accipitrid birds of prey.

Accipitrids exhibit a great variety of dietary specializations and include large scavengers, specialized insect eaters, and a few predominantly herbivorous forms. However, most accipitrid species forage for vertebrates, and the evolution of these birds is therefore likely to have been shaped by that of their prey, especially rodents and other small and medium-sized diurnal mammals. As in the case of falcons, the mid-Cenozoic spread of open landscapes may have favored the diversification of many accipitrid groups, as did the emergence of mammalian megafaunas, which provided food resources for large scavengers.

12 | The Cenozoic Radiation of Small Arboreal Birds

Trees provide numerous feeding opportunities for birds, either directly through fruits and seeds or indirectly via a plenitude of insect habitats. Trees also offer protected nesting sites in trunk cavities or difficult-to-reach places in the branches of the crown. As exemplified by the species-rich and diversified Enantiornithes, Mesozoic birds had already undergone a significant radiation in arboreal habitats. This diversification was probably accompanied by that of angiosperm plants and their insect pollinators, and stomach and crop contents suggest that birds acted as seed dispersers as early as in the Early Cretaceous.

However, early Mesozoic birds exhibit fewer morphological specializations for an arboreal way of life than extant tree-dwelling Neornithes, and the evolutionary success of arboreal crown group birds is probably due to both anatomical and behavioral characteristics. Compared to enantiornithines, arboreal neornithines show a much greater diversity of beak and foot morphologies, which is likely to be due to the higher degree of fusion of the bones involved and facilitated the exploitation of a greater variety of ecological niches. Arboreal birds are particularly vulnerable at their nests, and neornithine taxa evolved many sophisticated nesting strategies to reduce the risk of predation on their eggs and young. Together, these two factors – the high number of different feeding niches and the provision of safer nesting sites – probably account for the fact that the better part of extant birds belongs to one of the taxa discussed in the present chapter.

It is remarkable that no small arboreal neornithine birds are known from Cretaceous or even early Paleocene deposits. In the early Eocene, by contrast, stem group representatives of most extant arboreal lineages

Avian Evolution: The Fossil Record of Birds and its Paleobiological Significance,
First Edition. Gerald Mayr.
© 2017 John Wiley & Sons, Ltd. Published 2017 by John Wiley & Sons, Ltd.

were already present. It therefore seems very possible that a causal correlation existed between the extinction of the arboreal Enantiornithes at the end of the Mesozoic and the radiation of arboreal Neornithines thereafter, although this can only be evaluated once an improved fossil record from the latest Cretaceous and earliest Cenozoic is available.

At the beginning of the Cenozoic, the average temperatures were much higher than today, and most parts of the globe were covered with forests, which reached far into the northern latitudes. These relatively homogenous environments favored the evolution of insectivorous or frugivorous taxa with poor long-distance dispersal capabilities. The onset of global climatic cooling in the mid-Cenozoic and the formation of marked latitudinal temperature gradients restricted the distribution of these birds to tropical or subtropical latitudes. However, there are also other patterns in the past distribution of various taxa of the "arboreal landbirds" that cannot easily be explained by climatic factors. In virtually all of these cases, it is the improved understanding of the avian fossil record that sheds light on a sometimes complex biogeographic history.

The Courol and Mousebirds: Two African Relict Groups

As we have already seen, several avian groups that are today only found in Africa occurred outside the continent during the Cenozoic. However, although ostriches, turacos, and secretarybirds have a fossil record in the Miocene and Pliocene of Europe, these birds may have dispersed into Europe during periods of favorable climatic conditions and their distribution in Europe was restricted to the southern regions. Two other characteristic "African" taxa, by contrast, the Leptosomiformes (courols) and Coliiformes (mousebirds), clearly originated outside the continent.

Courols: Living fossils in Madagascar

Leptosomiformes include a single extant species, the Courol or "Cuckoo-roller" (*Leptosomus discolor*), a forest-dwelling bird of Madagascar and the Comoro islands, which feeds on chameleons and insects. The Courol is now known to be only distantly related

to rollers, to which it was long assigned, and resulted as the sister taxon of a clade including rollers, woodpeckers, and other small arboreal birds in molecular analyses (Hackett et al. 2008; Jarvis et al. 2014; Prum et al. 2015).

The Courol is among those Southern Hemispheric birds with a relictual extant distribution, and fossil stem group representatives of the Leptosomiformes are known from the early and middle Eocene of Europe and North America. These fossils were assigned to the taxon *Plesiocathartes*, whose skeletal morphology is remarkably similar to that of the extant Courol (Figure 12.1; Mayr 2008c). The *Plesiocathartes* fossils show that courols had a wide distribution over the Northern Hemisphere in the early Paleogene, and with five currently recognized species they were also more diversified by that time (Mayr 2009a).

All stratigraphically well-constrained courol fossils are from Eocene strata and these birds may have already become extinct on the Northern Hemisphere towards the

Figure 12.1 The stem group leptosomiform *Plesiocathartes* is one of the taxa that exemplify the great similarities between the early Eocene arboreal avifaunas of Europe and North America. Shown are skeletons of *Plesiocathartes* from (a, b) Messel (a: actual fossil with preserved feathering; b: specimen coated with ammonium chloride to enhance contrast of the bones) and (c) the Green River Formation. A comparison of (d, e) the coracoid, (f, g) the furcula, and (h, i) the tarsometatarsus of *Plesiocathartes* and the extant Courol (*Leptosomus*) illustrates the striking resemblances between this early Eocene leptosomiform and the single living species.

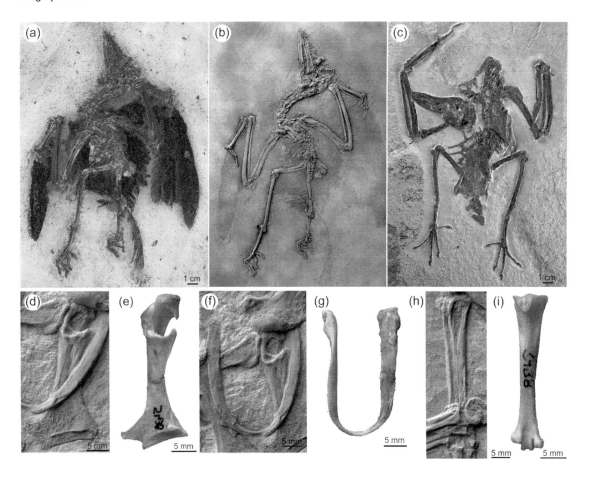

Oligocene. Leptosomiformes constitutes the only bird group with an extant distribution confined to the Madagascan region and a Cenozoic fossil record outside the island, and with its isolated phylogenetic position, restricted extant distribution, and morphological similarity to the Eocene stem group representatives, the extant Courol certainly fulfills the definition of a living fossil (Mayr 2008c).

Mousebirds: An Amazing Past Diversity

Africa south of the Sahara is home to the mousebirds (Coliiformes), distinctive sparrow-sized birds with a finch-like beak, which mainly feed on fruits but also take various other plant matter. Mousebirds today live in open woodlands, and the six extant species are classified in two taxa, *Colius* and *Urocolius*, which only differ in minor anatomical features. Mousebirds are the sole extant avian group with facultatively pamprodactyl and zygodactyl feet, and both the first and fourth toes can thus be moved forwards and backwards to enable agile locomotion in scrub and trees. Flight maneuvers in dense vegetation are facilitated by short and rounded wings and very long tail feathers, which are anchored in a large pygostyle with a characteristic derived morphology.

Coliiformes exhibit a derived arrangement of some tendons of the extensor muscles of the toes, which is otherwise only found in parrots. A sister group relationship to parrots is, however, only supported by some analyses with limited taxon sets (N. Wang et al. 2012), whereas analyses of more comprehensive molecular data either suggested closer affinities to strigiform birds (Ericson et al. 2006; Hackett et al. 2008) or placed Coliiformes between the branches leading to Strigiformes and Leptosomiformes (Jarvis et al. 2014; Prum et al. 2015).

Extant mousebirds are mere relics of a once much more widespread group of birds, and Coliiformes were diversified in the Cenozoic of the Northern Hemisphere. Pronounced differences in the bill and foot morphologies of the various species indicate that these fossil mousebirds occupied diverse ecological niches. Among the earliest and most basal stem group representatives are the Sandcoleidae, which occurred in the late Paleocene to middle Eocene of North America and Europe (e.g., *Sandcoleus, Anneavis, Eoglaucidium*; Houde and Olson 1992; Mayr and Peters 1998; Mayr 2009a). Some sandcoleid bones, especially the elongate humerus, show little similarity to the corresponding skeletal elements of modern mousebirds. The feet of sandcoleids appear to have been facultatively zygodactyl and pamprodactyl as in extant mousebirds, but the tarsometatarsus is shorter and stouter. Details in the morphology of the tarsometatarsal trochleae suggest a specialized grasping foot (Zelenkov and Dyke 2008). This is also supported by

the fact that, unlike in extant Coliiformes, the basal phalanges of all fore toes are greatly shortened and the pedal claws are very long and pointed. The beak of sandcoleids is of ordinary proportions, neither very long nor very short, and fossils with plumage remains show that they had long tail feathers (Plate 15a; Mayr and Peters 1998).

Other early Eocene Coliiformes more closely resemble extant mousebirds in their skeletal morphology, especially with regard to the shorter and stouter humerus, the narrower tarsometatarsus, and the presence of a large terminal disc on the pygostyle. This latter feature serves for the anchorage of the greatly elongated tail feathers, which in extant mousebirds also function as bracing devices of the feeding or resting bird. Presumably the earliest diverging of these more modern-type stem group Coliiformes is the early Eocene European *Selmes* (Mayr 2001; Ksepka and Clarke 2010b; Mayr 2015i), which overall resembles extant mousebirds in skeletal morphology, but has a longer beak (Figure 12.2b). As in sandcoleids, but unlike in extant mousebirds, the proximal phalanges of all three fore toes of *Selmes* are very short, which appears to be the primitive condition for Coliiformes and indicates a specialized foot use of the stem species of the clade.

Chascacocolius from the early Eocene of North America and Europe (Houde and Olson 1992; Mayr 2009a) is characterized by very long, blade-like retroarticular processes of the mandible (Figure 12.2b). Such processes increase the leverage of the muscles lowering the mandible and represent a gaping adaptation, which enables opening of the beak within substrate, such as fruit pulp, bark crevices, or soil. Long retroarticular processes are also present in *Masillacolius* from the early Eocene of Germany (Mayr 2015i), which has a greatly elongated tarsometatarsus and unusually short toes with robust and deep claws. *Masillacolius* appears

Figure 12.2 Mousebirds (Coliiformes) were very diversified in the early Cenozoic of Europe. Skeletons of (a) *Masillacolius* and (b) *Selmes*, two stem group Coliiformes from the early Eocene of Messel. Skulls of (c) *Chascacocolius* from the early Eocene of Messel and (d) *Oligocolius* from the late Oligocene of Germany in comparison to (e) the skull of an extant mousebird (*Urocolius*) and (f) a New World blackbird (*Amblyramphus*, Icteridae). Note the presence of greatly elongated retroarticular processes in (c) and (d), and the passeriform (f) blackbird (encircled), as well as the large seeds ingested by *Oligocolius*.

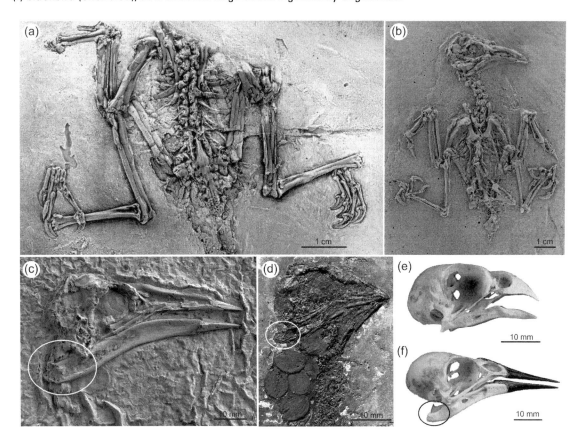

to have had fully pamprodactyl feet and its peculiar foot morphology was interpreted as an adaptation for clinging to vertical surfaces (Figure 12.2a; Mayr and Peters 1998). Current phylogenetic reconstructions show *Masillacolius* to be more closely related to crown group Coliiformes than are *Selmes* and *Chascacocolius* (Figure 12.3; Ksepka and Clarke 2010b; Mayr 2015i). Greatly elongated retroarticular processes of the mandible may therefore be plesiomorphic for Coliiformes, and these processes may have been secondarily reduced in the crown group.

Celericolius from the early Eocene Green River Formation of North America likewise resembles modern Coliiformes in overall skeletal morphology, but has proportionally longer wings (Ksepka and Clarke 2010b). Unlike in all other stem group Coliiformes of which the feet are known, the proximal phalanges of the second and third toes of *Celericolius* are not shortened. Unfortunately, the mandible of the only known *Celericolius* skeleton is too poorly preserved

Figure 12.3 Phylogenetic interrelationships and temporal occurrences of fossil mousebirds (Coliiformes; after Ksepka and Clarke 2010b and Mayr 2013c). The geographic occurrences of the taxa are indicated in parentheses (Afr: Africa, E: Europe, NA: North America).

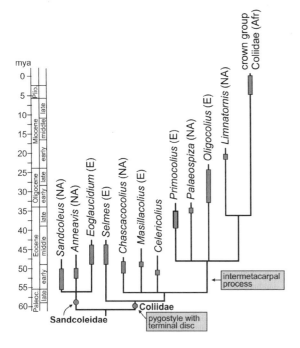

to assess whether retroarticular processes were present.

More advanced mousebirds belong to a clade that also includes the extant species and is characterized by a large intermetacarpal process of the carpometacarpus (Figure 12.3; Mayr 2001). The oldest of these more modern-type coliiform stem group representatives belong to *Primocolius* from the middle and late Eocene of France (Mourer-Chauviré 1988a). *Palaeospiza* is a *Primocolius*-like mousebird from the late Eocene of North America, which not only substantiates the close similarity of early Cenozoic European and North American avifaunas (see later in this chapter), but also represents the latest occurrence of a

mousebird in North America (Mayr 2001; Ksepka and Clarke 2009).

Among the known Paleogene stem group Coliiformes, the late Eocene *Primocolius* and *Palaeospiza* are most similar to extant mousebirds. A more aberrant coliiform taxon, however, still existed in the Oligocene of Germany. *Oligocolius* includes two species the size of extant mousebirds, each of which is known from a single skeleton (Mayr 2013c). Like the early Eocene *Celericolius*, *Oligocolius* has proportionally longer wings than extant mousebirds, which indicates that it was adapted to more sustained flight. A short beak, a distinct nasofrontal hinge, and a wide interorbital section lend its skull a parrot-like appearance, and like in *Chascacocolius* and *Masillacolius* the mandible exhibits long retroarticular processes (Figure 12.2a). In one of the *Oligocolius* skeletons large seeds are preserved in the area of the upper esophagus (Plate 15c, d; Mayr 2013c). This suggests that *Oligocolius* may have had a crop, which is absent in extant mousebirds and other predominantly frugivorous birds, and these Oligocene mousebirds therefore possibly fed on a higher proportion of less easily digestible plant material than their extant relatives. Seeds as stomach contents are also known from various early Eocene coliiform stem group representatives – that is, *Eoglaucidium* and *Selmes* (Plate 15b; Mayr and Peters 1998) – and document a long history of frugivory in Coliiformes.

Mousebirds were still widespread in the early and middle Miocene of Europe, and their latest European record is from the late Miocene, about 8 mya, of Austria (Mlíkovský 2002; Mayr 2010c, 2011d). These Miocene Coliiformes are assigned to the taxa *Limnatornis* and *Necrornis*, which exhibit essentially modern-type morphologies. The disappearance of mousebirds from Europe is likely to have been due to climatic cooling towards the late Miocene, as the emerging cold winter months did not allow the persistence of non-migratory frugivorous birds in northern latitudes.

In Africa, where Coliiformes only occur today, the earliest fossil record of mousebirds stems from the early Miocene of Namibia (Mourer-Chauviré 2008). All other African mousebird fossils are from the late Neogene and the only named extinct species is from the early Pliocene of South Africa (Rich and Haarhoff 1985).

The Long Evolutionary History of Owls

Extant owls include somewhat more than 200 species. These are classified in the sister taxa Tytonidae (barn owls), which include only the two taxa *Tyto* and *Phodilus*, and Strigidae (typical owls), which comprise all of the other living species. With very few exceptions, owls are nocturnal or crepuscular carnivores that feed on insects or vertebrates. The diet of most species consists of small mammals and the dietary activity patterns of owls therefore correspond with those of their main prey. Many of the anatomical characteristics of owls, such as the very large eyeballs and correlated changes in skull geometry, are due to their foraging patterns at times of low light intensity. Owls have strong feet with sharp talons and often stab their prey, a behavior that entailed characteristic modifications of the leg bones. The leg muscles are greatly developed, as are the tendons inserting on the toes, and as a consequence of this owls lost the supratendinal bridge of the tibiotarsus. Like in most diurnal birds of prey, the hypotarsus of the tarsometatarsus is formed by two widely separated crests. Most owls are arboreal perching birds, and to allow a firm grip of both prey and perches they have semizygodactyl feet, in which the fourth toe can be spread laterally.

The phylogenetic affinities of strigiform birds are still controversially resolved in analyses of different kinds of molecular data. Analyses of nuclear gene sequences resulted in a sister group relationship to either Coliiformes (Hackett et al. 2008) or to a clade including Coliiformes as well as "coraciiform" and piciform birds (Jarvis et al. 2014; Prum et al. 2015), whereas mitochondrial sequences indicated a sister group relationship to the Psittaciformes (Pacheco et al. 2011).

Strigiform stem group representatives are already known from Paleocene deposits and two distinctive taxa of that epoch were reported from North America and Europe, respectively. The North American *Ogygoptynx* (Ogygoptyngidae) is based on a long and slender tarsometatarsus from the late Paleocene of Colorado (Rich and Bohaska 1981). The tarsometatarsus of *Berruornis* from the late Paleocene of France and Germany, by contrast, is much stockier and differs from that of *Ogygoptynx* and other Strigiformes in that the skeletal characteristics associated with the development of a semizygodactyl foot are less pronounced (Mourer-Chauviré 1994; Mayr 2007). Apart from a tentatively referred beak, only hindlimb bones are known from *Berruornis*. The two species included in the taxon are large birds, the size of extant eagle owls (*Bubo*). Their tarsometatarsus departs from that of extant Strigiformes in some details, but other features of this bone as well as the morphology of the tibiotarsus support strigiform affinities of *Berruornis*.

The exact affinities of *Ogygoptynx* and *Berruornis* relative to each other and to other early Cenozoic Strigiformes are uncertain, but a position outside crown group Strigiformes is uncontested (Mayr 2009a). The pronounced differences in the tarsometatarsus morphology of these two taxa indicate that Strigiformes underwent a considerable radiation by the late Paleocene and suggest an earlier, possibly Late Cretaceous, origin of the group.

Another clade of stem group Strigiformes, the Protostrigidae, occurs in the Eocene of North America and Eurasia, and in the Oligocene of Europe. Protostrigid owls include the taxa *Eostrix*, *Oligostrix*, and *Minerva* ("*Protostrix*"), which are mainly known from hindlimb bones (Mayr 2009a; Kurochkin and Dyke 2011). The species of *Eostrix* from the early Eocene of North America and Mongolia are small birds, whereas the larger *Minerva* occurs in the middle and late Eocene of North America. The two species of *Oligostrix* stem from the early and late Oligocene of Europe (De Pietri et al. 2013b).

Protostrigid owls are distinguished from extant Strigiformes in the characteristic shape of the distal ends of the tibiotarsus and tarsometatarsus, and the hind toe of at least *Minerva* bears an unusually large claw (Mourer-Chauviré 1983b). Complete tarsometatarsi of *Eostrix* and *Minerva* have not been described, but in *Oligostrix* the bone is comparatively long and slender (De Pietri et al. 2013b). It is of particular interest that one of the lacrimal bones is preserved in an undescribed partial skeleton of a small protostrigid owl from the early Eocene British London Clay in a private collection. This bone exhibits a well-developed supraorbital process, whereas in extant Strigiformes the lacrimals are fused with the frontal bones, and supraorbital processes are either absent or vestigial. The London Clay fossil therefore strengthens ontogenetic evidence that these processes were reduced in the stem lineage of owls, possibly due to enlargement of the eyeballs in relation to nocturnal activity. Their presence in the London Clay strigiform may indicate that this bird had smaller eyes than modern owls and was less nocturnal, but a full appreciation of the significance of this fossil has to await its description.

The temporal origin of crown group Strigiformes is difficult to determine, mainly because no apomorphies have been identified that define a clade including the extant taxa. *Palaeoglaux* (Paleoglaucinae) from the late Eocene of the Quercy region in France shares pneumatic foramina in the acrocoracoid process of the coracoid with the Strigidae (Mourer-Chauviré 1987) and may therefore be within the crown group. Another species assigned to *Paleoglaux*, from the early Eocene of Messel in Germany, by contrast, lacks derived features of crown group Strigiformes (Peters 1992). Better knowledge of these fossils is required to assess whether the Messel and Quercy specimens did indeed belong to the same taxon and to establish their affinities within Strigiformes (Mayr 2009a).

Strigid owls, which comprise the majority of the extant strigiform species, are characterized by an ossified bony arch on the dorsal surface of the proximal end of the tarsometatarsus. This arch is absent in *Palaeoglaux* and first occurs in *Heterostrix* (Heterostrigidae) from the early Oligocene of Mongolia (Kurochkin and Dyke 2011). *Heterostrix* is distinguished from extant owls in the peculiar shape of the tarsometatarsal trochlea for the second toe, which indicates an unusual mobility of this toe and a nearly heterodactyl foot. Another, previously unrecognized heterostrigid owl, *Aurorornis*, stems from the late Eocene of the Crimea peninsula. Like *Heterostrix*, *Aurorornis* is only known from a tarsometatarsus and was originally assigned to the Protostrigidae (Panteleyev 2011). In *Aurorornis*, the bony arch on the proximal tarsometatarsus is only incompletely ossified, but otherwise the taxon exhibits the very same derived characteristics as *Heterostrix* and documents a wide distribution of small heterostrigid owls in the late Eocene and early Oligocene of Asia.

If strigid affinities of the Heterostrigidae are confirmed in future studies, the sister taxon of Strigidae, the Tytonidae, must have also existed by the late Eocene. Tytonid affinities have indeed been assumed for various modern-type Strigiformes from the late Eocene and Oligocene of the Quercy region. Some of these are known merely from a few bones and their taxonomic status and phylogenetic affinities can only be resolved with further material (Mayr 2009a). Of *Necrobyas*, however, which includes at least four species from late Eocene to late Oligocene sites of the Quercy region, all major limb bones have been described (Mourer-Chauviré 1987). *Necrobyas* and the very similar *Prosybris*, from the Oligocene and early Miocene of Europe, resemble the extant tytonid taxon *Phodilus* in overall morphology, especially with regard to the comparatively short tarsometatarsus. They are generally considered to be the earliest Tytonidae (Mourer-Chauviré 1987; Göhlich and Ballmann 2013), although some of the shared similarities may be plesiomorphic for Strigiformes.

The earliest definitive representative of the Tytonidae is *Miotyto* from the middle Miocene (15 mya) of Germany (Göhlich and Ballmann 2013), whose elongated tarsometatarsus suggests closer affinities to *Tyto* than to *Phodilus*. Even more modern-type Tytonidae, which were assigned to *Tyto*, are known from only slightly younger middle Miocene strata in France (Mlíkovský 2002).

Strigid owls occurred in the early Miocene of Europe and North America (e.g., Olson 1985a; Mlíkovský 2002). However, the exact affinities of these fossils remain uncertain, because all are based on a few fragmentary bones and the interrelationships of extant Strigidae are still incompletely understood. Fragmentary strigid remains were also reported from the early Miocene of Thailand (Cheneval et al. 1991). Identification of late Cenozoic strigid fossils is more straightforward, and some recent discoveries are of interest from a biogeographic point of view. The taxon *Surnia*, for example, is today restricted to the tundras of northern latitudes, but was reported from the late Pliocene of Hungary and Morocco (Mourer-Chauviré and Geraads 2010). Another unexpected biogeographic occurrence is the presence of the taxon *Athene* in Africa south of the Sahara in the early Pliocene, where it does not occur today (Pavia et al. 2015).

Early Cenozoic owls are only known from the Northern Hemisphere. The earliest South American strigiform fossils, by contrast, stem from the early or middle Miocene of Argentina (Chiappe 1991), and the oldest African ones are from the early Miocene of Kenya (Walker and Dyke 2006). This may be an artifact of the less complete fossil record of the Southern Hemisphere. However, because the early evolution of rodents, the predominant prey of extant owls, took place in the Northern Hemisphere, it may just as well indicate a true evolutionary pattern and a late dispersal of owls into the Southern Hemisphere.

Parrots and Passerines: An Unexpected Sister Group Relationship and Its Potential Evolutionary Implications

One of the unanticipated results of recent analyses of molecular data is the identification of parrots (Psittaciformes) as the closest extant relatives of passerines (Passeriformes). Psittacopasseres, the clade including both taxa, was so far only obtained in analyses of nuclear gene sequences, in which it is, however, robustly supported (Hackett et al. 2008; Suh et al. 2011; Wang et al. 2012; Yuri et al. 2013; Jarvis et al. 2014; Prum et al. 2015).

Parrots and passerines are very different in their anatomy and have not been considered as closely related in morphology-based studies. Earlier authors assumed that parrots show affinities to diurnal birds of prey, owls, or mousebirds, whereas passerines were often likened to woodpeckers and allies. Parrots have a characteristic bill morphology with a highly mobile upper beak and robust, zygodactyl feet. Passerines, by contrast, are long-legged anisodactyl birds, which differ distinctly from parrots in many anatomical features.

The recognition of close affinities between parrots and passerines is of particular interest, because Passeriformes have an extinct sister taxon with zygodactyl feet, the Zygodactylidae, the members of which closely resemble parrots in the derived morphology of the distal end of the tarsometatarsus (Figure 12.4 and later discussion). With two successive sister taxa of passerines therefore possessing zygodactyl feet, it is most parsimonious to assume that the passeriform stem species was also zygodactyl, and that the absence of this derived foot morphology in passerines represents a reversal into the plesiomorphic condition (Mayr 2009a). A zygodactyl foot is usually considered an adaptation for increased perching capabilities. If so, a secondary loss in passerines, by far the most diversified group of perching birds, would be difficult to understand. However, zygodactyl feet are likewise often associated with the habits of nesting in tree cavities, where the birds need to cling to vertical surfaces. Parrots breed in tree cavities and this may be the plesiomorphic breeding mode for Psittacopasseres, so that a loss of zygodactyl feet in passerines may have been related to a transition from cavity breeding to open nests (Mayr 2015e).

Recently a simple developmental mechanism was proposed to explain how the complex tarsometatarsal structures associated with a zygodactyl foot could have been secondarily lost (Botelho et al. 2014b). According to this model, the degeneration of one of the extensor muscles of the fourth toe results in asymmetric muscular forces, which lead to a reversion of this toe and, hence, to the formation of a zygodactyl

Figure 12.4 Tarsometatarsi of zygodactyl stem group representatives of passerines, in comparison to the tarsometatarsi of an extant parrot and an extant passerine. (a) An undescribed *Psittacopes*-like bird from the early Eocene London Clay. (b) The extant kea (*Nestor*, Psittacidae). (c) An undescribed zygodactylid from the London Clay. (d) An extant crow (*Corvus*, Passeriformes). The bones are from the left side and are shown in plantar view; the trochleae are numbered.

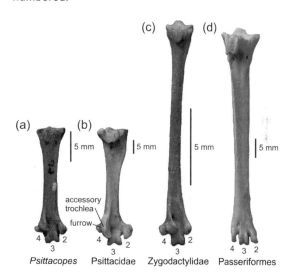

foot. It was experimentally shown that the changed orientation of the fourth toe affects early **osteogenesis** and results in the formation of the bone structures associated with zygodactyly (Botelho et al. 2014b). In Passeriformes, many of the foot muscles are greatly reduced, which restores muscular balance. It may be this peculiarity of the passerine anatomy that prevented the fourth toe from turning back, thereby eliminating the factors that in psittaciform birds trigger the formation of the skeletal correlates of a zygodactyl foot (Botelho et al. 2014b).

This model conflicts, however, with the fact that in zygodactylids the extensor muscle of the fourth toe appears to have been well developed, judging from the development of the correlated bone structures of its insertion site (Olson 1985a). The distal end of the tarsometatarsus of zygodactylids closely resembles that of parrots, and reduction of the extensor muscle of the fourth toe may therefore not be the only cause of the characteristic skeletal modifications associated with zygodactyl feet. Yet irrespective of the exact developmental mechanism of this toe arrangement, the fossil record, in concert with the results of current molecular analyses, suggests that the stem species of Psittacopasseres had zygodactyl feet. The recognition of this early origin of zygodactyl feet opens a new view on various fossil taxa that were initially considered to be stem group representatives of parrots.

Early Paleogene parrot-like birds

In fact, a fair number of morphologically diverse early Paleogene fossils were assigned to the Psittaciformes. Determination of the exact affinities of many of these taxa is not straightforward, nevertheless. This is particularly true for the Eocene Halcyornithidae and Messelasturidae.

Halcyornithids occurred in the early and middle Eocene of Europe and North America and include several very similar taxa, the exact taxonomic status of which still needs to be more thoroughly assessed (Figure 12.5a–e; Mayr 2009a; Ksepka et al. 2011). Messelasturids were found in the early Eocene of Europe and North America, and two taxa, *Tynskya* and *Messelastur*, are currently recognized (Figure 12.5d,e; Mayr 2009a, 2011e).

Halcyornithids have fully zygodactyl feet with a large accessory trochlea for the fourth toe, and compared with other zygodactyl birds the shape of their tarsometatarsus most closely resembles that of the Psittaciformes. In messelasturids, by contrast, the trochlea for the fourth toe merely bears a wing-like flange, and the feet may therefore only have been facultatively zygodactyl. Halcyornithids and messelasturids are small birds and are distinguished from extant parrots in many skeletal features (Mayr 2014a). Unlike in other Psittacopasseres, for example, the coracoid exhibits a foramen for the supracoracoideus nerve and the hypotarsus of the tarsometatarsus has only a single tendinal furrow. The skulls of halcyornithids and messelasturids furthermore feature large supraorbital processes of the lacrimal bones, similar to those of some diurnal birds of prey. The elongated and slender humeri of halcyornithids resemble those of owls, whereas messelasturids have very deep lower jaws and raptor-like pedal claws. These similarities to raptorial birds are of particular interest because analyses of molecular data either identified falcons as the closest extant relatives of Psittacopasseres (nuclear gene sequences; Hackett et al. 2008; Suh et al. 2011; Jarvis et al. 2014; Prum et al. 2015) or supported a sister group relationship between parrots and owls (mitochondrial data; Pacheco et al. 2011).

Another psittaciform-like bird is *Vastanavis* (Vastanavidae) from the early Eocene of India (Figure 12.6a). This taxon is well represented by most major limb bones, and *Vastanavis*-like birds also occurred in the early Eocene of the North American Green River Formation (*Avolatavis*) and the London Clay (Mayr et al. 2010, 2013d; Mayr 2015e; Ksepka and Clarke 2012). *Eurofluvioviridavis* from the Messel fossil site is a presumptive European relative of *Avolatavis* (Mayr 2015e). As in messelasturids, the tarsometatarsal trochlea for the fourth toe of *Vastanavis* and *Avolatavis* bears a plantar flange rather than a well-developed accessory trochlea, so that the foot was only semizygodactyl. Unlike in extant parrots, the coracoid of *Vastanavis* exhibits a deeply excavated, cup-like articulation facet for the scapula (the coracoid of *Avolatavis* is unknown). Both *Vastanavis* and *Avolatavis* have very long pedal claws, which show a resemblance to those of owls, falcons, and other birds of prey.

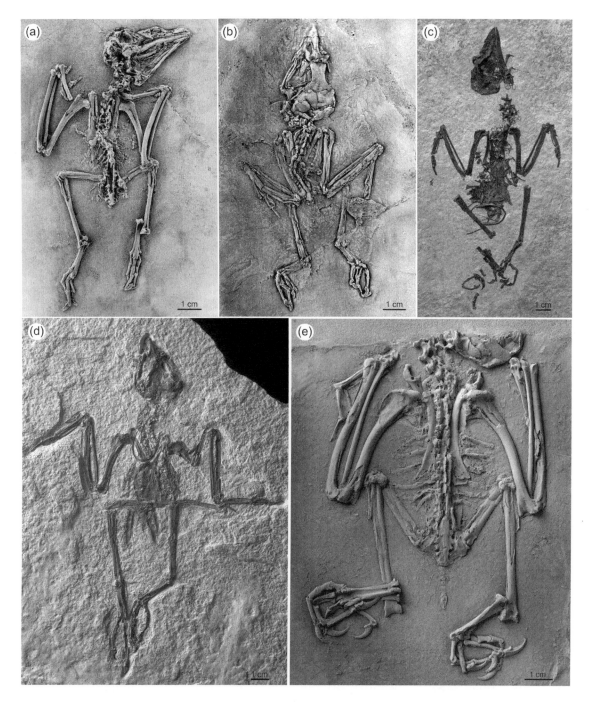

Figure 12.5 Skeletons of early Eocene representatives of Psittacopasseres, the clade including parrots and passerines. (a, b) The halcyornithid *Pseudasturides* from the Messel oil shale in Germany. (c) The very similar halcyornithid *Cyrilavis* from the North American Green River Formation (photograph by Lance Grande). (d) The messelasturid *Tynskya* from the Green River Formation. (e) *Messelastur*, a messelasturid from the Messel fossil site.

Figure 12.6 Humerus, coracoid, and tarsometatarsus of (a) the psittacopasserine *Vastanavis* (Vastanavidae) from the early Eocene of India, (b) *Quercypsitta* (Quercypsittidae) from the late Eocene of France (complete humeri of this taxon are unknown), and (c) an extant kea (*Nestor*, Psittacidae). Unlike in extant parrots, the coracoids of *Vastanavis* and *Quercypsitta* exhibit a plesiomorphic, cup-like articulation facet for the scapula.

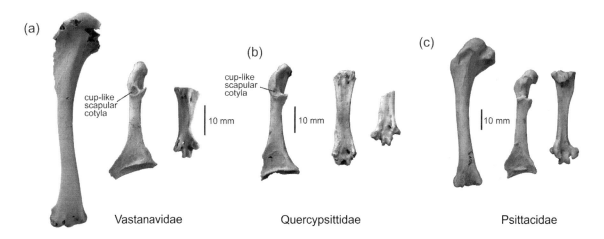

Quercypsitta (Quercypsittidae) is a psittaciform stem group taxon from the late Eocene of France (Figure 12.6b; Mourer-Chauviré 1992b), which is much more similar to crown group Psittaciformes than the aforementioned taxa. The tarsometatarsus of *Quercypsitta* exhibits a large accessory trochlea for the reversed fourth toe, which, like in extant parrots, is separated by a furrow from the main trochlea. The coracoid of *Quercypsitta*, however, resembles that of *Vastanavis* and still exhibits a plesiomorphic, cup-like scapular articulation facet. Flattening of the scapular facet of the coracoid in the evolution of parrots parallels that seen in Galliformes, and as in the latter it may have been related to the evolution of a crop in psittaciform birds. The large crop of crown group parrots also led to other modifications of their skeleton, such as a stout humerus with a well-developed deltopectoral crest, and the differences in the pectoral girdle morphology of stem group Psittaciformes indicate that these birds differed in their feeding ecology from modern parrots.

The origin of crown group parrots

Extant parrots are predominantly frugivorous or granivorous birds and are mainly found in the warmer tropical and subtropical regions, being most diverse in Australasia and South America. In the past few years a robust phylogenetic framework emerged from sequence-based analyses, and these studies found a clade formed by the two New Zealand taxa *Nestor* (Kea and kakas) and *Strigops* (Kakapo) to be the sister group of all other crown group Psittaciformes (Wright et al. 2008; Schweizer et al. 2011). The fact that the Australasian Cacatuini (cockatoos) are branching next suggests an Australasian origin of crown group Psittaciformes. The New World parrots (Arini) are monophyletic and their sister taxon is a clade including the African *Poicephalus* and *Psittacus* (Psittacini), which indicates that the colonization of the Americas by crown group parrots goes back to dispersal from the Old World.

The Australasian fossil record of parrots is meager, but from the early Miocene of New

Zealand three species of the taxon *Nelep-sittacus* were described and considered to be most closely related to the extant *Nestor* (Worthy et al. 2011a). A mandible fragment of a species of the Cacatuini was furthermore reported from the early or middle Miocene of Australia (Boles 1993b). These fossils document a divergence of crown group Psittacidae by the early Miocene at latest.

In Europe, three psittaciform taxa occurred in the early Miocene of Germany (*Mogontiacopsitta*), France (*Archaeopsittacus*), and the Czech Republic (*Xenopsitta*), and two more are known from the middle Miocene of Germany (*Bavaripsitta*) and France (*Pararallus*; Mlíkovský 2002; Mayr and Göhlich 2004; Mayr 2010c). Their exact phylogenetic affinities are uncertain, as the known bones exhibit no features that allow a well-founded assignment to any of the extant parrot groups. In overall morphology, however, *Archaeopsittacus*, *Mogontiacopsitta*, and *Bavaripsitta* resemble members of the Psittaculini (parakeets and allies; Figure 12.7), whereas the stouter tarsometatarsus of *Xenopsitta* is more similar to that of the African *Psittacus* and *Coracopsis*.

The earliest New World record of a modern-type parrot stems from the early Miocene of Nebraska and was assigned to the taxon *Conuropsis*, which also includes the recently extinct Carolina Parakeet (Olson 1985a). Because this fossil is, however, only represented by a humerus, a bone that shows little variation in psittaciform birds let alone New World Arini, its exact phylogenetic affinities are best considered unresolved. The earliest South American parrot fossil is a skull from the Pliocene of Argentina, which was referred to the extant taxon *Nandayus* (Tonni and Noriega 1996).

Compared to the Neotropic and Australasian regions, the extant psittaciform fauna of continental Africa is species poor and has a low diversity. The African fossil

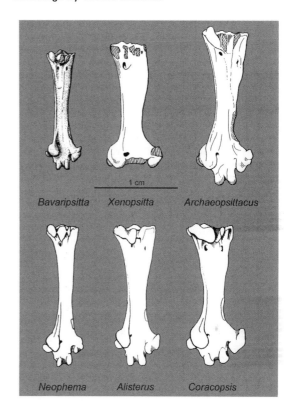

Figure 12.7 Tarsometatarsi of parrots from the Miocene of Germany (*Bavaripsitta*), the Czech Republic (*Xenopsitta*), and France (*Archaeopsittacus*) as well as those of extant Platycercini (*Neophema*), Psittaculini (*Alisterus*), and the Madagascan *Coracopsis*. Extant bones not to scale. Drawings by Ursula Göhlich.

record of crown group Psittaciformes is likewise scarce and the earliest fossils are from the Pliocene. Two taxa were distinguished in the early Pliocene of South Africa, with one of these, *Khwenena*, belonging to the Psittacini and the other being an extinct species of *Agapornis* (Manegold 2013). Another extinct *Agapornis* species was described from the late Pliocene of Morocco (Mourer-Chauviré and Geraads 2010). It is possible that parrots arrived late in Africa, but it is equally possible that their diversity never reached the level of Neotropic and

Australasian parrots (in this respect, it is notable that a negative correlation exists between the abundance of parrots and that of squirrels, which are more common in Africa than in the Neotropics and compete with parrots for ecological resources; Corlett and Primack 2006).

Earlier calibrations of gene sequence data indicated an initial diversification of crown group parrots in the Cretaceous (Wright et al. 2008), which is supported neither by the fossil record nor by subsequent molecular studies (Schweizer et al. 2011; Jarvis et al. 2014). Although parrots diverged from their sister taxon in the early Eocene at the latest, all of the Paleogene fossils belong to stem group representatives. When exactly crown group parrots dispersed out of the Australasian region is unknown, but the fossil record suggests that this was a comparatively late, mid-Cenozoic event. In any case, current data indicate that the diversification and biogeographic history of crown group parrots appear to have been remarkably similar to those of passerines, the crown group of which likewise seems to have originated in Australasia (see later discussion).

The Zygodactylidae

Zygodactylids, which are now recognized as the extinct sister group of passerines, were for a long time only known from partial hindlimb bones from the early and middle Miocene of Germany and France. These fossils were assigned to the taxon *Zygodactylus*, and it was noted that the tarsometatarsus of zygodactylids resembles that of parrots, in that the distal end exhibits a large accessory trochlea for a reversed fourth toe. However, because the bone otherwise shows an unusual character mosaic, the affinities of *Zygodactylus* remained unresolved.

Zygodactylids appear to have been among the most abundant arboreal birds in the early Cenozoic of the Northern Hemisphere, and in the past few years numerous remains of these birds, including complete and well-preserved skeletons, have been described from the early Eocene of Europe and North America (Figure 12.8; Mayr 1998, 2009a; Weidig 2010). These Eocene zygodactylids, which belong to the taxa *Primoscens*, *Primozygodactylus*, and *Eozygodactylus*, exhibit a somewhat less derived morphology than the later *Zygodactylus*, of which a complete skeleton from the early Oligocene of France has now been identified (Mayr 2008d). Several of the Eocene *Primozygodactylus* specimens are preserved with stomach contents consisting of grape seeds (Vitaceae), which document an at least partially frugivorous diet for these birds (Mayr 1998).

Zygodactylids are small birds with a long and slender tarsometatarsus; their beak has an average shape but unusually long nostrils. Apart from the presence of zygodactyl feet and the peculiar structure of the distal end of the tarsometatarsus, the skeleton of zygodactylids is strikingly similar to that of passerines. Shared derived features of zygodactylids and passerines include a well-developed intermetacarpal process on the carpometacarpus, very long legs, and a tarsometatarsal hypotarsus with two closed canals. Some zygodactylid species furthermore exhibit a derived humerus morphology that is otherwise exclusively known from passerines (Figure 12.8d; Mayr 2009a; Weidig 2010). Because this morphology is not present in all zygodactylids, its occurrence in some species probably represents a striking case of convergence. It should be noted, however, that if parrots are indeed the closest extant relatives of passerines and if the latter secondarily lost zygodactyl feet (see the discussion above), an alternative explanation of this character distribution is possible. In this case, zygodactylids may be paraphyletic, and the species showing the derived humerus morphology may be more closely related to passerines than those that lack this feature.

Figure 12.8 (a, b) Skeletons of the zygodactylid *Primozygodactylus* from the early Eocene of Messel in Germany. (c) Skeleton of *Zygodactylus* from the early Oligocene of France. (d) Bones of an undescribed small zygodactylid from the early Eocene London Clay. (e) Humerus and (f) hand skeleton of *Zygodactylus* (details of specimen shown in c). (g) Distal end of tarsometatarsus of *Primozygodactylus* (detail of specimen shown in b).

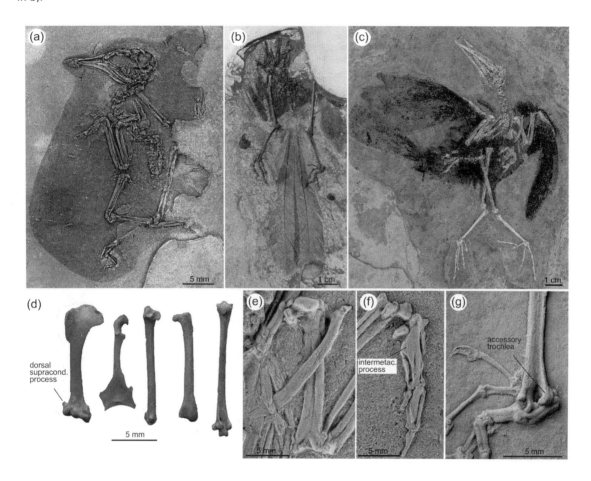

Psittacopes and allies: The most parrot-like stem group passerines?

Zygodactylids have a very parrot-like distal end of the tarsometatarsus, but the skeletal morphology of these birds is otherwise much more similar to passerines than to parrots. Until recently, no fossils were known that bridged the morphological gap between passerines and parrots. However, and as so often in paleontology, specimens of such birds already existed, but had just gone unrecognized because of their unusual character mosaic.

Psittacopes from the early Eocene of Europe was initially considered a psittaciform bird (Figure 12.9; Mayr and Daniels 1998; Mayr 2009a), but is now regarded as one of the more parrot-like zygodactyl stem group representatives of Passeriformes (Mayr 2015e). *Psittacopes* exhibits a similar overall morphology to parrots, but various differences were already noted in its original

Figure 12.9 Skeletons of (a) *Psittacopes* and (b) *Pumiliornis*, two putative zygodactyl stem group representatives of the Passeriformes from the early Eocene of Messel in Germany. Details of feet of (c) *Psittacopes* and (d) *Pumiliornis*. Skulls of (e) *Psittacopes* and (f) the extant *Agapornis* (Psittacidae). (g) Beak of *Psittacopes* (same specimen as in a and e, photo taken through the reverse of the transparent resin slab in which the fossil is embedded; matrix digitally removed).

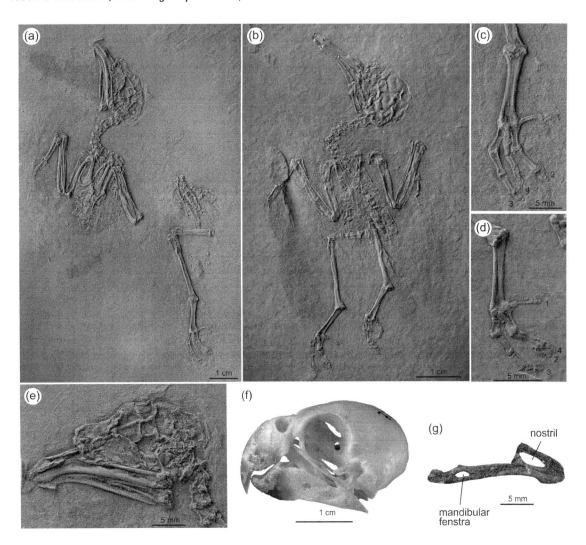

description. The beak of this taxon has very large nostrils and closely resembles that of some passerines (Figure 12.9e). Other salient features shared with passerines are a furcular apophysis, which is a characteristic of passerines but absent in psittaciform birds, and a very slender coracoid. Even though the distal end of the tarsometatarsus of *Psittacopes* closely matches that of crown group parrots, it is equally similar to the distal tarsometatarsus of zygodactylids (Figure 12.4), and these resemblances are likely to be plesiomorphic in case of a monophyletic Psittacopasseres (Mayr 2015e).

Another taxon that is now considered to be one of these early passerine ancestors is *Pumiliornis* from the early Eocene of Germany. This presumably nectarivorous bird (see Chapter 8) closely resembles *Psittacopes* in its postcranial anatomy, but has a much longer beak (Figure 12.9b). It is mainly the highly unusual mosaic combination of a long, rhynchokinetic beak with parrot-like feet that for a long time hindered a phylogenetic allocation of this small bird.

Whether *Morsoravis* from the early Eocene of Denmark, which is represented by an exceptionally well preserved partial skeleton (Plate 9c), is closely related to *Pumiliornis*, as previously suggested (Mayr 2009a), needs to be pursued further. Passeriform rather than psittaciform affinities of *Psittacopes* may also challenge the identification of a *Psittacopes*-like putative stem group psittaciform from the middle Eocene of Namibia (*Namapsitta*; Mourer-Chauviré et al. 2015).

Passerines: The most species-rich avian group

Passerines comprise more than half of all extant avian species and constitute the most widespread and most species-rich taxon of perching birds. The reasons for their diversification and evolutionary success remain poorly understood, but their ability to construct sophisticated nests may have played an important role and released passerines from competition for natural nesting cavities (Olson 2001). Not least because of the sheer number of species and their high degree of morphological similarity, the interrelationships of passerines have long remained elusive. Sequence-based analyses, however, resulted in a robust and congruently supported phylogeny for the major taxa. According to these studies, the Acanthisittidae (New Zealand wrens) are the sister taxon of all other passeriform birds, the Eupasseres, which can be divided in the sister taxa Suboscines and Oscines (Ericson et al. 2003; Barker et al. 2004).

Suboscines are mainly found in the New World, and by far the most species-rich taxon of extant Passeriformes are the Oscines (songbirds), to which most passerine species outside South and Central America belong. The earliest diverging Oscines are the Australian Menuridae (lyrebirds) and Atrichornithidae (scrub-birds), and the distribution of several subsequently branching oscine taxa is also confined to the Australasian region, where Oscines therefore probably originated (Barker et al. 2004). The most species-rich and most widely distributed group of Oscines is the Passerida, which contains most passerine taxa found in the Northern Hemisphere.

With Acanthisittidae occurring only in New Zealand and the basally diverging Oscines likewise inhabiting Australasia, the phylogenetic interrelationships of crown group passerines suggest their origin in the Australasian region. Passerine fossils are indeed unknown from early Cenozoic fossil sites of the Northern Hemisphere, and the earliest remains assigned to these birds are from the early Eocene of Australia (Boles 1995a, 1997c). The exact affinities of these fragmentary specimens cannot be determined, however, and even the most diagnostic bone among this material, an incomplete carpometacarpus, is distinguished from at least Eupasseres in some plesiomorphic characteristics (Mayr 2009a, 2014a).

After these fossils, there is a large gap in the passeriform fossil record of the Australasian region, and the next oldest Australian passeriform remains are logrunners (Orthonychidae) from late Oligocene strata of the Riversleigh fossil site (Nguyen et al. 2014). The early and middle Miocene localities of this site yielded extinct species of further distinctive taxa of Australian crown group Oscines, such as lyrebirds, treecreepers (Climacteridae), and butcherbirds (Cracticidae; Boles 1995b; Nguyen et al. 2013; Nguyen 2016). Crown group representatives of Passeriformes are also known from the early Miocene of New Zealand, where remains of Acanthisittidae (*Kuiornis*) and Cracticidae were identified (Worthy et al. 2007, 2010b).

Outside Australia, the earliest passerine fossils are from the early Oligocene of Central Europe. By that time, passerines already showed some morphological diversity, although the few described fossils, which belong to the taxa *Wieslochia*, *Jamna*, and *Resoviaornis*, were all small to very small birds (Mayr and Manegold 2006; Bocheński et al. 2011, 2013). The affinities of these fossils are not well constrained. *Wieslochia*, from the early Oligocene of Germany, lacks apomorphies of crown group Oscines and exhibits a suboscine-like skeletal morphology (Figure 12.10; Mayr and Manegold 2006). Regardless of their exact affinities, it is notable that early Oligocene European passerine fossils clearly do not belong to Oscines, the only extant passeriform taxon

in Europe. Definite oscine passerines first occurred in Europe in the late Oligocene and coexisted with suboscines at least into the early Miocene (Manegold 2008a).

The abrupt appearance of modern-type passerines in the European fossil record towards the earliest Oligocene temporally coincides with the early Oligocene closure of the Turgai Strait between Europe and Asia, and is in concordance with a passerine dispersal from Australia via Asia (Mayr 2009a). Unfortunately, the Asian passerine fossil record itself is still very limited, and the earliest fossil stems from the early or middle Miocene of Japan (Kakegawa and Hirao 2003).

Passerine remains are quite rare in the early Oligocene of Europe, whereas these birds constitute the dominant group of arboreal birds in some early and middle Miocene European fossil localities. By that time, passerines also seem to have exhibited

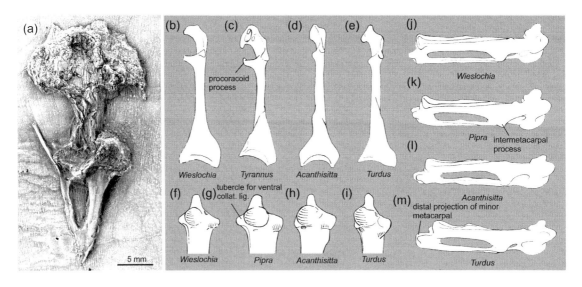

Figure 12.10 One of the earliest European passerines, *Wieslochia* from the early Oligocene of Germany. (a) Skull. (b–e) Coracoid, (f–i) proximal end of ulna, and (j–m) carpometacarpus of *Wieslochia* and representatives of the three extant passeriform subclades Acanthisittidae (*Acanthisitta*), Suboscines (*Tyrannus*, *Pipra*), and Oscines (*Turdus*). In all three bones, *Wieslochia* is clearly distinguished from oscine Passeriformes (see Mayr and Manegold 2006). b–m are not to scale.

a high diversity in Europe and even some representatives of oscine crown group taxa have been reported (e.g., Manegold 2008b). However, some middle Miocene passerines from Europe still show a plesiomorphic morphology of the hypotarsus of the tarsometatarsus, which may support their placement outside crown group Eupasseres (Manegold et al. 2004)

As yet, no Paleogene passerine remains are known from Africa and the Americas. The earliest African Passeriformes are from the early Miocene of Kenya (Mayr 2014b). Passerines are unknown from early Miocene avifaunas of Namibia, but occurred in the middle Miocene of Tunisia (Brunet 1971; Mourer-Chauviré 2003, 2008). Based on the current African fossil record, a passerine dispersal from Europe therefore seems likely.

The earliest passerine remains from the Americas are undescribed fossils from the early Miocene of Florida (Olson 1985a). The earliest South American record is from the early or middle Miocene of Argentina and may stem from a suboscine (Noriega and Chiappe 1993). Even though suboscines could have reached South America from Antarctica before glaciation of this continent started towards the late Eocene, the fossil record would be in better agreement with a dispersal from North America.

Passerines are among those avian taxa that best exemplify the disparate divergence dates derived from fossil and molecular data. Most calibrated gene sequence data yielded very early dates for the initial divergence of crown group passerines, which were reconstructed to have dated back into the Late Cretaceous, some 70–80 million years ago (Barker et al. 2004; Ericson et al. 2014). Because the skeletal morphology of all crown group passerines is very similar, such an early divergence would imply an unprecedented evolutionary stasis of passerines, as well as a geographic restriction of the crown group to the Southern Hemisphere for more than

40 million years, until the appearance of the first passerines in the early Oligocene of Europe (Mayr 2013a). More recent calibrated phylogenies, however, better conform to the fossil record and indicate an early Cenozoic diversification of crown group passerines (Jarvis et al. 2014; Prum et al. 2015).

Trogons, Rollers, and Woodpeckers: Cavity-Nesters with Diverse Foot Morphologies

Telluraves, the "arboreal landbirds," encompasses a clade formed by birds nesting in self-excavated cavities in trees, sandy banks, or soil. This telluravian subclade was termed Eucavitaves (Yuri et al. 2013) and includes trogons (Trogoniformes) as well as woodpeckers, hornbills, rollers, and allies (Piciformes and most of the paraphyletic "Coraciiformes"; Figure 12.11). Cavity nesting is likely to be an ancestral trait of these birds, and by reducing nest predation (e.g., Brightsmith 2005) it probably contributed to their evolutionary success. As detailed in the following sections, stem group representatives of most higher-level taxa of Eucavitaves already existed in the earliest Cenozoic.

Trogons: The only heterodactyl bird group

Trogons are colorful insectivorous or frugivorous birds of subtropical and tropical zones, and the long-tailed Central American quetzals (Pharomachrus) are the most widely known representatives of the group. The almost 40 extant species are the only birds with heterodactyl feet, in which the second toe is permanently reversed.

The first trogon fossils were described in the 19th century by the French paleornithologist Alphonse Milne-Edwards. These specimens, a few bones from the early Miocene of France, revealed that trogons once had a different distribution, and fossil finds of the past few years showed that the

Figure 12.11 Phylogenetic interrelationships and temporal occurrences of extant and fossil taxa of Eucavitaves, the clade including Trogoniformes (trogons), Alcediniformes (kingfishers and allies), Coraciiformes (rollers), and Piciformes (woodpeckers and allies).

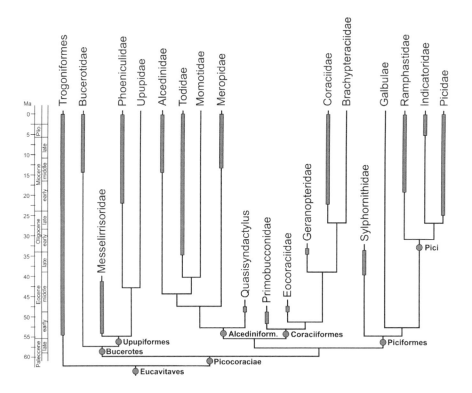

European range of stem group Trogoniformes even reached far north into Scandinavia. These new fossils also document that trogons are among those avian groups that underwent relatively few morphological changes during the past 50 million years.

Still undescribed specimens of heterodactyl stem group trogons were collected in the early Eocene British London Clay (Mayr 1999, 2009a). The oldest published fossil trogon is *Septentrogon* from the early Eocene of Denmark, which is based on a partial skull and constitutes the northernmost occurrence of the Trogoniformes (Kristoffersen 2002). *Masillatrogon* from the early Eocene German Messel oil shale (Figure 12.12) is represented by two skeletons that display the heterodactyl foot morphology but are distinguished from crown group Trogoniformes in some respects, including a narrower beak and plesiomorphic characters of the wing skeleton (Mayr 2009c). *Primotrogon* from the early Oligocene of France is another stem group representative of the Trogoniformes, which, like *Masillatrogon*, is smaller than the extant species and exhibits several plesiomorphic skeletal features (Mayr 1999). Trogon remains are known from various other early Oligocene localities in Europe (Mayr 2009a; Mayr and Smith 2013), and their latest occurrence in Europe is *Paratrogon* from the early Miocene (20–23 mya) of France (Mlíkovský 2002; Mayr 2011d).

Extant trogons have poor long-distance dispersal capabilities and, judging from the similar wing skeletons, this also seems to have been true for their extinct relatives.

Figure 12.12 (a) Skeleton of the stem group trogon *Masillatrogon* from the early Eocene Messel oil shale in Germany. Detail of (b) the hand skeleton and (d) the foot skeleton of *Masillatrogon* in comparison to (c, e) an extant trogon (*Harpactes*). A characteristic feature of trogons is a heterodactyl foot with a reversed second toe.

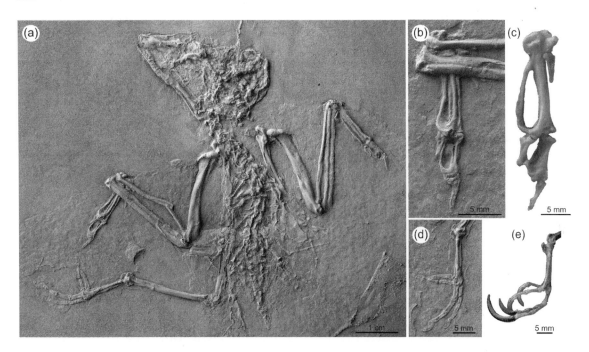

The disappearance of trogons from northern latitudes is therefore likely to have been due to the emergence of a marked climatic seasonality during the mid-Cenozoic, with the cold winters not providing enough food for insectivorous and frugivorous birds.

Outside Europe, trogon fossils are unknown from Cenozoic strata, although *Foshanornis* from the early Eocene of China shows some similarities to trogons and may be closely related (Zhao et al. 2015). Molecular evidence for the interrelationships of the crown group representatives is somewhat conflictive, but most analyses placed the African taxon *Apaloderma* as the sister group of a clade including the Asian and American species, which suggests an Old World origin of trogons (Johansson and Ericson 2005). When trogons dispersed into the New World remains unknown, but the fact that all Paleogene trogon fossils belong to stem group representatives indicates a dispersal date not before the Neogene.

Trogons do not occur on Madagascar and in the Australasian region. These birds have poor dispersal capabilities and at the time of their origin, let alone that of the crown group, Madagascar was already too far away from any continent for a dispersal of trogons. Whether their absence in the Australasian region – that is, east of Wallace's Line – is also due to the circumstance that seaways prevented dispersal, or whether it is due to ecological factors, is less clear and needs to be investigated further.

Bucerotes: Hoopoes, woodhoopoes, and hornbills

Bucerotes includes the African and Eurasian hoopoes (Upupidae; a single extant species) and the African woodhoopoes (Phoeniculidae; eight extant species), which together form the Upupiformes, as well as the hornbills (Bucerotiformes; nearly 60 extant species in Africa and Asia). These long-beaked birds feed on insects, small vertebrates, or – in the case of many hornbills – fruits.

The published fossil record of the Bucerotes stems from the Old World, but there is an undescribed skeleton from the Green River Formation, which may constitute a New World record of these birds (Mayr 2009a; Grande 2013). The oldest well-studied fossils belong to the Messelirrisoridae, which occurred in the Eocene of Europe and are particularly abundant in the Messel oil shale in Germany (Figure 12.13, Plate 12d; Mayr 1998, 2000, 2006a). Messelirrisorids are small birds and some species only reached the size of an average modern hummingbird. They exhibit a number of derived upupiform features, but likewise lack some of the apomorphies shared by Upupiformes and Bucerotiformes (Mayr 1998, 2006a). Even though there remains a possibility that these birds are actually stem group representatives of the more inclusive clade Bucerotes, their

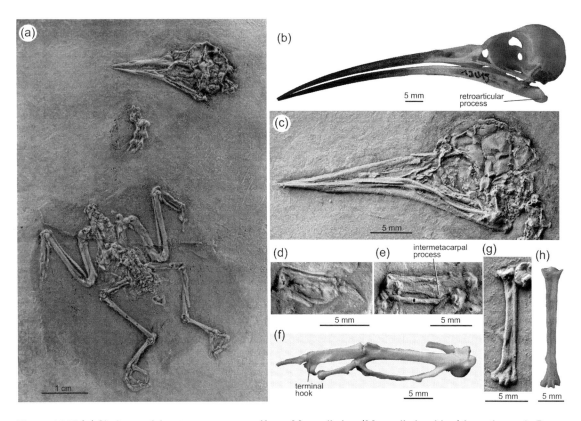

Figure 12.13 (a) Skeleton of the stem group upupiform *Messelirrisor* (Messelirrisoridae) from the early Eocene Messel oil shale in Germany. (a, b) Skull, (d–f) hand skeleton, and (g, h) tarsometatarsus of *Messelirrisor* and an extant hoopoe (*Upupa*). Note the different relative bill lengths of the specimens in (a) and (c), which may be due to sexual dimorphism. Among other plesiomorphic features, messelirrisorids differ from extant Upupiformes in the presence of an intermetacarpal process and the absence of a terminal hook of the proximal phalanx of the major wing digit.

identification as upupiform stem group representatives was supported by a phylogenetic analysis of a comprehensive data set (Mayr 2006a).

Like extant Upupiformes, messelirrisorids seem to have shown a marked sexual dimorphism in the length of their beaks. Judging from their foot morphology – that is, the short tarsometatarsus and the very long hind toe – they had perching habits, whereas crown group Upupiformes either forage on the ground (Upupidae) or are specialized for trunk climbing (Phoeniculidae). As evidenced by one particularly well-preserved *Messelirrisor* fossil from Messel, the tail feathers were distinctly barred like those of extant woodhoopoes (Plate 12d).

Small stem group Upupiformes were still present in the early Oligocene of Europe (Mayr 2009a; Mayr and Smith 2013). Smallness also characterizes the earliest representative of the Phoeniculidae, *Phirriculus* from the early Miocene of France and Germany (Mlíkovský and Göhlich 2000). Provided that this taxon was correctly assigned to woodhoopoes, whose extant distribution is confined to Africa, it documents a complex biogeographic history of upupiform birds. Phoeniculidae may have originated outside Africa, but it is equally possible that they dispersed into Europe from Africa towards the late Paleogene or earliest Neogene.

The other major group of Bucerotes, the hornbills, is divided into ground hornbills (Burcovinae, comprising the two extant species of *Bucorvus*) and a clade including all other species. Because their sister taxon, the Upupiformes, already occurs in the early Eocene, Bucerotiformes must have diverged by that time, too. The earliest hornbill records are, however, much younger and stem from the middle Miocene of Africa, where fossils were found in Morocco and Kenya. The Moroccan record was assigned to an extinct species of *Bucorvus* (Brunet 1971), but the much smaller Kenyan fossils exhibit a morphology resembling that of the extant hornbill taxon *Tockus* (Mayr 2014b). The split between ground hornbills and the other species therefore probably occurred before the middle Miocene and the fossils document a longer evolutionary history of hornbills, which may or may not have taken place in Africa. Bucerotiformes are among those taxa that seem to have dispersed into Europe during the Miocene Climatic Optimum, and an extinct taxon of putative Burcovinae was reported from the late Miocene of Bulgaria (*Euroceros*; Boev and Kovachev 2007).

The relictual Old World distribution of rollers

Rollers fall into two distinct but species-poor extant groups, the true rollers (Coraciidae) of Africa, Eurasia, and Australia, and the more terrestrial, long-legged Madagascan ground rollers (Brachypteraciidae). Extant rollers are carnivorous birds and today have an Old World distribution. The biogeographic history of these birds was complex, however, and fossils of at least two different lineages of stem group representatives were identified in the early Eocene of North America.

One of these is *Primobucco* (Primobucconidae), which is among the more abundant small avian taxa in the early Eocene of North America and Europe (Mayr et al. 2004; Ksepka and Clarke 2010c). All *Primobucco* species are significantly smaller than extant rollers and have a somewhat less robust beak. Otherwise, their skeletal features agree well with those of extant rollers and the legs are short, as in true rollers.

More closely related to extant rollers are two other stem group representatives; that is, *Paracoracias* from the Green River Formation and *Eocoracias* from the early Eocene of Messel (Figure 12.14b; Mayr et al. 2004; Clarke et al. 2009). Both taxa are quite similar, but a phylogenetic analysis suggested

Figure 12.14 (a) The stem group coraciiform *Eocoracias* and (b) the stem group alcediniform *Quasisyndactylus* from the early Eocene of Messel in Germany.

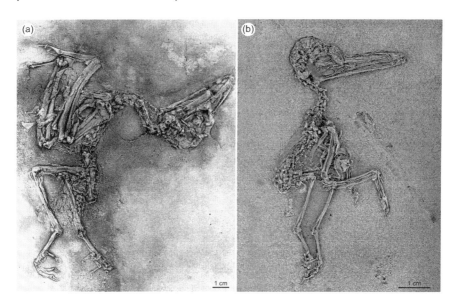

that *Paracoracias* is more closely related to crown group rollers than *Eocoracias* (Clarke et al. 2009).

Geranopterus, another taxon of stem group rollers, occurs in the late Eocene of France. It shares derived features with extant rollers that are absent in *Eocoracias* and *Paracoracias*, therefore indicating a closer relationship to the crown group (Mayr and Mourer-Chauviré 2000; Mayr 2009a). The earliest true roller (Coraciidae) is *Miocoracias* from the early Miocene of France (Mourer-Chauviré et al. 2013b).

Regardless of the skeletal similarities between even the oldest stem group rollers and their extant relatives, there appear to have been some significant shifts in the ecological preferences of these birds. Extant rollers, for example, mainly feed on invertebrates and small vertebrates, whereas their early stem group representatives seem to have been more opportunistic feeders, with seeds having been identified as stomach contents of *Primobucco* and *Eocoracias* fossils (Mayr and Mourer-Chauviré 2000;

Mayr et al. 2004). Some fossils of *Eocoracias* furthermore exhibit excellent feather preservation and, with regard to its rather short wings and long and graduated tail, the fossil taxon more closely agrees with the Madagascan Brachypteraciidae than with the Coraciidae. This suggests that the feathering of ground rollers, which is suitable for agile maneuvering in a forested environment, is plesiomorphic for rollers. The more elongated wings and forked tails of true rollers constitute an adaptation for more rapid flight in open landscapes, and evolved after true rollers left densely forested habitats (Mayr and Mourer-Chauviré 2000).

Why rollers disappeared from North America is unknown, but their extinction in the New World conforms to a pattern already outlined for other arboreal birds, such as mousebirds and courols. In the early Cenozoic, intermittent land corridors existed between Europe and North America. This absence of geographic barriers and the quite homogenous paleoenvironments resulted in great similarities in the composition of the arboreal avifaunas of both continents, which especially in the early Eocene had many avian taxa in common (Mayr 2009a). The reasons that led to the extinction of some of these arboreal taxa in the New World are not well understood, but an earlier disappearance of forested environments in North America may have played a role (Mayr 2009a).

The syndactyl Alcediniformes (bee-eaters, kingfishers, todies, and motmots)

The four extant groups of the Alcediniformes are the brightly colored kingfishers (Alcedinidae), bee-eaters (Meropidae), todies (Todidae), and motmots (Momotidae). Todidae and Momotidae today only occur in the New World. The distribution of the Meropidae, by contrast, is restricted to the Old World, and this is also true for all but six species of the Alcedinidae. Most alcediniform birds are small, long-beaked, and short-legged. All species nest in self-excavated burrows or tunnels. Because these are dug by use of the feet, the fore toes are connate over much of their length and form a so-called syndactyl foot, which is one of the alcediniform characteristics. Alcediniformes also share a characteristic morphology of the auditory ossicle, the columella (Feduccia 1999), as well as various other derived cranial and postcranial features. From a morphological point of view, an alcediniform clade is therefore fairly well supported.

However, although all recent molecular analyses obtained a clade including Alcedinidae, Momotidae, and Todidae, the affinities of the Meropidae were not congruently resolved, with bee-eaters resulting as the sister taxon of rollers in some analyses (Ericson et al. 2006; Hackett et al. 2008; Yuri et al. 2013; Prum et al. 2015). These studies furthermore did not support the traditionally assumed sister group relationship between Todidae and Momotidae, so that the extant New World distribution of these two taxa

is likely to go back to independent dispersal events from the Old World.

The early fossil record of alcediniform birds is mainly a European one. The oldest published specimens stem from the early Eocene of the Messel fossil site in Germany and belong to the taxon *Quasisyndactylus* (Figure 12.14). This comparatively abundant small bird has a long and flattened beak, which is remarkably similar to that of todies. However, *Quasisyndactylus* differs from the Todidae in a shorter tarsometatarsus and plesiomorphic postcranial features, some of which may even indicate a phylogenetic position outside crown group Alcediniformes (Mayr 1998, 2004c). As reflected by the taxon name, the feet of *Quasisyndactylus* appear to have been syndactyl, with the fore toes being closely adjoined in all of the known fossils.

Surprisingly, most Paleogene alcediniform fossils from Europe show affinities to New World Alcediniformes. *Protornis* from the early Oligocene of Switzerland, for example, shares with motmots and todies deep incisions in the caudal articular ends of the mandible (Olson 1976). Whether it is indeed a representative of Momotidae, which it also resembles in limb proportions and bill shape, has not been established beyond doubt (Mayr 2009a). However, if Momotidae and Todidae are not sister groups, as indicated by modern sequence-based analyses, an origin of either taxon outside the New World would be expected. That Momotidae once occurred outside their extant Neotropic range is also shown by a fossil from the late Miocene of Florida (Becker 1986b).

The extant distribution of the Todidae is restricted to the Greater Antilles, but stem group representatives were found in the early Cenozoic of North America and Europe. These fossils belong to the taxon *Palaeotodus*, which is known from the early Oligocene of Wyoming and the late Eocene and early Oligocene of France and Germany (Olson 1976; Mourer-Chauviré 1985;

Mayr and Micklich 2010). Two of the three known species of *Palaeotodus* distinctly exceed extant todies in size. The humerus of *Palaeotodus* is furthermore proportionally larger than that of extant Todidae, and Paleogene todies therefore probably had better-developed wings and greater dispersal capabilities than their extant relatives (Olson 1976).

The fossil record of the other alcediniform groups is surprisingly sparse, given the fact that Alcedinidae and Meropidae are today the most species-rich alcediniform clades, which also have the widest geographic distribution. The earliest published kingfisher remains are from the middle or late Miocene of Australia and the middle Miocene of Kenya, and show close resemblances to taxa of the extant sub-clades of the Alcedinidae (Boles 1997a; Mayr 2014b). A possible record of the Meropidae exists from the middle Miocene of Croatia (Mlíkovský 2002), but this fossil is too poorly preserved for a definitive identification.

The zygodactyl Piciformes (jacamars, toucans, woodpeckers, and allies)

Extant Piciformes comprise the Neotropic Galbulae (jacamars and puffbirds) and the Pici (barbets, toucans, honeyguides, wood-peckers, and allies), which have a nearly worldwide distribution but, like trogons, are absent from Australasia and Madagascar. All piciform birds are zygodactyl and therefore have a permanently reversed hind toe. Otherwise, Galbulae and Pici are very different in their anatomy. Overall, the Galbulae are much more similar to alcediniform birds, whereas the representatives of the Pici resemble passerines in skeletal morphology.

All Paleogene piciform birds belong to very small species and some were even smaller than the smallest extant ones. Small zygodactyl birds that resemble taxa of the Galbulae were reported from the early Eocene North American Green River Formation (Weidig 2010). Like extant Galbulae,

these birds show a close overall similarity to alcediniform birds and rollers, but their skeletal anatomy is too incompletely known for a well-founded classification. Whether the long-legged *Gracilitarsus* from the early Eocene of Germany is related to piciform birds (Mayr 2009a) is also far from certain.

Better supported are piciform affinities of the Sylphornithidae, which occurred in the late Eocene and early Oligocene of Europe and include the taxa *Sylphornis* and *Oligosylphe* (Mourer-Chauviré 1988a; Mayr 2009a). Sylphornithids are tiny birds with a long and slender tarsometatarsus. They had at least facultatively zygodactyl feet, and piciform affinities are supported by a large intermetacarpal process of the carpometacarpus and a derived morphology of the tarsometatarsal hypotarsus. A phylogenetic analysis found these birds to be the sister taxon of either Galbulae or all other crown group Piciformes (Mayr 2004e). As is the case with many other Paleogene fossil birds, however, more data on the skeletal morphology of these birds have to be gathered for a better-resolved phylogenetic placement.

Extant Galbulae breed in self-excavated burrows or termite mounds and a similar behavior is likely also to have been present in the piciform stem group representatives. More advanced nesting habits are found in the second clade of crown group Piciformes, the Pici. This clade includes the Ramphasti-dae (barbets and toucans), which are the sister taxon of a clade including Indica-toridae (honeyguides) and Picidae (piculets, wrynecks, and woodpeckers). Except for the brood-parasitic Indicatoridae, all Pici nest in tree cavities, which are self-excavated by all species other than toucans. All Pici are characterized by a derived morphology of the distal end of the tarsometatarsus, which bears a large and distally elongated accessory trochlea for the reversed fourth toe (Figure 12.15f) that is absent in the Galbulae (Figure 12.15g).

Figure 12.15 (a) Skeleton of the piciform *Rupelramphastoides* from the early Oligocene of Germany with interpretive drawing. (b) Hand skeleton of an extant barbet (*Psilopogon*; Ramphastidae). (c) Carpometacarpus of *Rupelramphastoides*. Distal ends of tarsometatarsus (plantar view) of (d) *Rupelramphastoides* and (e) an extant woodpecker (*Dendropicus*; Picidae). Distal ends of the tarsometatarsi (distal and plantar views) of (f) a barbet (*Lybius*; Ramphastidae, Pici) and (g) a puffbird (*Monasa*; Bucconidae, Galbulae).

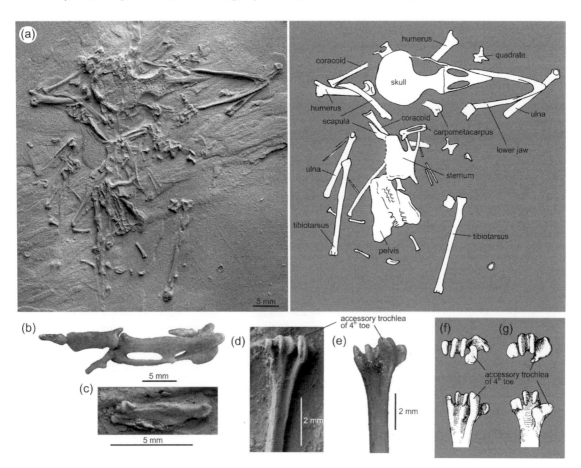

Miopico (Miopiconidae) from the middle Miocene of Morocco is a stem group representative of the Pici, whose tarsometatarsus differs that of extant Pici in presumably plesiomorphic features (Mayr 1998). Possibly in an even more basal phylogenetic position is *Picavus* (Picavidae) from the early Oligocene of the Czech Republic, although the skeletal morphology of this taxon is not well enough known for a strongly based classification (Mayr and Gregorová 2012).

The oldest uncontroversial modern-type representative of the Pici is the tiny *Rupelramphastoides* from the early Oligocene of southern Germany, which is also one of the smallest known members of the clade (Figure 12.15; Mayr 2005e, 2006d). All bones of *Rupelramphastoides*, including the highly distinctive distal end of the tarsometatarsus, exhibit morphologies that closely match those of extant Pici. The exact relationships of *Rupelramphastoides* remain to be established, although the long and slender tarsometatarsus suggests affinities to the Ramphastidae.

Fossils of several larger species of Pici from the early and middle Miocene of Europe were assigned to the taxon *Capitonides* and were classified in the Ramphastidae (Ballmann 1983). Again, it is difficult to support this placement with derived traits other than an elongated tarsometatarsus, which in some barbets and toucans is longer than in other crown group Pici. An occurrence of Ramphastidae by the early Miocene is, however, in concordance with the presence by that time of another piciform subclade, the Picidae.

Picidae form the most species-rich piciform clade which includes piculets (Picumninae), wrynecks (Jynginae), and true woodpeckers (Picinae). More than most other birds, these birds are adapted to trunk-climbing habits. This is especially the case in true woodpeckers, which evolved advanced attributes, including a stiffened support tail and derived modifications of the skull, which enable them to forage for insect larvae inside tree trunks and to excavate nesting cavities not just in brittle and decaying wood, but also in the trunks of living trees.

The earliest fossils assignable to the Picidae stem from the latest Oligocene and earliest Miocene of Europe, and *Piculoides* from the early Miocene of France was surmised to be a stem group representative of either the clade formed by Picumninae and Picinae, or of Picinae alone (De Pietri et al. 2011b). The earliest African record of the Picinae is *Australopicus* from the early Pliocene of South Africa (Manegold and Louchart 2012). In the New World, Picinae likewise do not occur before the Neogene, and the oldest skeletal remains are from the middle/late Miocene of Nebraska (*Palaeonerpes*) and the late Miocene of Kansas (*Pliopicus*; De Pietri et al. 2011b).

The only fossil Indicatoridae are from the early Pliocene of South Africa (Manegold et al. 2013), but as the sister taxon of the Picidae, honeyguides must also go back to the early Miocene at least. In any case, the fossil record as well as the interrelationships and geographic distribution of the extant species of crown group Pici suggest an Old World origin of the clade, and multiple dispersals into the New World occurred in Ramphastidae, Picumninae, and Picinae. The extant representatives of the Pici have limited capabilities for crossing the open sea, and as in the case of passerines, the sudden appearance of modern-type Pici in the early Oligocene of Europe may go back to dispersal from Asia, after closure of the Turgai Strait between Europe and Asia (Mayr 2009a).

13 Insular Avifaunas Now and Then, on Various Scales

Birds have a high dispersal potential and are found on even the most remote oceanic islands, on many of which they are the most diversified group of land vertebrates. Avian evolution on islands is different from that on continental land masses in several respects. This is due to various factors, not all of which relate to geographic isolation.

Many insular faunas are characterized by reduced predation pressure, owing to the absence or low abundance of terrestrial carnivores. As a result, islands often feature a disproportionately high number of flightless avian species. The limited ecological resources of small islands furthermore favor higher intraspecific competition and extravagant morphological specializations. The latter are also promoted by the fact that remote islands are usually colonized by small founding populations, which results in a reduced genetic variability and a higher degree of inbreeding.

At least one of these attributes, reduced predation pressure, does not only pertain to avifaunas of oceanic islands, but also occurred in some isolated continental regions. Because of plate tectonics and **marine transgressions**, the past geographies of the major land masses underwent significant changes in the course of avian evolution. Some of the southern continents were geographically isolated during most of the Cenozoic and served as refugia for various avian groups, which elsewhere succumbed to ecological competition or habitat changes. Such relict groups are also found on many of today's larger islands, and New Zealand and Madagascar in particular harbor several highly distinctive avian taxa, or at least did so before human arrival.

Avian Evolution: The Fossil Record of Birds and its Paleobiological Significance, First Edition. Gerald Mayr.

Islands and Isolated Continents as Refugia

New Zealand was home to a particularly high number of peculiar birds, which are unknown from anywhere else. Apart from the iconic moas (Chapter 6), the island housed the Aptornithidae (adzebills), of which two species were described from North and South Islands, respectively. Adzebills are mainly represented by Pleistocene and subfossil remains, but recently described early Miocene fossils show that they had a long evolutionary history on New Zealand (Worthy et al. 2007, 2011b). These flightless, turkey-sized birds have greatly reduced wings and a stoutly built skeleton with a fairly massive and curved beak (Figure 13.1). The phylogenetic affinities of adzebills are controversial, but most likely they were relicts of a group of birds that once also occurred outside New Zealand. Some skeletal features, such as a subtriangular sternum and a characteristic canal pattern of the hypotarsus of the tarsometatarsus, are shared with taxa of the Ralloidea, but some authors have also considered relationships between Aptornithidae and the Rhynochetidae (kagus) of New Caledonia.

A relictual distribution is also likely for the Rhynochetidae, which are the sister taxon of the South American Eurypygidae (sunbittern). The pre-Holocene evolutionary history of Rhynochetidae is unknown, but of Eurypygidae an undescribed putative record from the early Eocene of North America exists (see Mayr 2009a). If the identity of this fossil is confirmed in future studies, it would provide a geographic link between the disparate distributions of the two extant eurypygiform taxa.

The relict status of the Madagascan Leptosomiformes (cuckoo roller) has already been detailed in the previous chapter, and Mesitornithiformes (mesites) may constitute a further Madagascan relict taxon. This group of small, superficially rail-like birds of uncertain phylogenetic affinities includes three extant species and has no fossil record. Without doubt, however, the closest extant relatives of mesites are to be traced outside Madagascar, and they were affiliated with Columbiformes in the most recent molecular analyses (Hackett et al. 2008; Jarvis et al. 2014; Prum et al. 2015).

Relict taxa are not only found on islands, but also on continents that were geographically isolated during the Cenozoic. The fossil record shows that many of the extant avian groups, which were long considered Southern Hemispheric endemics, had stem group representatives in the Paleogene of the Northern Hemisphere (Mayr 2009a). This is especially true for birds of the Neotropic region, with Cariamiformes, Opisthocomiformes, Steatornithiformes, Nyctibiiformes, Trochilidae, and the alcediniform Todidae being among the groups that today only occur in South and Central America, but in the past had a distribution in the Northern Hemisphere. As documented by, for example, Galliformes, Anseriformes, core Gruiformes, and Strisores, another recurrent pattern in the evolution of several bird groups is the occurrence of early diverging taxa on southern continents that were geographically isolated during the Cenozoic, especially South America and Australia.

The geographic isolation of these southern continents especially aggravated the immigration of Northern Hemispheric placental mammals. Apart from rodents, Australia lacked terrestrial placental mammals before human arrival, and, until the formation of the Panamanian Isthmus initiated a major faunal exchange, only rodents and primates were potential predators and niche competitors of birds among the Neotropic placental mammals. It has been proposed that these differences in the mammalian faunas, as well as the absence of other biotic factors

Figure 13.1 Major skeletal elements of the flightless adzebill *Aptornis* (Aptornithidae) from the Quaternary of New Zealand (a: skull, b: scapula, c: coracoid, d: humerus, e: ulna, f: radius, g, h: tarsometatarsus with detail of hypotarsus, i: sternum, j: femur, and k: tibiotarsus). Note the reduced acrocoracoid process of the coracoid, the vestigial radius and ulna, and the absence of a sternal keel, which indicate a long history of flightlessness of aptornithids on New Zealand.

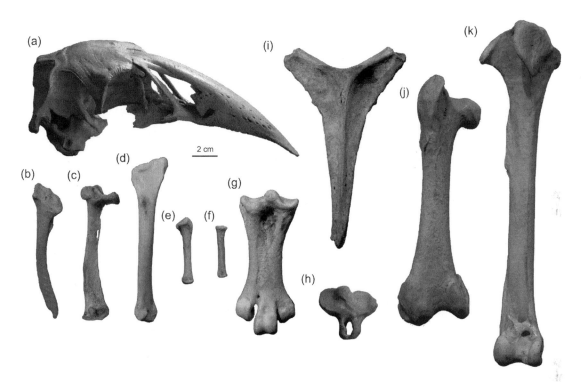

that impacted Northern Hemispheric habitats, contributed to the relictual "Southern Hemispheric pattern" of these taxa, which are likely to have succumbed to predation, competition, or habitat changes outside their extant Southern Hemispheric ranges (Mayr 2009a, 2011d).

The Evolution of Flightlessness in Predator-Free Environments

The initial evolution of flight capabilities in birds was favored by numerous selective advantages, including reduced predation pressure from terrestrial carnivores, increased dispersal potential, and the capabilities to exploit new food resources. These evolutionary benefits notwithstanding, flight is energetically costly, and in many cases natural selection led to secondary flightlessness in birds. Most often, this occurred in insular environments devoid of terrestrial predators.

Flightlessness usually evolves through heterochrony – that is, changes in the timing of developmental processes – and often paedomorphosis, the retention of juvenile features, is involved (Chapter 2; Feduccia 1999; Dove and Olson 2011). Flightlessness occurred in many avian lineages and resulted in diverse morphological specializations. With regard to locomotory adaptations, two extremes can be distinguished: the graviportal and the cursorial type. Graviportal birds are large, heavy, and stoutly built with comparatively short tarsometatarsi, which move rather slowly and often are herbivorous. Cursorial taxa, by contrast, have long hindlimbs and are fast runners. Most flightless predatory birds belong to the latter type, and only cursorial flightless birds have a chance of surviving the recurrence of predators in their habitats.

Flightless birds on islands

Carnivorous mammals are among the main predators of adult birds in extant terrestrial ecosystems. These have limited dispersal capabilities and are absent from remote oceanic islands where most extant flightless bird species occur. Although a strong correlation therefore exists between environments with reduced predation pressure and the evolution of flightlessness, several other factors have been suggested that may have contributed to the loss of flight in insular birds, such as cessation of the need for long-distance dispersal and the risk of being blown into the sea by strong winds (Feduccia 1999). However, flightlessness also evolved on large islands and in predator-free continental environments, where dispersal is an issue. Unlike in insects, the small population size and much longer generation periods of insular birds likewise do not render it likely that flightlessness would have become manifest if the main selective pressure was to prevent individuals from being blown into the sea (even more so as insular flightlessness most often occurs in medium-sized terrestrial birds, which are less prone to the effect of winds than are small aerial species). The absence of predation is therefore likely to be the prime cause for secondary flightlessness in birds.

New Zealand, which was devoid of any terrestrial mammals until the arrival of humans, is certainly among those geographic areas that held the highest number of coevally occurring flightless avian species.

The New Zealand fauna of flightless birds included nine species of moas (Dinornithiformes), two species of adzebills, one poorly flighted parrot (*Strigops*), a flightless goose with greatly reduced wings (*Cnemiornis*), two species of flightless teal (*Anas*), as well as various flightless Rallidae (Worthy and Holdaway 2002). At least for moas and adzebills, a long evolutionary history of flightlessness on New Zealand is documented, which goes back into the early Miocene.

A large number of flightless birds also existed on the Hawaiian archipelago, from where four species of the peculiar anseriform moa-nalos were reported (see later discussion), as well as other flightless ducks and geese, and two species of the flightless ibis *Apteribis* (Olson and James 1991; Sorenson et al. 1999; Dove and Olson 2011). Unlike New Zealand, the volcanic Hawaiian archipelago is of comparably recent origin, so that flightlessness in these taxa can be temporally constrained and must have evolved much more rapidly. Flightless ibises are also known from the Mascarene and West Indian islands (Longrich and Olson 2011; Parish 2012).

Flightless birds abounded on various Polynesian islands, where mainly species of the Megapodiidae, Rallidae, and Columbidae are involved (Steadman 2006). Among the most spectacular taxa of the latter is the giant pigeon *Natunaornis* from Viti Levu, which was considered to be most closely related to the extant New Guinean taxon *Goura* (Worthy 2001).

Columbidae gave also rise to insular flightless species on the Mascarene Islands, with the Dodo (*Raphus*; Figure 13.2) from Mauritius and the Solitaire (*Pezophaps*) from Rodriguez being some of the most iconic of all extinct birds (Parish 2012). Although both species are usually considered sister taxa (e.g., Shapiro et al. 2002), flightlessness must have evolved independently, because Mauritius and Rodriguez are, and always

Figure 13.2 Skulls of (a) the Dodo (*Raphus*) from the Holocene of Mauritius and (b) *Caloenas*, a presumably closely related extant columbiform. Not to scale. Dodo photograph courtesy of Hanneke Meijer and the Mauritius Museums Council.

(a)

(b)

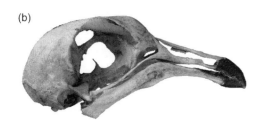

were, too widely separated for the dispersal of a flightless ancestor of these birds. This raises the question of whether certain birds are more prone to becoming flightless than others and, in the case of pigeons, arboreal species are probably less likely to lose their flight capabilities than terrestrial ones. Analyses of molecular data did indeed support a position of *Raphus* and *Pezophaps* within a clade including the Asian taxa *Caloenas* and *Goura*, which predominantly forage on the ground (Shapiro et al. 2002; Parish 2012).

Other than on remote oceanic islands, flightless birds also occurred in ephemeral insular areas that formed during marine transgressions. An example thereof is the late Miocene Gargano Island in Italy, where a large flightless bird of uncertain, possibly anseriform affinities occurred (*Garganornis*; Meijer 2014). Another late Miocene Italian island existed in the Montebamboli area in Tuscany, from where a putatively flightless duck was reported (*Bambolinessa*; Mayr and Pavia 2014).

Flightless birds in insular environments are often graviportal species, because there exist no selective advantages for the evolution of long-legged cursorial forms in the absence of predators and fast-running prey. Insular flightlessness occurred across many avian groups, but most flightless species are found among terrestrial taxa that forage on the ground. Flight loss is particularly abundant in taxa with great dispersal potential, as these constitute the majority of species that reached remote, predator-free islands. Rallidae in particular include a high number of flightless species on many oceanic islands (Olson 1977a).

At first sight, no selective advantage seems to exist for perching birds to lose their flight capabilities. In parrots (Psittacidae), however, this occurred at least twice: the large flightless *Lophopsittacus* lived on Mauritius and became extinct only in the 17th century (Hume 2007); and among extant parrots, the New Zealand *Strigops* is nearly flightless. In both cases, the evolution of flightlessness must have been facilitated by a plenitude of food items on ground level. Even among passerines, some insular species became flightless, such as some New Zealand wrens (Acanthisittidae) and an extinct species of buntings (Emberizidae) from the Canary Islands (Rando et al. 1999). Again, the ancestors of the species involved probably mainly foraged on the ground. This is also true for a large flightless hoopoe (Upupidae) that occurred on the South Atlantic island of Saint Helena (Olson 1985a).

Even more unexpected is the occurrence of flightlessness in insular raptorial birds. Owls in particular produced several flightless insular forms, most of which became very large. A well-known example concerns the giant *Ornimegalonyx* from the Pleistocene of Cuba, which reached a standing height of about 1 meter (Arredondo 1976). The evolution of flightlessness in diurnal birds of prey in insular environments, by contrast, appears to have been less common, although there is a record of a possibly flightless caracara (Falconidae) from the Holocene of Jamaica (Olson 2008).

Flightlessness in continental avifaunas

Flightless birds are very rare in extant continental avifaunas, and their occurrence is restricted to ostriches in Africa and rheas in South America (with the latter having more of an insular origin, owing to the long geographic isolation of the continent). This situation sharply contrasts with some past continental avifaunas, which featured a much higher diversity of flightless birds.

Even if the possibility is dismissed that *Caudipteryx* and other early theropods were secondarily flightless animals (Paul 2002), flightlessness already occurred in Mesozoic continental ecosystems. An uncontroversial case is *Patagopteryx* from the Late Cretaceous of Argentina. This taxon is particularly remarkable because it was not cursorial and coexisted with crocodilians and non-avian theropods (Chiappe 2002). The ecological conditions that facilitated flight loss in *Patagopteryx* are unknown, but so are the habitat preferences of the taxon, which must have lived in terrain inaccessible to potential predators. Further examples of Mesozoic flightless birds are *Gargantuavis* from the Late Cretaceous of France (Buffetaut 2010), if this taxon was indeed avian (Chapter 3), and the enantiornithine *Elsornis* from the Late Cretaceous of Mongolia, for which at least very weak flight capabilities have been assumed (Chiappe et al. 2007b).

Flightless birds abounded in the Paleocene and early Eocene of Europe, which was geographically isolated and devoid of large mammalian carnivores by that time. These include the Gastornithidae, Remiornithidae, Palaeotididae, some cariamiforms, and possibly other taxa as well (Mayr 2009a; Buffetaut and Angst 2014). The latest record of a flightless bird in the Paleogene of Europe

stems from the beginning of the Oligocene (Mayr 2011d) and therefore just postdates the "Grande Coupure," a faunal turnover that also involved the immigration of larger **carnivorans** from Asia after closure of the Turgai Strait (Mayr 2009a, 2011d). However, whether predation by newly arriving carnivorans did indeed terminate the existence of flightless birds in Europe can only be more rigorously evaluated once more data on the exact stratigraphic occurrence of both carnivorans and flightless birds have been gathered. No flightless birds are known from the late Oligocene and early Miocene of continental Europe, but during the Miocene Climatic Optimum cursorial ostriches dispersed into Southern Europe in the middle and late Pliocene.

The temporal distribution of flightless birds in North America and Asia is only incompletely known. The North American occurrence of gastornithids was apparently restricted to a comparatively short period in the early Eocene (Mayr 2009a; Buffetaut 2013), and presumably flightless bathornithids existed in the late Eocene and early Oligocene. Whether there were flightless birds between these periods and how long flightless birds persisted in North America remain to be studied.

In Asia, eogruids had a long evolutionary history, from the middle Eocene to the Pliocene. It is, however, debated whether the early representatives of these birds were already flightless, and the late forms were cursorial birds that could outrun potential predators. The same is true for Asian ostriches, which are first known from the early Miocene.

In Africa, flightless birds also seem to have been more widespread during the early Cenozoic. Although their fossil record in terms of skeletal remains is restricted to ostriches and the late Eocene *Eremopezus* and *Lavocatavis* (which may or may not represent the same or closely related taxa;

see Chapter 11), various types of eggshells of very large birds are common in localities throughout the Cenozoic (see Chapter 6). Again, the diversity of these birds is likely to be correlated with a comparatively late arrival of larger carnivorans on the African continent, which was geographically isolated during most of the early Paleogene (Smith et al. 1994). Placental carnivorans also dispersed late into other parts of the Southern Hemisphere; that is, South America, Antarctica, and Australia. At least in South America and Australia, flightless birds flourished throughout most of the Cenozoic to a similar degree as in insular Europe in the early Paleogene (Mayr 2009a). Antarctica was not only populated with penguins, but was also home to non-sphenisciform large flightless birds of uncertain affinities (Chapter 6).

In some respects, the ecological attributes of flightless continental birds differ from those of flightless birds on islands. Because flightless birds in continental ecosystems evolved from species of the native fauna, taxa with limited dispersal potential, such as cariamiforms, are also involved. Furthermore, whereas many flightless insular birds are predominantly herbivorous or omnivorous, continental avifaunas also produced numbers of flightless raptorial taxa, such as the South American phorusrhacids and the North American bathornithids.

The origin of flightlessness in continental ecosystems, however, is likely to have been due to the same reason that led to flight loss on oceanic islands; that is, reduced predation pressure. Most flightless lineages of continental birds appear to be very old ones, which attained flightlessness early in their evolutionary history, and the earliest records of penguins, flightless palaeognathous birds, gastornithids, and phorusrhacids date back to the Paleocene at least. The disappearance of larger terrestrial predators during the mass extinction events at the K/Pg boundary may

have been the critical factor that facilitated the evolution of flightlessness in these taxa, and it is to be hoped that analyses of an improved future fossil record will allow well-founded temporal correlations between the extinction of, for instance, non-avian dinosaurs and the first appearance of flightless birds in the latest Mesozoic and earliest Paleogene ecosystems.

Continental ecosystems were subjected to dramatic changes during the Cenozoic. Although there may have been predator-free periods in the earliest phases of flight loss, even the southern continents were not completely devoid of predators in the periods thereafter. One evolutionary strategy to reduce predation risk in terrestrial ecosystems is the attainment of a large size, which is also beneficial for dispersal over big home ranges. Contrary to insular faunas, no small flightless terrestrial birds are therefore known from continental ecosystems, with the smallest taxon being the turkey-sized Cretaceous *Patagopteryx* (penguins and other flightless aquatic birds are discussed in the following section). Another characteristic of flightless continental birds is that most species are cursorial, able to outrun potential predators. Examples thereof are the extant flightless palaeognathous birds in Africa and South America, as well as the extinct European palaeognathous birds, the Asian eogruids, and the South American phorusrhacids. Insular avifaunas, by contrast, mainly include graviportal flightless species. Although there were some graviportal flightless birds in the early Cenozoic of the Northern Hemisphere, in the long run only cursorial taxa, such as ostriches and eogruids, were able to coexist with large carnivorans.

Flightlessness in marine birds

Among aquatic birds, flightlessness occurs mainly in marine taxa, with the only notable examples of lacustrine diving birds without flight capabilities being some grebes (Podicipediformes) and the giant South American darters (Anhingidae). These latter groups are not included in the following section, which focuses on seabirds.

Even the most aquatic birds need to visit land in breeding periods, during which flightless taxa are particularly prone to predation. The existence of predator-free nesting grounds is therefore also one of the prerequisites for the evolution of flightlessness in seabirds.

Flightlessness evolved in several lineages of marine birds and over a large temporal period. The first seabirds without flight capabilities were the Cretaceous hesperornithiforms. Because the largest and most specialized of these birds were probably strongly restricted in their abilities of terrestrial locomotion, it has been surmised that they were migratory and may have bred in northern latitudes (but see Chapter 3).

Among neornithine (crown group) birds, flightless pelagic species are mainly found in the Anseriformes, Sphenisciformes, Plotopteridae, and Alcidae (the few flightless insular Phalacrocoracidae are not considered here). The origin of flightlessness in Sphenisciformes goes back to at least the Cretaceous–Paleogene boundary, and penguins appear to have lost their flight capabilities in the New Zealand or Antarctic region. Flightlessness in sphenisciforms is therefore likely to have initially been correlated with the end-Cretaceous extinction of potential terrestrial predators, and their Cenozoic diversification likewise took place largely in predator-free geographic areas.

There was a particularly high diversity of flightless birds in the Oligocene and Miocene of the North Pacific, where species of at least three different avian lineages lost their flight capabilities. All of these were reported from both sides of the Pacific; that is, from Japan and from fossil sites along the North American west coast, where their occurrence

was probably associated with predator-free breeding grounds on offshore volcanic islands (Goedert and Cornish 2002). Besides the Plotopteridae (Chapter 10) and mancalline auks (Chapter 9), a large, flightless diving anseriform occurred in the middle Miocene of California (*Megalodytes*; Howard 1992). This North American taxon is known only from a few limb bones, but a nearly complete skeleton of a closely related species was found in the middle Miocene of Japan; based on a high number of cervical vertebrae, it was identified as a flightless swan (Matsuoka et al. 2004).

The existence of plotopterids ceased in the early Miocene, but mancalline auks occurred until the late Pleistocene. In the Pleistocene and Holocene, there was a further large and flightless anseriform bird with greatly reduced wings along the Californian coast, where the offshore Channel Islands provided safe breeding grounds. *Chendytes* was considered to be a representative of the Mergini (sea ducks; Livezey 1993). It was hunted by indigenous North American tribes over a period of 8,000 years and became extinct some 2,400 years ago, presumably due to human overhunting (Jones et al. 2008). Again, a similar and possibly closely related taxon, *Shiriyanetta*, was reported from the Pleistocene of Japan (Watanabe and Matsuoka 2015).

Among the most widely distributed groups of flightless marine birds were the Cretaceous Hesperornithiformes. By contrast, only a few of the flightless Cenozoic marine bird groups had ranges that spanned different oceans, although the physical or ecological barriers that restricted their ranges remain poorly understood. Neither plotopterids nor mancalline auks, for example, are known from outside the North Pacific. Dispersal of mancalline auks into the Atlantic may have been prevented by the formation of the Isthmus of Panama, but plotopterids were already diversified in the late Eocene and Oligocene, by which time the Americas were not yet connected by a land bridge. After formation of a marked latitudinal temperature gradient in the mid-Cenozoic (e.g., Eldrett et al. 2009), the nutrient-poor tropical seas near the equator may have formed a barrier for diving seabirds and evidently they did so for penguins, whose distribution is restricted to the Southern Hemisphere. Oceanic currents likewise do not favor dispersal across the equator, and it may be no coincidence that the Cretaceous Hesperornithiformes lived at times with very different global oceanic circulation systems (e.g., Stille et al. 1996).

Even the earliest species of the Sphenisciformes and Plotopteridae were large in size, and giant species also evolved multiple times among the Hesperornithiformes. Like in flightless terrestrial taxa, a size increase in these marine groups was due to the release from aerodynamic constraints, and in seabirds a more favorable volume to surface ratio may have also played a role with regard to heat loss. In terrestrial ecosystems, a large size of flightless taxa contributes to reduced predation risk, but this may not necessarily be the case in marine environments.

The evolutionary factors that caused the extinction of these giant diving seabirds have been debated with some controversy. It is notable that, on a global scale, large flightless diving seabirds became extinct at similar times, towards the early Miocene. Most researchers agree that this demise of giant flightless seabirds was due to the rise of marine mammals, and two different scenarios have been discussed, which may not have been mutually exclusive and could have acted in concert. One of these is competition with gregarious pinnipeds for breeding sites. This hypothesis is in agreement with the fact that many extant penguins nest in earth tunnels or terrains, which do not allow large aggregations of pinnipeds; in some cases, a size decrease of penguins can furthermore be

correlated with the first local appearances of gregarious pinnipeds (Warheit and Lindberg 1988; Warheit 1992, 2002). On the other hand, the decreasing diversity of giant flightless seabirds temporally corresponds with the rise of odontocete whales, so that direct predation may have also played a role (Ando and Fordyce 2014), although plotopterids and odontocetes coexisted in the North Pacific from the late Eocene on (Goedert and Cornish 2002).

Insular Gigantism and Islands as Cradles of Unusual Morphologies

Insular gigantism in birds

It has long been known that some mammalian groups tend to become larger on islands (e.g., rodents), whereas others show a tendency towards dwarfism (e.g., carnivorans). Although the validity of this "island rule" is debated (e.g., Meiri et al. 2004 and references therein), some insular size trends have been observed in birds (Clegg and Owens 2002).

With regard to weight and standing height, insular avifaunas include some of the largest birds that ever lived. Particularly tall and heavy species are found among the Madagascan Aepyornithiformes, the New Zealand Dinornithiformes, and the Australian Dromornithidae, with the New Zealand moa *Dinornis*, the Madagascan elephant bird *Aepyornis*, and the Australian dromornithid *Bullockornis* rivaling for the weight record among birds (Worthy and Holdaway 2002; Murray and Vickers-Rich 2004). Gigantism in these birds is clearly related to their flightlessness, which released them from the physiological constraints that limit the size of volant birds. Equally large and heavy birds also occurred in insular South America, where the putative phorusrhacid *Brontornis* (see Chapter 11) and the phorusrhacid *Titanis* reached a size comparable to that of the aforementioned taxa. However, whereas virtually all giant birds on Madagascar, New Zealand, and Australia were omnivorous or herbivorous, the continental *Titanis* was a carnivore or scavenger.

Examples of extinct avian species from islands that were larger than closely related continental species are a Holocene species of *Leptoptilos* stork from Flores (Lesser Sunda Islands; Meijer and Due 2010) and a swan from the Pleistocene of the Mediterranean island of Malta (Northcote 1982), both of which probably had only limited flight capabilities. A size increase in volant insular birds mainly involved carnivores, such as the well-known extinct Haast's Eagle, *Harpagornis*, from New Zealand, which is among the largest eagles (Worthy and Holdaway 2002). Because this taxon is assumed to have preyed on moas, its large size is probably correlated with that of its predominant prey. Giant size also evolved among insular owls, such as *Ornimegalonyx* from the Pleistocene of Cuba (see further above in this chapter) and some *Tyto* species from the late Miocene to Pleistocene of various Mediterranean islands. Here, it is probably likewise related to a proportionally larger size of mammalian prey, which itself followed the "island rule" (Pavia 2004).

The evolution of unusual morphologies in island birds

Flightlessness in island birds has been extensively discussed in the literature, but another attribute of many insular avian species has received far less attention. This is the fact that island avifaunas produced some highly specialized birds with unusual morphologies, which are remarkably different from closely related continental species.

Textbook examples of such morphological outliers are the Dodo and the Solitaire from the Mascarene Islands of Mauritius and Rodriguez, which have already been mentioned. Despite being phylogenetically

nested within crown group Columbiformes (Shapiro et al. 2002), these flightless birds depart significantly from the usual columbiform morphology. Both species not only have much more massive beaks than putatively closely related pigeons (Figure 13.2). The solitaire, which appears to have been a highly territorial bird – its vernacular name refers to the circumstance that these birds were never seen in groups – also features unusual bony excrescences on the wing bones, which were used in combat fights of rival birds (Parish 2012). In their skeletal morphology, especially that of the skull, *Raphus* and *Pezophaps* are clearly distinguished from the presumedly closely related *Caloenas* and *Goura*, as well as all other extant Columbidae. That *Raphus* and *Pezophaps* show such close similarities and are so widely divergent from crown group Columbidae is particularly startling in light of the fact that these peculiar morphologies must have evolved convergently, if both taxa stem from a volant *Caloenas-* or *Goura*-like ancestor.

From a phylogenetic respect, the evolution of unusual skeletal morphologies appears to be unevenly distributed, and some taxa seem to have been more prone to evolving them than others. A particularly high number of grotesque adaptations are found among galloanserines, many insular forms of which show unusual bill shapes indicative of specialized feeding behaviors.

The galliform *Megavitiornis* from the Holocene of the Polynesian island Viti Levu, for example, has a very deep beak, which strikingly resembles that of gastornithids and dromornithids. This large, flightless bird may have been adapted to cracking large seeds and became extinct after the arrival of human settlers (Worthy 2000).

A very similar galloanserine bird is the flightless *Sylviornis* (Sylviornithidae) from the Holocene of New Caledonia and the nearby Île de Pins. This cassowary-sized bird has strongly reduced wings and a highly movable beak, which is almost completely separated from the neurocranium (Mourer-Chauviré and Balouet 2005). *Sylviornis* is characterized by several further extravagant features, including a large bony protuberance at the base of the upper beak. It is considered to be related to galliform birds, with which it shares fused thoracic vertebrae and a slender coracoid. *Sylviornis* was known as the "Du" by local people, and it has been surmised that it was the builder of large soil mounds, which were previously assumed to be human artifacts (Poplin et al. 1983). The peculiar bill morphology was interpreted as an adaptation for a carnivorous diet (Mourer-Chauviré and Balouet 2005), but may have equally served for the processing of hard vegetarian items. *Sylviornis* lived until at least 3,000 years ago, and its late Holocene extinction was attributed to overhunting by native people (Anderson et al. 2010).

Some highly derived morphologies also evolved among the Anseriformes. Striking examples are the goose-sized moa-nalos, which occurred on different islands of the Hawaiian archipelago during the Pleistocene and Holocene. Moa-nalos were flightless birds with greatly reduced wing bones and four species are known, which belong to the three taxa *Chelychelynechen*, *Thambetochen*, and *Ptaiochen* (Olson and James 1991). The beak of all moa-nalos is very short; in *Thambetochen* its cutting edges are serrated, thus being somewhat reminiscent of the bony pseudoteeth of pelagornithids. Irrespective of their goose-like beak, however, the presence of a syringeal bulla, a characteristic bulbous enlargement of the trachea of male Anatinae, indicates anatine affinities of moa-nalos, and analyses of molecular data have suggested that they are the sister taxon of dabbling ducks (Anatini; Sorenson et al. 1999). Moa-nalos were browsers with a hindgut fermentation (James and Burney 1997), and with regard to their feeding habits

they constitute an ecological equivalent of tortoises, which were absent on the Hawaiian archipelago. These birds colonized the Hawaiian archipelago at an early date, and their divergence from other Anatinae was estimated at 3.6 mya (Sorenson et al. 1999). Interestingly, they are absent on the main island of Hawaii itself, which emerged about 500,000 years ago, by which time moa-nalos were probably already flightless and could no longer disperse across sea barriers (Sorenson et al. 1999). Extinction of these birds is also hypothesized to have been due to human hunting pressure after the arrival of early Polynesian settlers.

Another bizarre anseriform is *Talpanas* from the Hawaiian island of Kauai. This taxon is known from a neurocranium and a few postcranial bones, which were found in 6,000-year-old lake deposits. Judging from its small orbits and reduced wings, *Talpanas* was probably nearly blind and flightless, and it is assumed to have been nocturnal. The Latin taxon name means "mole-duck," because the neurocranium remotely resembles that of a mole in its proportions. *Talpanas* has a very short tarsometatarsus and its relationships within Anatidae are uncertain (Iwaniuk et al. 2009).

The flightless ibis *Xenicibis* from Holocene cave deposits in Jamaica features particularly odd wing bones, with the unusually deformed and inflated carpometacarpus having thickened bone walls, and the distal wing phalanges being greatly reduced (Figure 13.3). This morphology resulted in "club-shaped" wings, which were probably used as weapons in combat fights (Longrich and Olson 2011).

Some of these birds must have evolved within a comparatively short time, since the islands housing them are not very old geologically, and several factors are likely to have played a role in the rapid establishment of a disproportionately high number of unusual adaptations among insular birds.

Figure 13.3 (a) Skeleton of the flightless ibis *Xenicibis xympethicus* from the Holocene of Jamaica. Wing bones of (b) *Xenicibis* and (c) the extant Scarlet Ibis, *Eudocimus ruber*. Note the short ulna and radius and the highly modified carpometacarpus of *Xenicibis*. Drawing and photographs by Nicholas Longrich.

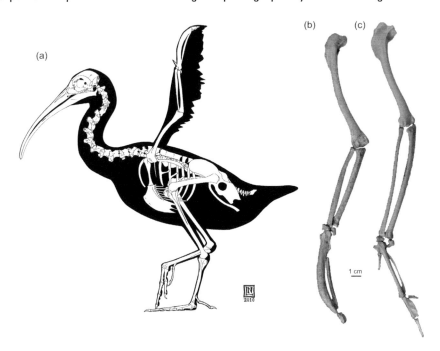

One of these is the founder effect; that is, the reduced genetic variability of the small initial populations (Mayr 1942), which in extreme cases may have consisted of a single gravid female. The founder effect, together with inbreeding, leads to the establishment of rare gene alleles and results in distinct phenotypic differences to the original populations. Often, birds arriving on remote oceanic islands also benefit from reduced competition, which allows the evolution of unusual structures that would have been negatively selected for under non-insular conditions. Compared to continental avifaunas, sexual selection is furthermore likely to be more pronounced on islands with only small populations, and the same may be true for competition among males for limited resources, be it territories or females, which would explain the evolution of the unusual combat structures in the solitaire and the Jamaican ibis.

Insular avifaunas offer critical insights into evolutionary processes beyond the textbook examples of speciation processes provided by the iconic Galapagos finches. As I hope to have shown in the present book, paleornithology as a whole likewise furthers our understanding of vertebrate evolution, be it through particularly demonstrative examples of transitional forms or unexpected biogeographic occurrences. Currently, much of this knowledge is restricted to a limited circle of specialists, especially where Cenozoic birds are concerned. I am confident, however, that future ornithological textbooks will provide a more balanced account of the overwhelmingly rich fossil record of birds, and hopefully this book has made a small contribution in that direction.

Glossary

apomorphy (derived character) an evolutionary novelty of a species and its descendants, which characterizes a clade.

Bergmann's Rule a biogeographic principle that postulates a larger body size of endothermic animals in colder climates.

calibrated molecular phylogeny an evolutionary "time tree" that shows the interrelationships and divergence times of organisms based on analyses of molecular data. Under the assumption of predictable nucleotide substitution rates, divergence times are calculated by the use of well-dated fossils or geographic events as calibration points.

Carnivora the clade of mammalian carnivores that includes cats, dogs, bears, and allies.

Cenozoic the geological era that encompasses the period from 66 million years ago until the present.

clade a natural phylogenetic unit, which consists of an ancestral species and all its descendants.

countershading a form of camouflage in which the upper side of the body of an animal is more pigmented (darker) than the under side.

crown group the clade encompassing the last common ancestor of the extant species of a certain taxon and all of its descendants (see Figure 1.1).

cursorial adapted for running.

derived character see "apomorphy."

distal a term referring to anatomical structures that are situated far away from the center of the body.

endemic native to a restricted geographic area.

epicontinental located on a continental shelf.

follicle a cellular aggregation in the ovary that contains the egg cell.

folivory a feeding ecology that involves a diet based mainly on leaves.

gastroliths ingested stones in the gastrointestinal tract (in birds usually the gizzard), which often serve to grind coarse food.

gizzard the specialized stomach of birds, which has strong muscular walls and assists in the processing of food.

granivory a feeding ecology that involves a diet based mainly on seeds.

homologous characters in different organisms are homologous if they have a common evolutionary origin.

Avian Evolution: The Fossil Record of Birds and its Paleobiological Significance,
First Edition. Gerald Mayr.
© 2017 John Wiley & Sons, Ltd. Published 2017 by John Wiley & Sons, Ltd.

homoplasy an umbrella term for a mosaic distribution of derived characters, which results either from convergent evolution or from a reversal into the primitive condition.

marine transgression a sea-level change that results in the flooding of land areas.

medullary bone a calcium reservoir for the formation of the eggshell that is deposited in the medullary cavity of the long bones of female birds shortly before or during egg laying.

megafauna a collective term for very large animals of a terrestrial habitat, which is mainly used for late Cenozoic mammalian herbivores.

melanosome a cell organelle involved in melanin synthesis.

Mesozoic the geological era that encompasses the period 252–66 million years ago.

monophyletic a taxonomic entity is monophyletic if it includes an ancestral species and all its descendants (see Figure 1.1); a monophyletic group is also termed a "clade."

Neogene the geological period that includes the Miocene and Pliocene epochs (23–2.6 million years ago).

niche segregation the separation of ecological niches of different species that allows their coexistence in the same area.

niche partitioning the process by which competing species are driven into different ecological niches through natural selection.

notarium an osseous structure that is formed by the fusion of thoracic vertebrae and occurs in some only distantly related bird groups.

ontogenetic a process or structure relating to the development of an organism from the embryo to the adult.

osteogenesis bone formation.

paedomorphosis the retention of juvenile traits in the adult.

Paleogene the geological period that includes the Paleocene, Eocene, and Oligocene epochs (66–23 million years ago). The Paleogene period is not to be confused with the Paleocene epoch.

paraphyletic a taxonomic entity is paraphyletic if it does not include all taxa that descended from its last common ancestor (see Figure 1.1).

parsimony-based analysis a phylogenetic analysis in which phylogenetic trees are constructed under the premise of a minimization of the amount of character homoplasy in the underlying data set.

pennaceous feather the "typical" vaned feather of modern birds, which consists of a shaft and serially arranged barbs.

phylogenetic systematics a method of reconstructing evolutionary trees that aims at identification of monophyletic groups (clades).

piscivorous a feeding ecology that involves a diet based mainly on fish.

plesiomorphy an ancestral (primitive) character that evolved in a species outside a clade.

polyphyletic a taxonomic unity is polyphyletic if it consists of only distantly related taxa (see Figure 1.1).

postcranial a term referring to all parts of the skeleton other than the skull.

preadaptation an evolutionary novelty that allows an organism to enter a new ecological zone.

primary feathers the long wing feathers that are attached to the hand section of the skeleton.

proximal a term referring to anatomical structures that are situated close to the center of the body.

retroarticular process a projection of the caudal end of the mandible that increases the leverage of the muscles lowering the mandible.

secondary feathers the long wing feathers that are attached to the ulna.

sister groups two taxa that have a common ancestor, which is not shared by other taxa.

stem group all species of a clade that are outside the crown group; the stem group is by definition a paraphyletic assemblage and only includes fossil taxa (see Figure 1.1).

stem lineage the evolutionary lineage of stem group representatives that leads directly to the crown group of a certain clade. Whereas the stem group may include coevally occurring species, the stem lineage represents a temporal sequence of species.

stem species the ancestral species of a clade, from which all other species developed.

taxon a group of organisms that is recognized as a systematic entity (plural: taxa). In this book, all taxonomic entities above the species level are denoted as taxa, because taxonomic categories, such as "genera," "families," or "orders," are arbitrary and have no equivalents in the real world, where only species can be distinguished.

transposable elements (transposons) DNA sequences that can change their position within the genome ("jumping genes").

trochlea a skeletal structure formed by the articular end of a bone.

Turgai Strait a shallow seaway that separated Europe and Asia until the early Oligocene.

Wallace's Line one of the various faunal boundaries that separate the Australasian region from Asia.

References

Abel, O. (1936) Kalman Lambrecht (1. Mai 1889–7. Januar 1936). *Paläontologische Zeitschrift* **18**, 11–17.

Acosta Hospitaleche, C. (2014) New giant penguin bones from Antarctica: Systematic and paleobiological significance. *Comptes Rendus Palevol* **13**, 555–560.

Acosta Hospitaleche, C., & Gelfo, J.N. (2015) New Antarctic findings of Upper Cretaceous and lower Eocene loons (Aves: Gaviiformes). *Annales de Paléontologie* **101**, 315–324.

Acosta Hospitaleche, C., & Reguero, M. (2014) *Palaeeudyptes klekowskii*, the best preserved penguin skeleton from the Eocene-Oligocene of Antarctica: Taxonomic and evolutionary remarks. *Geobios* **47**, 77–85.

Acosta Hospitaleche, C., & Tambussi, C. (2005) Phorusrhacidae Psilopterinae (Aves) en la Formación Sarmiento de la localidad de Gran Hondonada (Eoceno Superior), Patagonia, Argentina. *Revista Española de Paleontologia* **20**, 127–132.

Acosta Hospitaleche, C., Tambussi, C., Donato, M., & Cozzuol, M. (2007) A new Miocene penguin from Patagonia and its phylogenetic relationships. *Acta Palaeontologica Polonica* **52**, 299–314.

Acosta Hospitaleche, C., Reguero, M., & Scarano, A. (2013a) Main pathways in the evolution of the Paleogene Antarctic Sphenisciformes. *Journal of South American Earth Sciences* **43**, 101–111.

Acosta Hospitaleche, C., Griffin, M., Asensio, M., Cione, A. L., & Tambussi, C. (2013b) Restos de pingüinos del Cenozoico medio de la cordillera Patagónica. *Andean Geology* **40**, 409–503.

Agnolín, F.L. (2007) *Brontornis burmeisteri* Moreno & Mercerat, un Anseriformes (Aves) gigante del Mioceno Medio de Patagonia, Argentina. *Revista del Museo Argentino de Ciencias Naturales, nueva serie* **9**, 15–25.

Agnolín, F.L. (2009a) Una nueva especie del género *Megapaloelodus* (Aves: Phoenicopteridae: Palaelodinae) del Mioceno Superior del noroeste de Argentina. *Revista del Museo Argentino de Ciencias Naturales, nueva serie* **11**, 23–32.

Agnolín, F.L. (2009b) *Sistemática y filogenia de las aves fororracoideas (Gruiformes: Cariamae).* Buenos Aires: Fundación de Historia Natural Félix de Azara.

Agnolín, F.L. (2010) An avian coracoid from the Upper Cretaceous of Patagonia, Argentina. *Studia Geologica Salmanticensia* **46**, 99–119.

Agnolín, F. (2013) La posición sistemática de *Hermosiornis* (Aves, Phororhacoidea) y sus implicancias filogenéticas. *Revista del Museo Argentino de Ciencias Naturales, nueva serie* **15**, 39–60.

Agnolín, F.L., & Martinelli, A.G. (2009) Fossil birds from the Late Cretaceous Los Alamitos Formation, Río Negro Province, Argentina. *Journal of South American Earth Sciences* **27**, 42–49.

Agnolín, F.L., & Novas, F.E. (2011) Unenlagiid theropods: Are they members of the Dromaeosauridae (Theropoda, Maniraptora)?

Anais da Academia Brasileira de Ciências **83**, 117–162.

Agnolín, F.L., & Novas, F.E. (2012) A carpometacarpus from the upper cretaceous of patagonia sheds light on the Ornithurine bird radiation. *Paläontologische Zeitschrift* **86**, 85–89.

Agnolín, F.L., & Novas, F.E. (2013) *Avian Ancestors: A Review of the Phylogenetic Relationships of the Theropods Unenlagiidae, Microraptoria, Anchiornis and Scansoriopterygidae*. Dordrecht: Springer.

Agnolín, F.L., Novas, F.E., & Lio, G. (2006) Neornithine bird coracoid from the Upper Cretaceous of Patagonia. *Ameghiniana* **43**, 245–248.

Agnolín, F.L., Tomassini, R.L., & Contreras, V.H. (2016) Oldest record of Thinocoridae (Aves, Charadriiformes) from South America. *Annales de Paléontologie* **102**, 1–6.

Alegret, L., Thomas, E., & Lohmann, K.C. (2012) End-Cretaceous marine mass extinction not caused by productivity collapse. *Proceedings of the National Academy of Sciences* **109**, 728–732.

Alexander, D.E., Gong, E., Martin, L.D., Burnham, D.A., & Falk, A.R. (2010) Model tests of gliding with different hindwing configurations in the four-winged dromaeosaurid *Microraptor gui*. *Proceedings of the National Academy of Sciences USA* **107**, 2972–2976.

Allen, V., Bates, K.T., Li, Z., & Hutchinson, J.R. (2013) Linking the evolution of body shape and locomotor biomechanics in bird-line archosaurs. *Nature* **497**, 104–107.

Altamirano-Sierra, A. (2013) Primer registro de pelícano (Aves: Pelecanidae) para el Mioceno tardío de la formación Pisco, Perú. *Bulletin de l'Institut Français d'Études Andines* **42**, 1–12.

Alvarenga, H.M.F. (1983) Uma ave ratitae do Paleoceno Brasileiro: Bacia calcária de Itaboraí, Estado do Rio de Janeiro, Brasil. *Boletim do Museu Nacional, Geologia* **41**, 1–8.

Alvarenga, H.M.F. (1985a) Um novo Psilopteridae (Aves: Gruiformes) dos sedimentos Terciários de Itaboraí, Rio de Janeiro, Brasil. In: D.A. Campos, C.S. Ferreira, I.M. Brito, & C.F. Viana (eds.), Anais do VIII Congresso Brasileiro de Paleontologia. MME-DNPM, Série Geologia 27, Paleontologia, Estratigrafia 2, 17–20.

Alvarenga, H.M.F. (1985b) Notas sobre os Cathartidae (Aves) e descrição de um novo gênero do Cenozóico Brasileiro. *Anais da Academia Brasileira de Ciências* **57**, 349–357.

Alvarenga, H.M.F. (1990) Flamingos fósseis da Bacia de Taubaté, estado de São Paulo, Brasil: Descrição de nova espécie. *Anais da Academia Brasileira de Ciências* **62**, 335–345.

Alvarenga, H.M.F. (1995) A large and probably flightless anhinga from the Miocene of Chile. *Courier Forschungsinstitut Senckenberg* **181**, 149–161.

Alvarenga, H.M.F. (1999) A fossil screamer (Anseriformes: Anhimidae) from the middle Tertiary of southeastern Brazil. *Smithsonian Contributions to Paleobiology* **89**, 223–230.

Alvarenga, H.M.F., & Guilherme, E. (2003) The darters (Aves: Anhingidae) from the Upper Tertiary (Miocene-Pliocene) of Southwestern Amazonia. *Journal of Vertebrate Paleontology* **23**, 614–621.

Alvarenga, H.M.F., & Höfling, E. (2003) Systematic revision of the Phorusrhacidae (Aves: Ralliformes). *Papéis Avulsos de Zoologia* **43**, 55–91.

Alvarenga, H., Jones, W., & Rinderknecht, A. (2010) The youngest record of phorusrhacid birds (Aves: Phorusrhacidae) from the late Pleistocene of Uruguay. *Neues Jahrbuch für Geologie und Paläontologie, Abhandlungen* **256**, 229–234.

Alvarenga, H., Chiappe, L., & Bertelli, S. (2011) Phorusrhacids: The terror birds. In: G. Dyke & G. Kaiser (eds.), *Living Dinosaurs: The Evolutionary History of Modern Birds*. Chichester: John Wiley & Sons, pp. 187–208.

Anderson, A., Sand, C., Petchey, F., & Worthy, T.H. (2010) Faunal extinction and human habitation in New Caledonia: Initial results and implications of new research at the Pindai Caves. *Journal of Pacific Archaeology* **1**, 89–109.

Ando, T., & Fordyce, R.E. (2014) Evolutionary drivers for flightless, wing-propelled divers in the Northern and Southern Hemispheres. *Palaeogeography, Palaeoclimatology, Palaeoecology* **400**, 50–61.

Andors, A. (1992) Reappraisal of the Eocene groundbird *Diatryma* (Aves: Anserimorphae). *Natural History Museum of Los Angeles County, Science Series* **36**, 109–125.

Angst, D., Buffetaut, E., Lécuyer, C., & Amiot, R. (2013) "Terror Birds" (Phorusrhacidae) from the

Eocene of Europe imply trans-Tethys dispersal. *PLoS ONE* **8**, e80357.

Angst, D., Lécuyer, C., Amiot, R., Buffetaut, E., Fourel, F., Martineau, F., Legendre, S., Abourachid, A., & Herrel, A. (2014) Isotopic and anatomical evidence of an herbivorous diet in the Early Tertiary giant bird *Gastornis*. Implications for the structure of Paleocene terrestrial ecosystems. Naturwissenschaften **101**, 313–322.

Angst, D., Buffetaut, E., Lécuyer, C., Amiot, R., Smektala, F., Giner, S., Méchin, A., Méchin, P., Amoros, A., Leroy, L., Guiomar, M., Tomg, H., & Martinez, A. (2015) Fossil avian eggs from the Palaeogene of southern France: New size estimates and a possible taxonomic identification of the egg-layer. *Geological Magazine* **152**, 70–79.

Areta, J.I., Noriega, J.I., & Agnolín, F. (2007) A giant darter (Pelecaniformes: Anhingidae) from the Upper Miocene of Argentina and weight calculation of fossil Anhingidae. *Neues Jahrbuch für Geologie und Paläontologie, Abhandlungen* **243**, 343–350.

Arredondo, O. (1976) The great predatory birds of the Pleistocene of Cuba. *Smithsonian Contributions to Paleobiology* **27**, 169–187.

Baier, D.B., Gatesy, S.M., & Jenkins, F.A. (2006) A critical ligamentous mechanism in the evolution of avian flight. *Nature* **445**, 307–310.

Baird, R.F., & Vickers-Rich, P. (1998) *Palaelodus* (Aves: Palaelodidae) from the Middle to Late Cainozoic of Australia. *Alcheringa* **22**, 135–151.

Baker, A.J., Pereira, S.L., & Paton, T.A. (2007) Phylogenetic relationships and divergence times of Charadriiformes genera: Multigene evidence for the Cretaceous origin of at least 14 clades of shorebirds. *Biology Letters* **3**, 205–209.

Baker, A.J., Haddrath, O., McPherson, J.D., & Cloutier, A. (2014) Genomic support for a moa-tinamou clade and adaptive morphological convergence in flightless ratites. *Molecular Biology and Evolution* **31**, 1686–1696.

Balanoff, A.M., & Rowe, T. (2007) Osteological description of an embryonic skeleton of the extinct elephant bird, *Aepyornis* (Palaeognathae: Ratitae). *Journal of Vertebrate Paleontology* **27** (Supplement), 1–53.

Balanoff, A.M., Xu, X., Kobayashi, Y., Matsufune, Y., & Norell, M.A. (2009) Cranial osteology of the theropod dinosaur *Incisivosaurus gauthieri* (Theropoda: Oviraptorosauria). *American Museum Novitates* **3651**, 1–35.

Balanoff, A.M., Bever, G.S., Rowe, T.B., & Norell, M.A. (2013) Evolutionary origins of the avian brain. *Nature* **501**, 93–96.

Balanoff, A.M., Bever, G.S., & Norell, M.A. (2014) Reconsidering the avian nature of the oviraptorosaur brain (Dinosauria: Theropoda). *PLoS ONE* **9**, e113559.

Ballmann, P. (1979) Fossile Glareolidae aus dem Miozän des Nördlinger Ries (Aves: Charadriiformes). *Bonner zoologische Beiträge* **30**, 51–101.

Ballmann, P. (1983) A new species of fossil barbet (Aves: Piciformes) from the Middle Miocene of the Nördlinger Ries (Southern Germany). *Journal of Vertebrate Paleontology* **3**, 43–48.

Ballmann, P. (2004) Fossil Calidridinae (Aves: Charadriiformes) from the Middle Miocene of the Nördlinger Ries. *Bonner zoologische Beiträge* **52**, 101–114.

Barker, F.K., Cibois, A., Schikler, P., Feinstein, J., & Cracraft, J. (2004) Phylogeny and diversification of the largest avian radiation. *Proceedings of the National Academy of Sciences USA* **101**, 11040–11045.

Becker, J.J. (1985) *Pandion lovensis*, a new species of osprey from the late Miocene of Florida. *Proceedings of the Biological Society of Washington* **98**, 314–320.

Becker, J.J. (1986a) Re-identification of "*Phalacrocorax*" *subvolans* Brodkorb as the earliest record of Anhingidae. *The Auk* **103**, 804–808.

Becker, J.J. (1986b) A fossil motmot (Aves: Momotidae) from the late Miocene of Florida. *The Condor* **88**, 478–482.

Becker, J.J. (1987a) Additional material of *Anhinga grandis* Martin and Mengel (Aves: Anhingidae) from the late Miocene of Florida. *Proceedings of the Biological Society of Washington* **100**, 358–363.

Becker, J.J. (1987b) Revision of "*Falco*" *ramenta* Wetmore and the Neogene evolution of Falconidae. *The Auk* **104**, 270–276.

Bell, A., & Chiappe, L.M. (2015) Identification of a new hesperornithiform from the Cretaceous Niobrara Chalk and implications for ecologic diversity among early diving birds. *PloS ONE* **10**, e0141690.

Bell, A.K., & Chiappe, L.M. (2016) A species-level phylogeny of the Cretaceous Hesperornithiformes (Aves: Ornithuromorpha): Implications for body size evolution amongst the earliest diving birds. *Journal of Systematic Palaeontology* **14**, 239–251.

Bell, A.K., Chiappe, L.M., Erickson, G.M., Suzuki, S., Watabe, M., Barsbold, R., & Tsogtbaatar, K. (2010) Description and ecologic analysis of *Hollanda luceria*, a Late Cretaceous bird from the Gobi Desert (Mongolia). *Cretaceous Research* **31**, 16–26.

Bertelli, S., Giannini, N.P., & Ksepka, D.T. (2006) Redescription and phylogenetic position of the early Miocene Penguin *Paraptenodytes antarcticus* from Patagonia. *American Museum Novitates* **3525**, 1–36.

Bertelli, S., Chiappe, L.M., & Tambussi, C. (2007) A new phorusrhacid (Aves: Cariamae) from the middle Miocene of Patagonia, Argentina. *Journal of Vertebrate Paleontology* **27**, 409–419.

Bertelli, S., Chiappe, L.M., & Mayr, G. (2011) A new Messel rail from the Early Eocene Fur Formation of Denmark (Aves, Messelornithidae). *Journal of Systematic Palaeontology* **9**, 551–562.

Bertelli, S., Lindow, B.E.K., Dyke, G.J., & Mayr, G. (2013) Another charadriiform-like bird from the lower Eocene of Denmark. *Paleontological Journal* **47**, 1282–1301.

Bertelli, S., Chiappe, L.M., & Mayr, G. (2014) Phylogenetic interrelationships of living and extinct Tinamidae, volant palaeognathous birds from the New World. *Zoological Journal of the Linnean Society* **172**, 145–184.

Bever, G.S., Gauthier, J.A., & Wagner, G.P. (2011) Finding the frame shift: Digit loss, developmental variability, and the origin of the avian hand. *Evolution & Development* **13**, 269–279.

Bhullar, B.-A.S., Marugán-Lobón, J., Racimo, F., Bever, G.S., Rowe, T.B., Norell, M.A., & Abzhanov, A. (2012) Birds have paedomorphic dinosaur skulls. *Nature* **487**, 223–226.

Bibi, F., Shabel, A.B., Kraatz, B.P., & Stidham, T.A. (2006) New fossil ratite (Aves: Palaeognathae) eggshell discoveries from the Late Miocene Baynunah Formation of the United Arab Emirates, Arabian Peninsula. *Palaeontologia Electronica* **9.1.2A**, 1–13.

Bickart, K.J. (1990) The birds of the Late Miocene–Early Pliocene Big Sandy Formation, Mohave County, Arizona. *Ornithological Monographs* **44**, 1–72.

Blackburn, D.G., & Evans, H.E. (1986) Why are there no viviparous birds? *American Naturalist* **128**, 165–190.

Blaszyk, P. (1935) Untersuchungen über die Stammesgeschichte der Vogelschuppen und Federn und über die Abhängigkeit ihrer Ausbildung am Vogelfuß von der Funktion. *Morphologisches Jahrbuch* **75**, 483–567.

Bleiweiss, R. (1998) Tempo and mode of hummingbird evolution. *Biological Journal of the Linnean Society* **65**, 63–76.

Bocheński, Z., & Bocheński, Z.M. (2008) An Old World hummingbird from the Oligocene: A new fossil from Polish Carpathians. *Journal of Ornithology* **149**, 211–216.

Bocheński, Z.M., Tomek, T., Bujoczek, M., & Wertz, K. (2011) A new passerine bird from the early Oligocene of Poland. *Journal of Ornithology* **152**, 1045–1053.

Bochenski, Z.M., Tomek, T., Wertz, K., & Swidnicka, E. (2013) The third nearly complete passerine bird from the early Oligocene of Europe. *Journal of Ornithology* **154**, 923–931.

Boessenecker, R.W., & Smith, N.A. (2011) Latest Pacific Basin record of a bony-toothed bird (Aves, Pelagornithidae) from the Pliocene Purisima Formation of California, U.S.A. *Journal of Vertebrate Paleontology* **31**, 652–657.

Boev, Z. (2002) Fossil record and disappearance of peafowl (*Pavo* Linnaeus) from the Balkan Peninsula and Europe (Aves: Phasianidae). *Historia Naturalis Bulgarica* **14**, 109–115.

Boev, Z. (2011) *Falco bulgaricus* sp. n. (Aves: Falconiformes) from the Late Miocene of Hadzhidimovo (SW Bulgaria). *Acta Zoologica Bulgarica* **63**, 17–35.

Boev, Z., & Kovachev, D. (2007) *Euroceros bulgaricus* gen. nov., sp. nov. from Hadzhidimovo (SW Bulgaria) (Late Miocene) – the first European record of hornbills (Aves: Coraciiformes). *Geobios* **40**, 39–49.

Boev, Z., & Spassov, N. (2009) First record of ostriches (Aves, Struthioniformes, Struthionidae) from the late Miocene of Bulgaria with taxonomic and zoogeographic discussion. *Geodiversitas* **31**, 493–507.

Boev, Z., Lazaridis, G., & Tsoukala, E. (2013) *Otis hellenica* sp. nov., a new Turolian bustard (Aves: Otididae) from Kryopigi (Chalkidiki, Greece). *Geologica Balcanica* **42**, 59–65.

Boles, W.E. (1993a) *Pengana robertbolesi*, a peculiar bird of prey from the Tertiary of Riversleigh, northwestern Queensland, Australia. *Alcheringa* **17**, 19–25.

Boles, W.E. (1993b) A new cockatoo (Psittaciformes: Cacatuidae) from the Tertiary of Riversleigh, northwestern Queensland, and an evaluation of rostral characters in the systematics of parrots. *The Ibis* **135**, 8–18.

Boles, W.E. (1995a) The world's oldest songbird. *Nature* **374**, 21–22.

Boles, W.E. (1995b) A preliminary analysis of the Passeriformes from Riversleigh, N.W. Queensland, Australia, with the description of a new species of lyrebird. *Courier Forschungsinstitut Senckenberg* **181**, 163–170.

Boles, W.E. (1997a) A kingfisher (Halcyonidae) from the Miocene of Riversleigh, northwestern Queensland, with comments on the evolution of kingfishers in Australo-Papua. *Memoirs of the Queensland Museum* **41**, 229–234.

Boles, W.E. (1997b) Hindlimb proportions and locomotion of *Emuarius gidju* (Patterson & Rich, 1987) (Aves: Casuariidae). *Memoirs of the Queensland Museum* **41**, 235–240.

Boles, W.E. (1997c) Fossil songbirds (Passeriformes) from the early Eocene of Australia. *Emu* **97**, 43–50.

Boles, W.E. (2001) A swiftlet (Apodidae: Collocaliini) from the Oligo-Miocene of Riversleigh, northwestern Queensland. *Memoir of the Association of Australasian Palaeontologists* **25**, 45–52.

Boles, W.E. (2005) A review of the Australian fossil storks of the genus *Ciconia* (Aves: Ciconiidae), with the description of a new species. *Records of the Australian Museum* **57**, 165–178.

Boles, W.E. (2008) Systematics of the fossil Australian giant megapodes *Progura* (Aves: Megapodiidae). *Orytos* **7**, 195–215.

Boles, W.E., & Ivison, T.J. (1999) A new dwarf megapode (Galliformes: Megapodiidae) from the late Oligocene of Central Australia. *Smithsonian Contributions to Paleobiology* **89**, 199–206.

Bond, W.J., & Silander, J.A. (2007) Springs and wire plants: Anachronistic defences against Madagascar's extinct elephant birds. *Proceedings of the Royal Society B: Biological Sciences* **274**, 1985–1992.

Botelho, J.F., Ossa-Fuentes, L., Soto-Acuña, S., Smith-Paredes, D., Nuñez-León, D., Salinas-Saavedra, M., Ruiz-Flores, M., & Vargas, A.O. (2014a) New developmental evidence clarifies the evolution of wrist bones in the dinosaur-bird transition. *PLoS Biology* **12**, e1001957.

Botelho, J.F., Smith-Paredes, D., Nuñez-Leon, D., Soto-Acuña, S., & Vargas, A.O. (2014b) The developmental origin of zygodactyl feet and its possible loss in the evolution of Passeriformes. *Proceedings of the Royal Society of London, Series B* **281**, 20140765.

Bourdon, E. (2005) Osteological evidence for sister group relationship between pseudo-toothed birds (Aves: Odontopterygiformes) and waterfowls (Anseriformes). *Naturwissenschaften* **92**, 586–591.

Bourdon, E., & Cappetta, H. (2012) Pseudo-toothed birds (Aves, Odontopterygiformes) from the Eocene phosphate deposits of Togo, Africa. *Journal of Vertebrate Paleontology* **32**, 965–970.

Bourdon, E., Bouya, B., & Iarochène, M. (2005) Earliest African neornithine bird: A new species of Prophaethontidae (Aves) from the Paleocene of Morocco. *Journal of Vertebrate Paleontology* **25**, 157–170.

Bourdon, E., Mourer-Chauviré, C., Amaghzaz, M., & Bouya, B. (2008a) New specimens of *Lithoptila abdounensis* (Aves, Prophaethontidae) from the Lower Paleogene of Morocco. *Journal of Vertebrate Paleontology* **28**, 751–761.

Bourdon, E., Amaghzaz, M., & Bouya, B. (2008b) A new seabird (Aves, cf. Phaethontidae) from the Lower Eocene phosphates of Morocco. *Geobios* **41**, 455–459.

Bourdon, E., Castanet, J., De Ricqles, A., Scofield, P., Tennyson, A., Lamrous, H., & Cubo, J. (2009a) Bone growth marks reveal protracted growth in New Zealand kiwi (Aves, Apterygidae). *Biology Letters* **5**, 639–642.

Bourdon, E., De Ricqles, A., & Cubo, J. (2009b) A new transantarctic relationship: Morphological evidence for a Rheidae-Dromaiidae-Casuariidae clade (Aves, Palaeognathae, Ratitae). *Zoological Journal of the Linnean Society* **156**, 641–663.

Bourdon, E., Amaghzaz, M., & Bouya, B. (2010) Pseudotoothed birds (Aves, Odontopterygiformes) from the early Tertiary of Morocco. *American Museum Novitates* **3704**, 1–71.

Bourdon, E., Mourer-Chauviré, C., & Laurent, Y. (2016) Early Eocene birds from La Borie, southern France. *Acta Palaeontologica Polonica* **61**, 175–190.

Brennan, P.L.R., Birkhead, T.R., Zyskowski, K., van der Waag, J., & Prum, R.O. (2008) Independent evolutionary reductions of the phallus in basal birds. *Journal of Avian Biology* **39**, 487–492.

Brightsmith, D.J. (2005) Competition, predation and nest niche shifts among tropical cavity nesters: Phylogeny and natural history evolution of parrots (Psittaciformes) and trogons (Trogoniformes). *Journal of Avian Biology* **36**, 64–73.

Brikiatis, L. (2014) The De Geer, Thulean and Beringia routes: Key concepts for understanding early Cenozoic biogeography. *Journal of Biogeography* **41**, 1036–1054.

Brunet, J. (1971) Oiseaux miocènes de Beni Mellal (Maroc); un complément à leur étude. *Notes et Mémoires du Service Géologique du Maroc* **31**, 109–111.

Buffetaut, E. (2010) *Gargantuavis philoinos*: Giant bird or giant pterosaur? *Annales de Paléontologie* **96**, 135–141.

Buffetaut, E. (2013) The giant bird *Gastornis* in Asia: A revision of *Zhongyuanus xichuanensis* Hou, 1980, from the early Eocene of China. *Paleontological Journal* **47**, 1302–1307.

Buffetaut, E., & Angst, D. (2013) New evidence of a giant bird from the Late Cretaceous of France. *Geological Magazine* **150**, 173–176.

Buffetaut, E., & Angst, D. (2014) Stratigraphic distribution of large flightless birds in the Palaeogene of Europe and its palaeobiological and palaeogeographical implications. *Earth-Science Reviews* **138**, 394–408.

Bunce, M., Worthy, T.H., Ford, T., Hoppitt, W., Willerslev, E., Drummond, A., & Cooper, A. (2003) Extreme reversed sexual size dimorphism in the extinct New Zealand moa *Dinornis*. *Nature* **425**, 172–175.

Burgers, P., & Chiappe, L.M. (1999) The wing of *Archaeopteryx* as a primary thrust generator. *Nature* **399**, 60–62.

Burnham, D.A. (2004) New information on *Bambiraptor feinbergi* from the Late Cretaceous of Montana. In: P.J. Currie, E.B. Koppelhus, M.A. Shugar, & J.L. Wright (eds.), *Feathered Dragons: Studies on the Transition from Dinosaurs to Birds*. Bloomington: Indiana University Press, pp. 67–111.

Campbell, K.E. (1996) A new species of giant anhinga (Aves: Pelecaniformes: Anhingidae) from the upper Miocene (Huayquerian) of Amazonian Peru. *Natural History Museum of Los Angeles County, Contributions in Science* **460**, 1–9.

Campbell, K.E., & Tonni, E.P. (1980) A new genus of teratorn from the Huayquerian of Argentina (Aves: Teratornithidae). *Natural History Museum of Los Angeles County, Contributions in Science* **330**, 59–68.

Campbell, K.E., & Tonni, E.P. (1983) Size and locomotion in teratorns (Aves: Teratornithidae). *The Auk* **100**, 390–403.

Carney, R.M., Vinther, J., Shawkey, M.D., D'Alba, L., & Ackermann, J. (2012) New evidence on the colour and nature of the isolated *Archaeopteryx* feather. *Nature Communications* **3**, 637.

Cau, A., Brougham, T., & Naish, D. (2015) The phylogenetic affinities of the bizarre Late Cretaceous Romanian theropod *Balaur bondoc* (Dinosauria, Maniraptora): Dromaeosaurid or flightless bird? *PeerJ* **3**, e1032.

Cenizo, M.M. (2012) Review of the putative Phorusrhacidae from the Cretaceous and Paleogene of Antarctica: New records of ratites and pelagornithid birds. *Polish Polar Research* **3**, 225–244.

Cenizo, M.M., & Agnolín, F.L. (2010) The southernmost records of Anhingidae and a new basal species of Anatidae (Aves) from the lower–middle Miocene of Patagonia, Argentina. *Alcheringa* **34**, 493–514.

Cenizo, M.M., Tambussi, C.P., & Montalvo, C.I. (2012) Late Miocene continental birds from the Cerro Azul Formation in the Pampean region (central-southern Argentina). *Alcheringa* **36**, 47–68.

Cenizo, M.M., Noriega, J.I., & Reguero, M.A. (2016) A stem falconid bird from the Lower Eocene of Antarctica and the early southern radiation of the falcons. *Journal of Ornithology* **157**, 885–894.

Chandler, R.M., & Parmley, D. (2003) The earliest North American record of auk (Aves: Alcidae) from the Late Eocene of Central Georgia. *Oriole* **68**, 7–9.

Chatterjee, S. (2002) The morphology and systematics of *Polarornis*, a Cretaceous loon (Aves: Gaviidae) from Antarctica. In: Z.-H. Zhou & F.-Z. Zhang (eds.), *Proceedings of the 5th Symposium of the Society of Avian Paleontology and Evolution, 1-4 June 2000.* Beijing: Science Press, pp. 125–155.

Chatterjee, S. (2015) *The Rise of Birds: 225 Million Years of Evolution*, 2nd edn. Baltimore, MD: Johns Hopkins University Press.

Chatterjee, S., & Templin, R.J. (2003) The flight of *Archaeopteryx. Naturwissenschaften* **90**, 27–32.

Chatterjee, S., & Templin, R.J. (2007) Biplane wing planform and flight performance of the feathered dinosaur *Microraptor gui. Proceedings of the National Academy of Sciences USA* **104**, 1576–1580.

Chatterjee, S.R., Templin J., & Campbell, K.E. (2007) The aerodynamics of *Argentavis*, the world's largest flying bird from the Miocene of Argentina. *Proceedings of the National Academy of Sciences USA* **104**, 12398–12403.

Cheneval, J. (1984) Les oiseaux aquatiques (Gaviiformes à Ansériformes) du gisement aquitanien de Saint-Gérand-le-Puy (Allier, France): Révision systématique. *Palaeovertebrata* **14**, 33–115.

Cheneval, J. (2000) L'avifaune de Sansan. *Mémoires du Muséum national d'histoire naturelle* **183**, 321–388.

Cheneval, J., & Escuillié, F. (1992) New data concerning *Palaelodus ambiguus* (Aves: Phoenicopteriformes: Palaelodidae): Ecological and evolutionary interpretations. *Natural History Museum of Los Angeles County, Science Series* **36**, 208–224.

Cheneval, J., Ginsburg, L., Mourer-Chauviré, C., & Ratanasthien, B. (1991) The Miocene avifauna of the Li Mae Long locality, Thailand: Systematics and paleoecology. *Journal of Southeast Asian Earth Sciences* **6**, 117–126.

Chiappe, L.M. (1991) Fossil birds from the Miocene Pinturas Formation of southern Argentina. *Journal of Vertebrate Paleontology* **11**, 21–22.

Chiappe, L.M. (2002) Osteology of the flightless *Patagopteryx deferrariisi* from the Late Cretaceous of Patagonia (Argentina). In: L.M. Chiappe & L.M. Witmer (eds.), *Mesozoic Birds: Above the Heads of Dinosaurs.* Berkeley: University of California Press, pp. 281–316.

Chiappe, L.M. (2007) *Glorified Dinosaurs: The Origin and Early Evolution of Birds.* Hoboken, NJ: John Wiley & Sons.

Chiappe, L.M., & Walker, C.A. (2002) Skeletal morphology and systematics of the Cretaceous Euenantiornithes (Ornithothoraces: Enantiornithes). In: L.M. Chiappe & L.M. Witmer (eds.), *Mesozoic Birds: Above the Heads of Dinosaurs.* Berkeley: University of California Press, pp. 240–267.

Chiappe, L.M., Ji, S.-A., Ji, Q., & Norell, M.A. (1999) Anatomy and systematics of the Confuciusornithidae (Theropoda, Aves) from the late Mesozoic of northeastern China. *Bulletin of the American Museum of Natural History* **242**, 1–89.

Chiappe, L.M., Norell, M., & Clark, J. (2001) A new skull of *Gobipteryx minuta* (Aves: Enantiornithes) from the Cretaceous of the Gobi Desert. *American Museum Novitates* **3346**, 1–15.

Chiappe, L.M., Ji, S.-A., & Ji, Q. (2007a) Juvenile birds from the Early Cretaceous of China: Implications for enantiornithine ontogeny. *American Museum Novitates* **3594**, 1–46.

Chiappe, L.M., Suzuki, S., Dyke, G.J., Watabe, M., Tsogtbaatar, K., & Barsbold, R. (2007b) A new enantiornithine bird from the Late Cretaceous of the Gobi Desert. *Journal of Systematic Palaeontology* **5**, 193–208.

Chiappe, L.M., Marugán-Lobón, J., & Zhou, Z. (2008) Life history of a basal bird: Morphometrics of the Early Cretaceous *Confuciusornis. Biology Letters* **4**, 719–723.

Chiappe, L.M., Zhao, B., O'Connor, J.K., Chunling, G., Wang, X., Habib, M., Marugan-Lobon, J., Meng, Q., & Cheng, X. (2014) A new specimen of the Early Cretaceous bird *Hongshanornis longicresta*: Insights into the aerodynamics and diet of a basal ornithuromorph. *PeerJ* **2**, e234.

Chinsamy, A., & Elzanowski, A. (2001) Bone histology: Evolution of growth pattern in birds. *Nature* **412**, 402–403.

Chinsamy, A., Chiappe, L.M., & Dodson, P. (1994) Growth rings in Mesozoic birds. *Nature* **368**, 196–197.

Chinsamy, A., Chiappe, L.M., Marugán-Lobón, J., Gao, C., & Zhang, F. (2013) Gender identification of the Mesozoic bird *Confuciusornis sanctus*. *Nature Communications* **4**, 1381.

Claessens, L.P. (2004) Dinosaur gastralia; origin, morphology, and function. *Journal of Vertebrate Paleontology* **24**, 89–106.

Claramunt, S., & Cracraft, J. (2015) A new time tree reveals Earth history's imprint on the evolution of modern birds. *Science Advances* **1**, e1501005.

Clarke, J.A. (2004) Morphology, phylogenetic taxonomy, and systematics of *Ichthyornis* and *Apatornis* (Avialae: Ornithurae). *Bulletin of the American Museum of Natural History* **286**, 1–179.

Clarke, J.A., & Norell, M.A. (2002) The morphology and phylogenetic position of *Apsaravis ukhaana* from the Late Cretaceous of Mongolia. *American Museum Novitates* **3387**, 1–46.

Clarke, J.A., Tambussi, C.P., Noriega, J.I., Erickson, G.M., & Ketcham, R.A. (2005a) Definitive fossil evidence for the extant avian radiation in the Cretaceous. *Nature* **433**, 305–308.

Clarke, J.A., Norell, M.A., & Dashzeveg, D. (2005b) New avian remains from the Eocene of Mongolia and the phylogenetic position of the Eogruidae (Aves, Gruoidea). *American Museum Novitates* **3494**, 1–17.

Clarke, J.A., Zhou, Z., & Zhang, F. (2006) Insight into the evolution of avian flight from a new clade of Early Cretaceous ornithurines from China and the morphology of *Yixianornis grabaui*. *Journal of Anatomy* **208**, 287–308.

Clarke, J.A., Ksepka, D.T., Stucchi, M., Urbina, M., Giannini, N., Bertelli, S., Narváez, Y., & Boyd, C.A. (2007) Paleogene equatorial penguins challenge the proposed relationship between penguin biogeography, diversity, and Cenozoic climate change. *Proceedings of the National Academy of Sciences USA* **104**, 11545–11550.

Clarke, J.A., Ksepka, D.T., Smith N.A., & Norell, M.A. (2009) Combined phylogenetic analysis of a new North American fossil species confirms widespread Eocene distribution rollers (Aves, Coracii). *Zoological Journal of the Linnean Society* **157**, 586–611.

Clarke, J.A., Ksepka, D.T., Salas-Gismondi, R., Altamirano, A.J., Shawkey, M.D., D'Alba, L., Vinther, J., DeVries, T.J., & Baby, P. (2010) Fossil evidence for evolution of the shape and color of penguin feathers. *Science* **330**, 954–957.

Clarke, S.J., Miller, G.H., Fogel, M.L., Chivas, A.R., & Murray-Wallace, C.V. (2006) The amino acid and stable isotope biogeochemistry of elephant bird (*Aepyornis*) eggshells from southern Madagascar. *Quaternary Science Reviews* **25**, 2343–2356.

Clegg, S.M., & Owens, P.F. (2002) The 'island rule' in birds: Medium body size and its ecological explanation. *Proceedings of the Royal Society of London. Series B: Biological Sciences* **269**, 1359–1365.

Close, R.A., & Rayfield, E.J. (2012) Functional morphometric analysis of the furcula in Mesozoic birds. *PLoS ONE* **7**, e36664.

Close, R.A., Vickers-Rich, P., Trusler, P., Chiappe, L.M., O'Connor, J., Rich, T.H., Kool, L., & Komarower, P. (2009) Earliest Gondwanan bird from the Cretaceous of southeastern Australia. *Journal of Vertebrate Paleontology* **29**, 616–619.

Cooper, A., Lalueza-Fox, C., Anderson, S., Rambaut, A., Austin, J., & Ward, R. (2001) Complete mitochondrial genome sequences of two extinct moas clarify ratite evolution. *Nature* **409**, 704–707.

Corlett, R.T., & Primack, R.B. (2006) Tropical rainforests and the need for cross-continental comparisons. *Trends in Ecology & Evolution* **21**, 104–110.

Cracraft, J. (1968) A review of the Bathornithidae (Aves, Gruiformes), with remarks on the relationships of the suborder Cariamae. *American Museum Novitates* **2326**, 1–46.

Cracraft, J. (2001) Avian evolution, Gondwana biogeography and the Cretaceous-Tertiary mass extinction event. *Proceedings of the Royal Society of London, Series B* **268**, 459–469.

Czerkas, S.A., & Ji, Q. (2002) A preliminary report on an omnivorous volant bird from northeast China. *Dinosaur Museum Journal* **1**, 127–135.

Davis, P.C. (2003) The oldest record of the genus *Diomedea, Diomedea tanakai* sp. nov. (Procellariiformes: Diomedeidae): An albatross from the Miocene of Japan. *Bulletin of the*

National Science Museum, Tokyo, Series C **29**, 39–48.

De Pietri, V.L. (2013) Interrelationships of the Threskiornithidae and the phylogenetic position of the Miocene ibis "*Plegadis*" *paganus* from the Saint-Gérand-le-Puy area in central France. *The Ibis* **155**, 544–560.

De Pietri, V.L., & Mayr, G. (2012) An assessment of the diversity of early Miocene Scolopaci (Aves, Charadriiformes) from Saint-Gérand-le-Puy (Allier, France). *Palaeontology* **55**, 1177–1197.

De Pietri, V.L., & Mayr, G. (2014a) The phylogenetic relationships of the Early Miocene stork *Grallavis edwardsi*, with comments on the interrelationships of living Ciconiidae (Aves). *Zoologica Scripta* **43**, 576–585.

De Pietri, V.L., & Mayr, G. (2014b) Reappraisal of early Miocene rails (Aves, Rallidae) from central France: Diversity and character evolution. *Journal of Zoological Systematics and Evolutionary Research* **52**, 312–322.

De Pietri, V.L., & Mayr, G. (2014c) The enigmatic *Ibidopodia* from the early Miocene of France – the first Neogene record of Cariamiformes (Aves) in Europe. *Journal of Vertebrate Paleontology* **34**, 1470–1475.

De Pietri, V.L., & Scofield, R.P. (2014) The earliest European record of a Stone-curlew (Charadriiformes, Burhinidae) from the late Oligocene of France. *Journal of Ornithology* **155**, 421–426.

De Pietri, V.L., Berger, J.-P., Pirkenseer, C., Scherler, L., & Mayr, G. (2010) New skeleton from the early Oligocene of Germany indicates a stem-group position of diomedeoidid birds. *Acta Palaeontologica Polonica* **55**, 23–34.

De Pietri, V.L., Costeur, L., Güntert, M., & Mayr, G. (2011a) A revision of the Lari (Aves: Charadriiformes) from the early Miocene of Saint-Gérand-le-Puy (Allier, France). *Journal of Vertebrate Paleontology* **31**, 812–828.

De Pietri, V.L., Manegold, A., Costeur, L., & Mayr, G. (2011b) A new species of woodpecker (Aves; Picidae) from the early Miocene of Saulcet (Allier, France). *Swiss Journal of Palaeontology* **130**, 307–314.

De Pietri, V.L., Güntert, M., & Mayr, G. (2013a) A *Haematopus*-like skull and other remains of Charadrii (Aves, Charadriiformes) from the early Miocene of Saint-Gérand-le-Puy (Allier, France).

In: U.B. Göhlich & A. Kroh (eds.), *Paleornithological Research 2013 – Proceedings of the 8th International Meeting of the Society of Avian Paleontology and Evolution*. Vienna: Natural History Museum Vienna, pp. 93–101.

De Pietri, V.L., Mourer-Chauviré, C., Menkveld-Gfeller, U., Meyer, C.A., & Costeur, L. (2013b) An assessment of the Cenozoic avifauna of Switzerland, with a description of two fossil owls (Aves, Strigiformes). *Swiss Journal of Geosciences* **106**, 187–197.

De Pietri, V.L., Camens, A.B., & Worthy, T.H. (2015) A Plains-wanderer (Pedionomidae) that did not wander plains: A new species from the Oligocene of South Australia. *The Ibis* **157**, 68–74.

De Pietri, V.L., Scofield, R.P., Tennyson, A.J., Hand, S.J., & Worthy, T.H. (2016) Wading a lost southern connection: Miocene fossils from New Zealand reveal a new lineage of shorebirds (Charadriiformes) linking Gondwanan avifaunas. *Journal of Systematic Palaeontology* **14**, 603–616.

de Souza Carvalho, I., Novas, F.E., Agnolín, F.L., Isasi, M.P., Freitas, F.I., & Andrade, J.A. (2015) A Mesozoic bird from Gondwana preserving feathers. *Nature Communications* **6**, 7141.

Dececchi, T.A., & Larsson, H.C. (2011) Assessing arboreal adaptations of bird antecedents: Testing the ecological setting of the origin of the avian flight stroke. *PLoS ONE* **6**, e22292.

Deeming, D.C., & Ruta, M. (2014) Egg shape changes at the theropod–bird transition, and a morphometric study of amniote eggs. *Royal Society Open Science* **1**, 140311.

Degrange, F.J., & Tambussi, C.P. (2011) Re-examination of *Psilopterus lemoinei* (Aves, Phorusrhacidae), a late early Miocene little terror bird from Patagonia (Argentina). *Journal of Vertebrate Paleontology* **31**, 1080–1092.

Degrange, F.J., Tambussi, C.P., Moreno, K., Witmer, L.M., & Wroe, S. (2010) Mechanical analysis of feeding behavior in the extinct "terror bird" *Andalgalornis steulleti* (Gruiformes: Phorusrhacidae). *PLoS ONE* **5**, e11856.

Degrange, F.J, Noriega, J.I., & Areta, J.I. (2012) Diversity and paleobiology of the Santacrucian birds. In: S.F. Vizcaino, R.F. Kay, & M.S. Bargo (eds.), *Early Miocene Paleobiology in Patagonia: High-Latitude Paleocommunities of the Santa*

Cruz Formation. Cambridge: Cambridge University Press, pp. 138–155.

Degrange, F.J., Tambussi, C.P., Taglioretti, M.L., Dondas, A., & Scaglia, F. (2015a) A new Mesembriornithinae (Aves, Phorusrhacidae) provides new insights into the phylogeny and sensory capabilities of terror birds. *Journal of Vertebrate Paleontology* **35**, e912656.

Degrange, F.J., Noriega, J.I., & Vizcaíno, S.F. (2015b) Morphology of the forelimb of *Psilopterus bachmanni* (Aves, Cariamiformes) (early Miocene of Patagonia). *Paläontologische Zeitschrift* **89**, 1087–1096.

Dial, K.P. (2003a) Wing-assisted incline running and the evolution of flight. *Science* **299**, 402–404.

Dial, K.P. (2003b) Evolution of avian locomotion: Correlates of flight style, locomotor modules, nesting biology, body size, development, and the origin of flapping flight. *The Auk* **120**, 941–952.

Domínguez Alonso, P.D., Milner, A.C., Ketcham, R.A., Cookson, M.J., & Rowe, T.B. (2004) The avian nature of the brain and inner ear of *Archaeopteryx. Nature* **430**, 666–669.

Donne-Goussé, C., Laudet, V., & Hänni, C. (2002) A molecular phylogeny of anseriformes based on mitochondrial DNA analysis. *Molecular Phylogenetics and Evolution* **23**, 339–356.

Dove, C.J., & Olson, S.L. (2011) Fossil feathers from the Hawaiian flightless ibis (*Apteribis* sp.): Plumage coloration and systematics of a prehistorically extinct bird. *Journal of Paleontology* **85**, 892–897.

Dwyer, G.S., & Chandler, M.A. (2009) Mid-Pliocene sea level and continental ice volume based on coupled benthic Mg/Ca palaeotemperatures and oxygen isotopes. *Philosophical Transactions of the Royal Society A: Mathematical, Physical and Engineering Sciences* **367**, 157–168.

Dyke, G.J., & Kaiser, G.W. (2010) Cracking a developmental constraint: Egg size and bird evolution. *Records of the Australian Museum* **62**, 207–216.

Dyke, G.J., & Mayr, G. (1999) Did parrots exist in the Cretaceous period? *Nature* **399**, 317–318.

Dyke, G.J., & Walker, C.A. (2008) New records of fossil "waterbirds" from the Miocene of Kenya. *American Museum Novitates* **3610**, 1–12.

Dyke, G.J., Malakhov, D.V., & Chiappe, L.M. (2006) A re-analysis of the marine bird *Asiahesperornis* from northern Kazakhstan. *Cretaceous Research* **27**, 947–953.

Dyke, G.J., Nudds, R.L., & Walker, C.A. (2007) The Pliocene *Phoebastria* ("*Diomedea*") *anglica*: Lydekker's English fossil albatross. *The Ibis* **149**, 626–631.

Dyke, G.J., Wang, X., & Habib, M.B. (2011) Fossil plotopterid seabirds from the Eo-Oligocene of the Olympic Peninsula (Washington State, USA): Descriptions and functional morphology. *PLoS ONE* **6**, e25672.

Dyke, G., de Kat, R., Palmer, C., van der Kindere, J., Naish, D., & Ganapathisubramani, B. (2013) Aerodynamic performance of the feathered dinosaur *Microraptor* and the evolution of feathered flight. *Nature Communications* **4**, 2489.

Eberle, J.J., & Greenwood, D.R. (2012) Life at the top of the greenhouse Eocene world: A review of the Eocene flora and vertebrate fauna from Canada's High Arctic. *GSA Bulletin* **124**, 3–23.

Eldrett, J.S., Greenwood, D.R., Harding, I.C., & Huber, M. (2009) Increased seasonality through the Eocene to Oligocene transition in northern high latitudes. *Nature* **459**, 969–973.

Elzanowski, A. (1991) New observations on the skull of *Hesperornis* with reconstructions of the bony palate and otic region. *Postilla* **207**, 1–20.

Elzanowski, A. (1999) A comparison of the jaw skeleton in theropods and birds, with a description of the palate in the Oviraptoridae. *Smithsonian Contributions to Paleobiology* **89**, 311–323.

Elzanowski, A. (2002) Archaeopterygidae (Upper Jurassic of Germany). In: L.M. Chiappe & L.M. Witmer (eds.), *Mesozoic Birds: Above the Heads of Dinosaurs.* Berkeley: University of California Press, pp. 129–159.

Elzanowski, A., & Boles, W.E. (2012) Australia's oldest anseriform fossil: A quadrate from the Early Eocene Tingamarra Fauna. *Palaeontology* **55**, 903–911.

Elzanowski, A., & Galton, P.M. (1991) Braincase of *Enaliornis*, an early Cretaceous bird from England. *Journal of Vertebrate Paleontology* **11**, 90–107.

Elzanowski, A., & Stidham, T.A. (2010) Morphology of the quadrate in the Eocene anseriform *Presbyornis* and extant galloanserine birds. *Journal of Morphology* **271**, 305–323.

Elzanowski, A., & Stidham, T.A. (2011) A galloanserine quadrate from the Late Cretaceous Lance Formation of Wyoming. *The Auk* **128**, 138–145.

Elzanowski, A., & Wellnhofer, P. (1996) Cranial morphology of *Archaeopteryx*: Evidence from the seventh skeleton. *Journal of Vertebrate Paleontology* **16**, 81–94.

Emslie, S.D. (1988) An early condor-like vulture from North America. *The Auk* **105**, 529–535.

Erickson, G.M., Rauhut, O.W., Zhou, Z., Turner, A.H., Inouye, B.D., Hu, D., & Norell, M.A. (2009) Was dinosaurian physiology inherited by birds? Reconciling slow growth in *Archaeopteryx*. *PLoS ONE* **4**, e7390.

Ericson, P.G.P. (1997) Systematic relationships of the palaeogene family Presbyornithidae (Aves: Anseriformes). *Zoological Journal of the Linnean Society* **121**, 429–483.

Ericson, P.G.P. (2000) Systematic revision, skeletal anatomy, and paleoecology of the New World early Tertiary Presbyornithidae (Aves: Anseriformes). *PaleoBios* **20**, 1–23.

Ericson, P.G.P., Irestedt, M., & Johansson, U.S. (2003) Evolution, biogeography, and patterns of diversification in passerine birds. *Journal of Avian Biology* **34**, 3–15.

Ericson, P.G.P., Anderson, C.L., Britton, T., Elzanowski, A., Johansson, U.S., Källersjö, M., Ohlson, J.I., Parsons, T.J., Zuccon, D., & Mayr, G. (2006) Diversification of Neoaves: Integration of molecular sequence data and fossils. *Biology Letters* **2**, 543–547.

Ericson, P.G.P., Klopfstein, S., Irestedt, M., Nguyen, J.M.T., & Nylander, J.A.A. (2014) Dating the diversification of the major lineages of Passeriformes (Aves). *BMC Evolutionary Biology* **14**, 8.

Fain, M.G., & Houde, P. (2004) Parallel radiations in the primary clades of birds. *Evolution* **58**, 2558–2573.

Fain, M.G., & Houde, P. (2007) Multilocus perspectives on the monophyly and phylogeny of the order Charadriiformes (Aves). *BMC Evolutionary Biology* **7**, 35.

Feduccia, A. (1993) Evidence from claw geometry indicating arboreal habits of Archaeopteryx. *Science* **259**, 790–793.

Feduccia, A. (1999) *The Origin and Evolution of Birds*, 2nd edn. New Haven, CT: Yale University Press.

Feduccia, A. (2003) "Big bang" for Tertiary birds? *Trends in Ecology and Evolution* **18**, 172–176.

Feduccia, A. (2012) *Riddle of the Feathered Dragons: Hidden Birds of China*. New Haven, CT: Yale University Press.

Feduccia, A. (2014) Avian extinction at the end of the Cretaceous: Assessing the magnitude and subsequent explosive radiation. *Cretaceous Research* **50**, 1–15.

Feduccia, A., & Czerkas, S.A. (2015) Testing the neoflightless hypothesis: Propatagium reveals flying ancestry of oviraptorosaurs. *Journal of Ornithology* **156**, 1067–1074.

Feduccia, A., & Voorhies, M.R. (1989) Miocene hawk converges on Secretarybird. *The Ibis* **131**, 349–354.

Feduccia, A., & Voorhies, M.R. (1992) Crowned cranes (Gruidae: *Balearica*) in the Miocene of Nebraska. *Natural History Museum of Los Angeles County, Science Series* **36**, 239–248.

Feo, T.J., Field, D.J., & Prum, R.O. (2015) Barb geometry of asymmetrical feathers reveals a transitional morphology in the evolution of avian flight. *Proceedings of the Royal Society of London B: Biological Sciences* **282**, 20142864.

Fitzgerald, E.M.G., Park, T., & Worthy, T.H. (2012) First giant bony-toothed bird (Pelagornithidae) from Australia. *Journal of Vertebrate Paleontology* **32**, 971–974.

Fordyce, R.E., & Thomas, D. (2011) *Kaiika maxwelli*, a new Early Eocene archaic penguin (Sphenisciformes, Aves) from Waihao Valley, South Canterbury, New Zealand. *New Zealand Journal of Geology and Geophysics* **54**, 43–51.

Forster, C.A., Sampson, S.D., Chiappe, L.M., & Krause, D.W. (1998) The theropod ancestry of birds: New evidence from the Late Cretaceous of Madagascar. *Science* **279**, 1915–1919.

Forster, C.A., Chiappe, L.M., Krause, D.W., & Sampson, S.D. (2002) *Vorona berivotrensis*, a primitive bird from the Late Cretaceous of Madagascar. In: L.M. Chiappe & L.M. Witmer (eds.), *Mesozoic Birds: Above the Heads of Dinosaurs*. Berkeley: University of California Press, pp. 268–280.

Foth, C., Tischlinger, H., & Rauhut, O. (2014) New specimen of *Archaeopteryx* provides insights into the evolution of pennaceous feathers. *Nature* **511**, 79–82.

Fowler, D.W., Freedman, E.A., Scannella, J.B., & Kambic, R.E. (2011) The predatory ecology of

Deinonychus and the origin of flapping in birds. *PLoS ONE* **6**, e28964.

Friis, E.M., Crane, P.R., & Pedersen, K.R. (2011) *Early Flowers and Angiosperm Evolution.* Cambridge: Cambridge University Press.

Fuchs, J., Johnson, J.A., & Mindell, D.P. (2015) Rapid diversification of falcons (Aves: Falconidae) due to expansion of open habitats in the Late Miocene. *Molecular Phylogenetics and Evolution* **82**, 166–182.

Gaff, P., & Boles, W.E. (2010) A new eagle (Aves: Accipitridae) from the mid Miocene Bullock Creek Fauna of Northern Australia. *Records of the Australian Museum* **62**, 71–76.

Galton, P.M., & Martin, L.D. (2002) *Enaliornis*, an Early Cretaceous hesperornithiform bird from England, with comments on other Hesperornithiformes. In: L.M. Chiappe & L.M. Witmer (eds.), *Mesozoic Birds: Above the Heads of Dinosaurs.* Berkeley: University of California Press, pp. 317–338.

Gao, C., Chiappe, L.M., Meng, Q., O'Connor, J.K., Wang, X., Cheng, X., & Liu, J. (2008) A new basal lineage of Early Cretaceous birds from China and its implications on the evolution of the avian tail. *Palaeontology* **51**, 775–791.

Gao, C., Chiappe, L.M., Zhang, F., Pomeroy, D.L., Shen, C., Chinsamy, A., & Walsh, M.O. (2012) A subadult specimen of the Early Cretaceous bird *Sapeornis chaoyangensis* and a taxonomic reassessment of sapeornithids. *Journal of Vertebrate Paleontology* **32**, 1103–1112.

García-R, J.C., Gibb, G.C., & Trewick, S.A. (2014) Deep global evolutionary radiation in birds: Diversification and trait evolution in the cosmopolitan bird family Rallidae. *Molecular Phylogenetics and Evolution* **81**, 96–108.

García-Talavera, I. (1990) Aves gigantes en el Mioceno de Famara (Lanzarote). *Revista de la Academia Canaria de Ciencas* **2**, 71–79.

Gatesy, S.M., & Dial, K.P. (1996) Locomotor modules and the evolution of avian flight. *Evolution* **50**, 331–340.

Gauthier, J. (1986) Saurischian monophyly and the origin of birds. *Memoirs of the California Academy of Sciences* **8**, 1–55.

Godefroit, P., Golovneva, L., Shchepetov, S., Garcia, G., & Alekseev, P. (2009) The last polar dinosaurs: High diversity of latest Cretaceous arctic dinosaurs in Russia. *Naturwissenschaften* **96**, 495–501.

Godefroit, P., Cau, A., Hu, D.-Y., Escuillié, F., Wu, W., & Dyke, G. (2013a) A Jurassic avialan dinosaur from China resolves the early phylogenetic history of birds. *Nature* **498**, 359–362.

Godefroit, P., Demuynck, H., Dyke, G., Hu, D., Escuillié, F., & Claeys, P. (2013b) Reduced plumage and flight ability of a new Jurassic paravian theropod from China. *Nature Communications* **4**, 1394.

Godefroit, P., Sinitsa, S.M., Dhouailly, D., Bolotsky, Y.L., Sizov, A.V., McNamara, M.E., Benton, M.J., & Spagna, P. (2014) A Jurassic ornithischian dinosaur from Siberia with both feathers and scales. *Nature* **345**, 451–455.

Goedert, J.L. (1988) A new late Eocene species of Plotopteridae (Aves: Pelecaniformes) from northwestern Oregon. *Proceedings of the California Academy of Sciences* **45**, 97–102.

Goedert, J.L., & Cornish, J. (2002) A preliminary report on the diversity and stratigraphic distribution of the Plotopteridae (Pelecaniformes) in Paleogene rocks of Washington State, USA. In: Z.-H. Zhou & F.-Z. Zhang (eds.), *Proceedings of the 5th Symposium of the Society of Avian Paleontology and Evolution, 1-4 June 2000.* Beijing: Science Press, pp. 63–76.

Göhlich, U.B. (2003a) A new crane (Aves: Gruidae) from the Miocene of Germany. *Journal of Vertebrate Paleontology* **23**, 387–393.

Göhlich, U.B. (2003b) The avifauna of the Grund Beds (Middle Miocene, Early Badenian, northern Austria). *Annalen des Naturhistorischen Museums in Wien* **104**, 237–249.

Göhlich, U.B. (2007) The earliest record of the extant genus *Spheniscus* (Aves: Spheniscidae) – a new species from the Miocene of the Pisco Formation, Peru. *Acta Palaeontologica Polonica* **52**, 285–298.

Göhlich, U.B., & Ballmann, P. (2013) A new barn owl (Aves: Strigiformes: Tytonidae) from the Middle Miocene of the Nördlinger Ries (Germany) with remarks on the history of the owls. In: U.B. Göhlich & A. Kroh (eds.), *Paleornithological Research 2013 – Proceedings of the 8th International Meeting of the Society of Avian Paleontology and Evolution.* Vienna: Natural History Museum Vienna, pp. 103–122.

Göhlich, U.B., & Mourer-Chauviré, C. (2005) Revision of the phasianids Aves: Galliformes) from the lower Miocene of Saint-Gérand-le-Puy (Allier, France). *Palaeontology* **48**, 1331–1350.

Göhlich, U.B., & Pavia, M. (2008) A new species of *Palaeortyx* (Aves: Galliformes: Phasianidae) from the Neogene of Gargano, Italy. *Oryctos* **7**, 95–108.

Goodman, S.M., & Jungers, W.L. (2014) *Extinct Madagascar: Picturing the Island's Past.* Chicago, IL: University of Chicago Press.

Grande, L. (2013) *The Lost World of Fossil Lake: Snapshots from Deep Time.* University of Chicago, IL: University of Chicago Press.

Grellet-Tinner, G., Murelaga, X., Larrasoaña, J.C., Silveira, L.F., Olivares, M., Ortega, L.A., Trimby, P.W., & Pascual, A. (2012) The first occurrence in the fossil record of an aquatic avian twig-nest with Phoenicopteriformes eggs: Evolutionary implications. *PLoS ONE* **7**, e46972.

Grellet-Tinner, G., Spooner, N.A., & Worthy, T.H. (2016) Is the "*Genyornis*" egg of a mihirung or another extinct bird from the Australian dreamtime? *Quaternary Science Reviews* **133**, 147–164.

Gunn, R.G., Douglas, L.C., & Whear, R.L. (2011) What bird is that? Identifying a probable painting of *Genyornis newtoni* in Western Arnhem land. *Australian Archaeology* **73**, 1–12.

Hackett, S.J., Kimball, R.T., Reddy, S., Bowie, R.C.K., Braun, E.L., Braun, M.J., Chojnowski, J.L., Cox, W.A., Han, K.-L., Harshman, J., Huddleston, C.J., Marks, B.D., Miglia, K.J., Moore, W.S., Sheldon, F.H., Steadman, D.W., Witt, C.C., & Yuri, T. (2008) A phylogenomic study of birds reveals their evolutionary history. *Science* **320**, 1763–1767.

Haddrath, O., & Baker, A.J. (2012) Multiple nuclear genes and retroposons support vicariance and dispersal of the palaeognaths, and an Early Cretaceous origin of modern birds. *Proceedings of the Royal Society of London, Series B* **279**, 4617–4625.

Haidr, N., & Acosta Hospitaleche, C. (2012) Feeding habits of Antarctic Eocene penguins from a morphofunctional perspective. *Neues Jahrbuch für Geologie und Paläontologie, Abhandlungen* **263**, 125–131.

Han, G., Chiappe, L.M., Ji, S.-A., Habib, M., Turner, A.H., Chinsamy, A., Liu, X., & Han, L. (2014) A new raptorial dinosaur with exceptionally long feathering provides insights into dromaeosaurid flight performance. *Nature Communications* **5**, 4382.

Han, K.L., Braun, E.L., Kimball, R.T., Reddy, S., Bowie, R.C., Braun, M.J., Chojnowski, J.L., Hackett, S.J., Harshman, J., Huddleston, C.J., Marks, B.D., Miglia, K.J., Moore, W.S., Sheldon, F.H., Steadman, D.W., Witt, C.C., & Yuri, T. (2011) Are transposable element insertions homoplasy free? An examination using the avian Tree of Life. *Systematic Biology* **60**, 375–386.

Harrison, C.J.O., & Walker, C.A. (1976) A reappraisal of *Prophaethon shrubsolei* Andrews (Aves). *Bulletin of the British Museum (Natural History)* **27**, 1–30.

Harrison, T., & Msuya, C.P. (2005) Fossil struthionid eggshells from Laetoli, Tanzania: Taxonomic and biostratigraphic significance. *Journal of African Earth Sciences* **41**, 303–315.

Harshman, J., Braun, E.L., Braun, M.J., Huddleston, C.J., Bowie, R.C.K., Chojnowski, J.L., Hackett, S.J., Han, K.-L., Kimball, R.T., Marks, B.D., Miglia, K.J., Moore, W.S., Reddy, S., Sheldon, F.H., Steadman, D.W., Steppan, S.J., Witt, C.C., & Yuri, T. (2008) Phylogenomic evidence for multiple losses of flight in ratite birds. *Proceedings of the National Academy of Sciences USA* **36**, 13462–13467.

Haug, G.H., & Tiedemann, R. (1998) Effect of the formation of the Isthmus of Panama on Atlantic Ocean thermohaline circulation. *Nature* **393**, 673–676.

Heers, A.M., Dial, K.P., & Tobalske, B.W. (2014) From baby birds to feathered dinosaurs: Incipient wings and the evolution of flight. *Paleobiology* **40**, 459–476.

Hertel, F., & Campbell, K.E. (2007) The antitrochanter of birds: Form and function in balance. *The Auk* **124**, 789–805.

Hesse, A. (1990) Die Beschreibung der Messelornithidae (Aves: Gruiformes: Rhynocheti) aus dem Alttertiär Europas und Nordamerikas. *Courier Forschungsinstitut Senckenberg* **128**, 1–176.

Hills, L.V., Nicholls, E.L., Núñez-Betelu, L.K.M., & McIntyre, D.J. (1999) *Hesperornis* (Aves) from Ellesmere Island and palynological correlation of known Canadian localities. *Canadian Journal of Earth Sciences* **36**, 1583–1588.

Holdaway, R.N., & Jacomb, C. (2000) Rapid extinction of the moas (Aves:

Dinornithiformes): Model, test, and implications. *Science* **287**, 2250–2254.

Hope, S. (2002) The Mesozoic radiation of Neornithes. In: L.M. Chiappe & L.M. Witmer (eds.), *Mesozoic Birds: Above the Heads of Dinosaurs*. Berkeley: University of California Press, pp. 339–388.

Hou, L.-H. (1999) New hesperornithid (Aves) from the Canadian Arctic. *Vertebrata PalAsiatica* **37**, 228–233.

Hou, L., Zhou, Z., Zhang, F., & Wang, Z. (2005) A Miocene ostrich fossil from Gansu Province, northwest China. *Chinese Science Bulletin* **50**, 1808–1810.

Houde, P. (1988) Paleognathous birds from the early Tertiary of the Northern Hemisphere. *Publications of the Nuttall Ornithological Club* **22**, 1–148.

Houde, P., & Haubold, H. (1987) *Palaeotis weigelti* restudied: A small Middle Eocene ostrich (Aves: Struthioniformes). *Palaeovertebrata* **17**, 27–42.

Houde, P., & Olson, S.L. (1992) A radiation of coly-like birds from the early Eocene of North America (Aves: Sandcoleiformes new order). *Natural History Museum of Los Angeles County, Science Series* **36**, 137–160.

Howard, H. (1955) A new wading bird from the Eocene of Patagonia. *American Museum Novitates* **1710**, 1–25.

Howard, H. (1957) A gigantic "toothed" marine bird from the Miocene of California. *Santa Barbara Museum of Natural History Bulletin (Geology Department)* **1**, 1–23.

Howard, H. (1969) A new avian fossil from Kern County, California. *The Condor* **71**, 68–69.

Howard, H. (1992) New records of Middle Miocene anseriform birds from Kern County, California. *Natural History Museum of Los Angeles County, Science Series* **36**, 231–237.

Hu, D., Hou, L., Zhang, L., & Xu, X. (2009) A pre-*Archaeopteryx* troodontid theropod from China with long feathers on the metatarsus. *Nature* **461**, 640–643.

Hu, D., Li, L., Hou, L., & Xu, X. (2010) A new sapeornithid bird from China and its implication for early avian evolution. *Acta Geologica Sinica (English edition)* **84**, 472–482.

Hu, H., Zhou, Z.-H., & O'Connor, J.K. (2014) A subadult specimen of *Pengornis* and character evolution in Enantiornithes. *Vertebrata PalAsiatica* **52**, 77–97.

Hu, H., O'Connor, J.K., & Zhou, Z. (2015) A new species of Pengornithidae (Aves: Enantiornithes) from the Lower Cretaceous of China suggests a specialized scansorial habitat previously unknown in early birds. *PLoS ONE* **10**, e0126791.

Hume, J.P. (2007) Reappraisal of the parrots (Aves: Psittacidae) from the Mascarene Islands, with comments on their ecology, morphology, and affinities. *Zootaxa* **1513**, 1–75.

Hutchinson, J.R. (2001a) The evolution of pelvic osteology and soft tissues on the line to extant birds (Neornithes). *Zoological Journal of the Linnean Society* **131**, 123–168.

Hutchinson, J.R. (2001b). The evolution of femoral osteology and soft tissues on the line to extant birds (Neornithes). *Zoological Journal of the Linnean Society* **131**, 169–197.

Hutchinson, J.R., & Allen, V. (2009) The evolutionary continuum of limb function from early theropods to birds. *Naturwissenschaften* **96**, 423–448.

Hwang, S.H., Mayr, G., & Minjin, B. (2010) The earliest record of a galliform bird in Asia, from the late Paleocene/early Eocene of the Gobi Desert, Mongolia. *Journal of Vertebrate Paleontology* **30**, 1642–1644.

Iwaniuk A.L., Olson S.L., & James H.F. (2009) Extraordinary cranial specialization in a new genus of extinct duck (Aves: Anseriformes) from Kauai, Hawaiian Islands. *Zootaxa* **2296**, 47–67.

Jacobs, B., Kingston, J.D., & Jacobs, L. (1999) The origin of grass-dominated ecosystems. *Annals of the Missouri Botanical Garden* **399**, 673–676.

Jadwiszczak, P. (2001) Body size of Eocene Antarctic penguins. *Polish Polar Research* **22**, 147–158.

Jadwiszczak, P. (2006) Eocene penguins of Seymour Island, Antarctica: Taxonomy. *Polish Polar Research* **27**, 3–62.

Jadwiszczak, P. (2009) Penguin past: The current state of knowledge. *Polish Polar Research* **30**, 3–28.

Jadwiszczak, P. (2012) Partial limb skeleton of a "giant penguin" Anthropornis from the Eocene of Antarctic Peninsula. *Polish Polar Research* **33**, 259–274.

Jadwiszczak, P. (2013) Taxonomic diversity of Eocene Antarctic penguins: A changing picture.

Geological Society, London, Special Publications **381**, 129–138.

Jadwiszczak, P., & Chapman, S.D. (2011) The earliest fossil record of a medium-sized penguin. *Polish Polar Research* **32**, 269–277.

Jadwiszczak, P., Acosta Hospitaleche, C., & Reguero, M. (2013) Redescription of *Crossvallia unienviella*: The only Paleocene Antarctic penguin. *Ameghiniana* **50**, 545–553.

James, H.F., & Burney, D.A. (1997) The diet and ecology of Hawaii's extinct flightless waterfowl: Evidence from coprolites. *Biological Journal of the Linnean Society* **62**, 279–297.

Jarvis, E.D. et al. (104 coauthors) (2014) Whole-genome analyses resolve early branches in the tree of life of modern birds. *Science* **346**, 1320–1331.

Jenkyns, H.C. (2003) Evidence for rapid climate change in the Mesozoic–Palaeogene greenhouse world. *Philosophical Transactions of the Royal Society of London. Series A* **361**, 1885–1916.

Jetz, W., Thomas, G.H., Joy, J.B., Hartmann, K., & Mooers, A.O. (2012) The global diversity of birds in space and time. *Nature* **491**, 444–448.

Ji, Q., Currie, P.J., Norell, M.A., & Ji, S.-A. (1998) Two feathered dinosaurs from northeastern China. *Nature* **393**, 753–761.

Ji, S., & Ji, Q. (2007) *Jinfengopteryx* compared to *Archaeopteryx*, with comments on the mosaic evolution of long-tailed avialan birds. *Acta Geologica Sinica* **81**, 337–343.

Jin, F., Zhang, F., Li, Z., Zhang, J., Li, C., & Zhou, Z. (2008) On the horizon of *Protopteryx* and the early vertebrate fossil assemblages of the Jehol Biota. *Chinese Science Bulletin* **53**, 2820–2827.

Johansson, U.S., & Ericson, P.G.P. (2005) A re-evaluation of basal phylogenetic relationships within trogons (Aves: Trogonidae) based on nuclear DNA sequences. *Journal of Zoological Systematics and Evolutionary Research* **43**, 166–173.

Johnston, P. (2011) New morphological evidence supports congruent phylogenies and Gondwana vicariance for palaeognathous birds. *Zoological Journal of the Linnean Society* **163**, 959–982.

Jones, T.L., Porcasi, J.F., Erlandson, J.M., Dallas, H., Wake, T.A., & Schwaderer, R. (2008) The protracted Holocene extinction of California's flightless sea duck (*Chendytes lawi*) and its implications for the Pleistocene overkill

hypothesis. *Proceedings of the National Academy of Sciences USA* **105**, 4105–4108.

Kaiser, G., Watanabe, J., & Johns, M. (2015) A new member of the family Plotopteridae (Aves) from the late Oligocene of British Columbia, Canada. *Palaeontologia Electronica* **18.3.52A**, 1–18.

Kakegawa, J., & Hirao, K. (2003) A Miocene passeriform bird from the Iwami Formation, Tottori Group, Tottori, Japan. *Bulletin of the National Science Museum, Series C* **29**, 33–37.

Karhu, A. (1988) [A new family of swift-like birds from the Paleogene of Europe]. *Paleontological Journal* **3**, 78–88 [in Russian].

Karhu, A. (1997) A new species of *Urmiornis* (Gruiformes: Ergilornithidae) from the early Miocene of western Kazakhstan. *Paleontological Journal* **31**, 102–107.

Karhu, A. (1999) A new genus and species of the family Jungornithidae (Apodiformes) from the Late Eocene of the Northern Caucasus, with comments on the ancestry of hummingbirds. *Smithsonian Contributions to Paleobiology* **89**, 207–216.

Kawabe, S., Ando, T., & Endo, H. (2014) Enigmatic affinity in the brain morphology between plotopterids and penguins, with a comprehensive comparison among water birds. *Zoological Journal of the Linnean Society* **170**, 467–493.

Kinsky, F.C. (1971) The consistent presence of paired ovaries in the Kiwi (*Apteryx*) with some discussion of this condition in other birds. *Journal für Ornithologie* **112**, 334–357.

Koschowitz, M.C., Fischer, C., & Sander, M. (2014) Beyond the rainbow. *Science* **346**, 416–418.

Kristoffersen, A.V. (2002) An early Paleogene trogon (Aves: Trogoniformes) from the Fur Formation, Denmark. *Journal of Vertebrate Paleontology* **22**, 661–666.

Ksepka, D.T. (2014) Flight performance of the largest volant bird. *Proceedings of the National Academy of Sciences USA* **111**, 10624–10629.

Ksepka, D.T., & Ando, T. (2011) Penguins past, present, and future: Trends in the evolution of the Sphenisciformes. In: G. Dyke & G. Kaiser (eds.), *Living Dinosaurs: The Evolutionary History of Modern Birds*. Chichester: John Wiley & Sons, pp. 155–186.

Ksepka, D.T., & Clarke, J.A. (2009) Affinities of *Palaeospiza bella* and the phylogeny and biogeography of mousebirds (Coliiformes). *The Auk* **126**, 245–259.

Ksepka, D.T., & Clarke, J.A. (2010a) The basal penguin (Aves: Sphenisciformes) *Perudyptes devriesi* and a phylogenetic evaluation of the penguin fossil record. *Bulletin of the American Museum of Natural History* **337**, 1–77.

Ksepka, D.T., & Clarke, J.A. (2010b) New fossil mousebird (Aves: Coliiformes) with feather preservation provides insight into the ecological diversity of an Eocene North American avifauna. *Zoological Journal of the Linnean Society* **160**, 685–706.

Ksepka, D.T., & Clarke, J.A. (2010c) *Primobucco mcgrewi* (Aves: Coracii) from the Eocene Green River Formation: New anatomical data from the earliest definitive record of stem rollers. *Journal of Vertebrate Paleontology* **30**, 215–225.

Ksepka, D.T., & Clarke, J.A. (2012) A new stem parrot from the Green River Formation and the complex evolution of the grasping foot in Pan-Psittaciformes. *Journal of Vertebrate Paleontology* **32**, 395–406.

Ksepka, D.T., & Thomas, D.B. (2012) Multiple cenozoic invasions of Africa by penguins (Aves, Sphenisciformes). *Proceedings of the Royal Society, B, Biological Sciences* **279**, 1027–1032.

Ksepka, D.T., Clarke, J.A., DeVries, T.J., & Urbina, M. (2008) Osteology of *Icadyptes salasi*, a giant penguin from the Eocene of Peru. *Journal of Anatomy* **213**, 131–147.

Ksepka, D.T., Clarke, J.A., & Grande, L. (2011) Stem parrots (Aves, Halcyornithidae) from the Green River Formation and a combined phylogeny of Pan-Psittaciformes. *Journal of Paleontology* **85**, 835–852.

Ksepka, D.T., Fordyce, R.E., Ando, T., & Jones, C.M. (2012) New fossil penguins (Aves, Sphenisciformes) from the Oligocene of New Zealand reveal the skeletal plan of stem penguins. *Journal of Vertebrate Paleontology* **32**, 235–254.

Ksepka, D.T., Balanoff, A.M., Bell, M.A., & Houseman, M.D. (2013a) Fossil grebes from the Truckee Formation (Miocene) of Nevada and a new phylogenetic analysis of Podicipediformes (Aves). *Palaeontology* **56**, 1149–1169.

Ksepka, D.T., Clarke, J.A., Nesbitt, S.J., Kulp, F.B., & Grande, L. (2013b) Fossil evidence of wing shape in a stem relative of swifts and hummingbirds (Aves, Pan-Apodiformes).

Proceedings of the Royal Society B: Biological Sciences **280**, 20130580.

Ksepka, D.T., Ware, J.L., & Lamm, K.S. (2014) Flying rocks and flying clocks: Disparity in fossil and molecular dates for birds. *Proceedings of the Royal Society B: Biological Sciences* **281**, 20140677.

Kurochkin, E.N. (1976) A survey of the Paleogene birds of Asia. *Smithsonian Contributions to Paleobiology* **27**, 75–86.

Kurochkin, E.N. (1999) The relationships of the Early Cretaceous *Ambiortus* and *Otogornis* (Aves: Ambiortiformes). *Smithsonian Contributions to Paleobiology* **89**, 275–284.

Kurochkin, E.N., & Dyke, G.J. (2010) A large collection of *Presbyornis* (Aves, Anseriformes, Presbyornithidae) from the late Paleocene and early Eocene of Mongolia. *Geological Journal* **45**, 375–387.

Kurochkin, E.N., & Dyke, G.J. (2011) The first fossil owls (Aves: Strigiformes) from the Paleogene of Asia and a review of the fossil record of Strigiformes. *Paleontological Journal* **45**, 445–458.

Kurochkin, E.N., Dyke, G.J., & Karhu, A.A. (2002) A new presbyornithid bird (Aves, Anseriformes) from the Late Cretaceous of Southern Mongolia. *American Museum Novitates* **3386**, 1–11.

Kurochkin, E.N., Zelenkov, N.V., Averianov, A.O., & Leshchinskiy, S.V. (2011) A new taxon of birds (Aves) from the Early Cretaceous of Western Siberia, Russia. *Journal of Systematic Palaeontology* **9**, 109–117.

Kurochkin, E.N., Chatterjee, S., & Mikhailov, K.E. (2013) An embryonic enantiornithine bird and associated eggs from the Cretaceous of Mongolia. *Paleontological Journal* **47**, 1252–1269.

Lambrecht, K. (1933) *Handbuch der Palaeornithologie*. Berlin: Gebrüder Borntraeger.

Lambrecht, K. (1938) *Urmiornis abeli* n. sp., eine pliozäne Vogelfährte aus Persien. *Palaeobiologica* **6**, 242–245.

Lee, M.S., Cau, A., Naish, D., & Dyke, G.J. (2014) Sustained miniaturization and anatomical innovation in the dinosaurian ancestors of birds. *Science* **345**, 562–566.

Lenz, O.K., Wilde, V., Mertz, D.F., & Riegel, W. (2015) New palynology-based astronomical and revised ^{40}Ar/^{39}Ar ages for the Eocene maar lake of Messel (Germany). *International Journal of Earth Sciences* **104**, 873–889.

Leonard, L., Dyke, G.J., & Walker, C.A. (2006) New specimens of a fossil ostrich from the Miocene of Kenya. *Journal of African Earth Sciences* **45**, 391–394.

Lerner, H.R.L., & Mindell, D.P. (2005) Phylogeny of eagles, Old World vultures, and other Accipitridae based on nuclear and mitochondrial DNA. *Molecular Phylogenetics and Evolution* **37**, 327–346.

Li, Q., Gao, K.-Q., Vinther, J., Shawkey, M.D., Clarke, J.A., D'Alba, L., Neng, Q., Briggs, D.E.G., & Prum, R.O. (2010) Plumage color patterns of an extinct dinosaur. *Science* **327**, 1369–1372.

Li, Q., Gao, K.Q., Meng, Q., Clarke, J.A., Shawkey, M.D., D'Alba, L., Pei, R., Ellison, M., Norell, M.A., & Vinther, J. (2012) Reconstruction of *Microraptor* and the evolution of iridescent plumage. *Science* **335**, 1215–1219.

Li, Z., Zhou, Z., Wang, M., & Clarke, J.A. (2014a) A new specimen of large-bodied basal enantiornithine *Bohaiornis* from the Early Cretaceous of China and the inference of feeding ecology in Mesozoic Birds. *Journal of Paleontology* **88**, 99–108.

Li, Z., Zhou, Z., Deng, T., Li, Q., & Clarke, J.A. (2014b) A falconid from the Late Miocene of northwestern China yields further evidence of transition in Late Neogene steppe communities. *The Auk* **131**, 335–350.

Liu, D., Chiappe, L.M., Zhang, Y., Bell, A., Meng, Q., Ji, Q., & Wang, X. (2014) An advanced, new long-legged bird from the Early Cretaceous of the Jehol Group (northeastern China): Insights into the temporal divergence of modern birds. *Zootaxa* **3884**, 253–266.

Liu, Y.-Q., Kuang, H.-W., Jiang, X.-J., Peng, N., Xu, H., & Sun, H.-Y. (2012) Timing of the earliest known feathered dinosaurs and transitional pterosaurs older than the Jehol Biota. *Palaeogeography, Palaeoclimatology, Palaeoecology* **323**, 1–12.

Livezey, B.C. (1993) Morphology of flightlessness in *Chendytes*, fossil seaducks (Anatidae: Mergini) of coastal California. *Journal of Vertebrate Paleontology* **13**, 185–199.

Livezey, B.C. (1997) A phylogenetic analysis of basal Anseriformes, the fossil *Presbyornis*, and the interordinal relationships of waterfowl. *Zoological Journal of the Linnean Society* **121**, 361–428.

Livezey, B.C., & Martin, L.D. (1988) The systematic position of the Miocene anatid *Anas*[?] *blanchardi* Milne-Edwards. *Journal of Vertebrate Paleontology* **8**, 196–211.

Livezey, B.C., & Zusi, R.L. (2007) Higher-order phylogeny of modern birds (Theropoda, Aves: Neornithes) based on comparative anatomy: II. – Analysis and discussion. *Zoological Journal of the Linnean Society* **149**, 1–94.

Long, J.A., Vichers-Rich, P., Hirsch, K., Bray, E., & Tuniz, C. (1998) The Cervantes egg: An early Malagasy tourist to Australia. *Records of the Western Australian Museum* **19**, 39–46.

Longrich, N. (2009) Structure and function of hindlimb feathers in *Archaeopteryx lithographica*. *Paleobiology* **32**, 417–431.

Longrich, N.R., & Olson, S.L. (2011) The bizarre wing of the Jamaican flightless ibis *Xenicibis xympithecus*: A unique vertebrate adaptation. *Proceedings of the Royal Society B: Biological Sciences* **278**, 2333–2337.

Longrich, N.R., Tokaryk, T., & Field, D.J. (2011) Mass extinction of birds at the Cretaceous-Paleogene (K–Pg) boundary. *Proceedings of the National Academy of Sciences USA* **108**, 15253–15257.

Longrich, N.R., Vinther, J., Meng, Q., Li, Q., & Russell, A.P. (2012) Primitive wing feather arrangement in *Archaeopteryx lithographica* and *Anchiornis huxleyi*. *Current Biology* **22**, 2262–2267.

Louchart, A. (2003) A true peafowl in Africa. *South African Journal of Science* **99**, 368–371.

Louchart, A., & Viriot, L. (2011) From snout to beak: The loss of teeth in birds. *Trends in Ecology and Evolution* **26**, 663–673.

Louchart, A., Vignaud, P., Likius, A., MacKaye, H.T., & Brunet, M. (2005a) A new swan (Aves: Anatidae) in Africa, from the latest Miocene of Chad and Libya. *Journal of Vertebrate Paleontology* **25**, 384–392.

Louchart, A., Mourer-Chauviré, C., Vignaud, P., Mackaye, H.T., & Brunet, M. (2005b) A finfoot from the Late Miocene of Toros Menalla (Chad, Africa): Palaeobiogeographical and palaeoecological implications. *Palaeogeography, Palaeoclimatology, Palaeoecology* **222**, 1–9.

Louchart, A., Vignaud, P., Likius, A., Brunet, M., & White, T.D. (2005c) A large extinct marabou stork in African Pliocene hominid sites, and a review of the fossil species of *Leptoptilos*. *Acta Palaeontologica Polonica* **50**, 549–563.

Louchart, A., Tourment, N., Carrier, J., Roux, T., & Mourer-Chauviré, C. (2008a) Hummingbird with modern feathering: An exceptionally well-preserved Oligocene fossil from southern France. *Naturwissenschaften* **95**, 171–175.

Louchart, A., Haile-Selassie, Y., Vignaud, P., Likius, A., & Brunet, N. (2008b) Fossil birds from the Late Miocene of Chad and Ethiopia and zoogeographical implications. *Oryctos* **7**, 147–167.

Louchart, A., Tourment, N., & Carrier, J. (2011) The earliest known pelican reveals 30 million years of evolutionary stasis in beak morphology. *Journal of Ornithology* **152**, 15–20.

Louchart, A., Sire, J.-Y., Mourer-Chauviré, C., Geraads, D., Viriot, L., & de Buffrénil, V. (2013) Structure and growth pattern of pseudoteeth in *Pelagornis mauretanicus* (Aves, Odontopterygiformes, Pelagornithidae). *PLoS ONE* **8**, e80372.

Lü, J., & Brusatte, S.L. (2015) A large, short-armed, winged dromaeosaurid (Dinosauria: Theropoda) from the Early Cretaceous of China and its implications for feather evolution. *Scientific Reports* **5**, 11775.

MacFadden, B.J., Labs-Hochstein, J., Hulbert, R.C., & Baskin, J.A. (2007) Revised age of the late Neogene terror bird (*Titanis*) in North America during the Great American Interchange. *Geology* **35**, 123–126.

Mackness, B. (1995) *Anhinga malagurala*, a new pygmy darter from the Early Pliocene Bluff Downs local fauna, north-eastern Queensland. *Emu* **95**, 265–271.

Makovicky, P.J., & Norell, M.A. (2004) Troodontidae. In: D.B. Weishampel, P. Dodson, & H. Osmólska (eds.), *The Dinosauria*, 2nd edn. Berkeley: University of California Press, pp. 184–195.

Makovicky, P.J., & Zanno, L.E. (2011) Theropod diversity and the refinement of avian characteristics. In: G. Dyke & G. Kaiser (eds.), *Living Dinosaurs: The Evolutionary History of Modern Birds*. Chichester: John Wiley & Sons, pp. 9–29.

Makovicky, P.J., Apesteguía, S., & Agnolín, F.L. (2005) The earliest dromaeosaurid theropod from South America. *Nature* **437**, 1007–1011.

Manegold, A. (2008a) Passerine diversity in the late Oligocene of Germany: Earliest evidence for the sympatric coexistence of Suboscines and Oscines. *The Ibis* **150**, 377–387.

Manegold, A. (2008b) Earliest fossil record of the Certhioidea (treecreepers and allies) from the early Miocene of Germany. *Journal of Ornithology* **149**, 223–228.

Manegold, A. (2013) Two new parrot species (Psittaciformes) from the early Pliocene of Langebaanweg, South Africa, and their palaeoecological implications. *The Ibis* **155**, 127–139.

Manegold, A., & Louchart, A. (2012) Biogeographic and paleoenvironmental implications of a new woodpecker species (Aves, Picidae) from the early Pliocene of South Africa. *Journal of Vertebrate Paleontology* **32**, 926–938.

Manegold, A., Mayr, G., & Mourer-Chauviré, C. (2004) Miocene songbirds and the composition of the European passeriform avifauna. *The Auk* **121**, 1155–1160.

Manegold, A., Louchart, A., Carrier, J., & Elzanowski, A. (2013) The Early Pliocene avifauna of Langebaanweg (South Africa): A review and update. In: U.B. Göhlich & A. Kroh (eds.), *Paleornithological Research 2013 – Proceedings of the 8th International Meeting of the Society of Avian Paleontology and Evolution*. Vienna: Natural History Museum Vienna, pp. 135–152.

Manegold, A., Pavia, M., & Haarhoff, P. (2014) A new species of *Aegypius* vulture (Aegypiinae, Accipitridae) from the early Pliocene of South Africa. *Journal of Vertebrate Paleontology* **34**, 1394–1407.

Manning, P.L., Payne, D., Pennicott, J., Barrett, P.M., & Ennos, R.A. (2006) Dinosaur killer claws or climbing crampons? *Biology Letters* **2**, 110–112.

Marsh, O.C. (1880) *Odontornithes: A Monograph on the Extinct Toothed Birds of North America*. Washington: US Government Printing Office.

Marshall, L.G., Webb, S.D., Sepkoski, J.J., & Raup, D.M. (1982) Mammalian evolution and the great American interchange. *Science* **215**, 1351–1357.

Martin, L.D. (1992) The status of the Late Paleocene birds *Gastornis* and *Remiornis*.

Natural History Museum of Los Angeles County, Science Series **36**, 97–108.

Martin, L.D., & Lim, J.-D. (2002) New information on the hesperornithiform radiation. In: Z.-H. Zhou & F.-Z. Zhang (eds.), *Proceedings of the 5th Symposium of the Society of Avian Paleontology and Evolution, 1–4 June 2000.* Beijing: Science Press, pp. 113–124.

Martin, L.D., & Tate, J. (1976) The skeleton of *Baptornis advenus* (Aves: Hesperornithiformes). *Smithsonian Contributions to Paleobiology* **27**, 35–66.

Martin, L.D., Kurochkin, E.N., & Tokaryk, T.T. (2012) A new evolutionary lineage of diving birds from the Late Cretaceous of North America and Asia. *Palaeoworld* **21**, 59–63.

Maryańska, T., Osmólska, H., & Wolsan, M. (2002) Avialan status for Oviraptorosauria. *Acta Palaeontologica Polonica* **47**, 97–116.

Matsuoka, H., Hasegawa, Y., & Takakuwa, Y. (2004) Osteological note on the completely prepared fossil "Annaka Short-winged Swan" from the Miocene Tomioka Group, Japan. *Bulletin of Gunma Museum of Natural History* **8**, 35–56.

Matthew, W.D., & Granger, W. (1917) The skeleton of *Diatryma*, a gigantic bird from the Lower Eocene of Wyoming. *Bulletin of the American Museum of Natural History* **37**, 307–326.

Mayr, E. (1942) *Systematics and the Origin of Species, from the Viewpoint of a Zoologist.* New York: Columbia University Press.

Mayr, G. (1998) "Coraciiforme" und "piciforme" Kleinvögel aus dem Mittel-Eozän der Grube Messel (Hessen, Deutschland). *Courier Forschungsinstitut Senckenberg* **205**, 1–101.

Mayr, G. (1999) A new trogon from the Middle Oligocene of Céreste, France. *The Auk* **116**, 427–434.

Mayr, G. (2000) Tiny hoopoe-like birds from the Middle Eocene of Messel (Germany). *The Auk* **117**, 968–974.

Mayr, G. (2001) New specimens of the Middle Eocene fossil mousebird *Selmes absurdipes* Peters 1999. *The Ibis* **143**, 427–434.

Mayr, G. (2002) Osteological evidence for paraphyly of the avian order Caprimulgiformes (nightjars and allies). *Journal für Ornithologie* **143**, 82–97.

Mayr, G. (2003a) Phylogeny of early Tertiary swifts and hummingbirds (Aves: Apodiformes). *The Auk* **120**, 145–151.

Mayr, G. (2003b) A new Eocene swift-like bird with a peculiar feathering. *The Ibis* **145**, 382–391.

Mayr, G. (2004a) A partial skeleton of a new fossil loon (Aves, Gaviiformes) from the early Oligocene of Germany with preserved stomach content. *Journal of Ornithology* **145**, 281–286.

Mayr, G. (2004b) Morphological evidence for sister group relationship between flamingos (Aves: Phoenicopteridae) and grebes (Podicipedidae). *Zoological Journal of the Linnean Society* **140**, 157–169.

Mayr, G. (2004c) New specimens of *Hassiavis laticauda* (Aves: Cypselomorphae) and *Quasisyndactylus longibrachis* (Aves: Alcediniformes) from the Middle Eocene of Messel, Germany. *Courier Forschungsinstitut Senckenberg* **252**, 23–28.

Mayr, G. (2004d) Old World fossil record of modern-type hummingbirds. *Science* **304**, 861–864.

Mayr, G. (2004e) The phylogenetic relationships of the early Tertiary Primoscenidae and Sylphornithidae and the sister taxon of crown group piciform birds. *Journal of Ornithology* **145**, 188–198.

Mayr, G. (2005a) A new cypselomorph bird from the Middle Eocene of Germany and the early diversification of avian aerial insectivores. *The Condor* **107**, 342–352.

Mayr, G. (2005b) The Palaeogene Old World potoo *Paraprefica* Mayr, 1999 (Aves, Nyctibiidae): Its osteology and affinities to the New World Preficinae Olson, 1987. *Journal of Systematic Palaeontology* **3**, 359–370.

Mayr, G. (2005c) Tertiary plotopterids (Aves, Plotopteridae) and a novel hypothesis on the phylogenetic relationships of penguins (Spheniscidae). *Journal of Zoological Systematics and Evolutionary Research* **43**, 61–71.

Mayr, G. (2005d) "Old World phorusrhacids" (Aves, Phorusrhacidae): A new look at *Strigogyps* ("*Aenigmavis*") *sapea* (Peters 1987). *PaleoBios* **25**, 11–16.

Mayr, G. (2005e) A tiny barbet-like bird from the Lower Oligocene of Germany: The smallest species and earliest substantial fossil record of the Pici (woodpeckers and allies). *The Auk* **122**, 1055–1063.

Mayr, G. (2006a) New specimens of the Eocene Messelirrisoridae (Aves: Bucerotes), with comments on the preservation of uropygial gland waxes in fossil birds from Messel and the phylogenetic affinities of Bucerotes. *Paläontologische Zeitschrift* **80**, 405–420.

Mayr, G. (2006b) New specimens of the early Eocene stem group galliform *Paraortygoides* (Gallinuloididae), with comments on the evolution of a crop in the stem lineage of Galliformes. *Journal of Ornithology* **147**, 31–37.

Mayr, G. (2006c) A new raptorial bird from the Middle Eocene of Messel, Germany. *Historical Biology* **18**, 95–102.

Mayr, G. (2006d) First fossil skull of a Paleogene representative of the Pici (woodpeckers and allies) and its evolutionary implications. *The Ibis* **148**, 824–827.

Mayr, G. (2007) The birds from the Paleocene fissure filling of Walbeck (Germany). *Journal of Vertebrate Paleontology* **27**, 394–408.

Mayr, G. (2008a) Avian higher-level phylogeny: Well-supported clades and what we can learn from a phylogenetic analysis of 2954 morphological characters. *Journal of Zoological Systematics and Evolutionary Research* **46**, 63–72.

Mayr, G. (2008b) Phylogenetic affinities and morphology of the late Eocene anseriform bird *Romainvillia stehlini* Lebedinsky, 1927. *Neues Jahrbuch für Geologie und Paläontologie, Abhandlungen* **248**, 365–380.

Mayr, G. (2008c) The Madagascan "cuckoo-roller" (Aves: Leptosomidae) is not a roller – notes on the phylogenetic affinities and evolutionary history of a "living fossil." *Acta Ornithologica* **43**, 226–230.

Mayr, G. (2008d) Phylogenetic affinities of the enigmatic avian taxon *Zygodactylus* based on new material from the early Oligocene of France. *Journal of Systematic Palaeontology* **6**, 333–344.

Mayr, G. (2009a) *Paleogene Fossil Birds*. Heidelberg: Springer.

Mayr, G. (2009b) A well-preserved skull of the "falconiform" bird *Masillaraptor* from the middle Eocene of Messel (Germany). *Palaeodiversity* **2**, 315–320.

Mayr, G. (2009c) A well-preserved second trogon skeleton (Aves, Trogonidae) from the middle Eocene of Messel, Germany. *Palaeobiodiversity and Palaeoenvironments* **89**, 1–6.

Mayr, G. (2010a) Phylogenetic relationships of the paraphyletic "caprimulgiform" birds (nightjars and allies). *Journal of Zoological Systematics and Evolutionary Research* **48**, 126–137.

Mayr, G. (2010b) Reappraisal of *Eocypselus* – a stem group representative of apodiform birds from the early Eocene of Northern Europe. *Palaeobiodiversity and Palaeoenvironments* **90**, 395–403.

Mayr, G. (2010c) Mousebirds (Coliiformes), parrots (Psittaciformes), and other small birds from the late Oligocene/early Miocene of the Mainz Basin, Germany. *Neues Jahrbuch für Geologie und Paläontologie, Abhandlungen* **258**, 129–144.

Mayr, G. (2011a) Metaves, Mirandornithes, Strisores and other novelties – a critical review of the higher-level phylogeny of neornithine birds. *Journal of Zoological Systematics and Evolutionary Research* **49**, 58–76.

Mayr, G. (2011b) The phylogeny of charadriiform birds (shorebirds and allies) – reassessing the conflict between morphology and molecules. *Zoological Journal of the Linnean Society* **161**, 916–934.

Mayr, G. (2011c) Cenozoic mystery birds – on the phylogenetic affinities of bony-toothed birds (Pelagornithidae). *Zoologica Scripta* **40**, 448–467.

Mayr, G. (2011d) Two-phase extinction of "Southern Hemispheric" birds in the Cenozoic of Europe and the origin of the Neotropic avifauna. *Palaeobiodiversity and Palaeoenvironments* **91**, 325–333.

Mayr, G. (2011e) Well-preserved new skeleton of the Middle Eocene *Messelastur* substantiates sister group relationship between Messelasturidae and Halcyornithidae (Aves, ?Pan-Psittaciformes). *Journal of Systematic Palaeontology* **9**, 159–171.

Mayr, G. (2013a) The age of the crown group of passerine birds and its evolutionary significance – molecular calibrations versus the fossil record. *Systematics and Biodiversity* **11**, 7–13.

Mayr, G. (2013b) Parvigruidae (Aves, core-Gruiformes) from the early Oligocene of

Belgium. *Palaeobiodiversity and Palaeoenvironments* **93**, 77–89.

Mayr, G. (2013c) Late Oligocene mousebird converges on parrots in skull morphology. *The Ibis* **155**, 384–396.

Mayr, G. (2014a) The origins of crown group birds: Molecules and fossils. *Palaeontology* **57**, 231–242.

Mayr, G. (2014b) On the middle Miocene avifauna of Maboko Island, Kenya. *Geobios* **47**, 133–146.

Mayr, G. (2014c) A hoatzin fossil from the middle Miocene of Kenya documents the past occurrence of modern-type Opisthocomiformes in Africa. *The Auk* **131**, 55–60.

Mayr, G. (2014d) The Eocene *Juncitarsus* – its phylogenetic position and significance for the evolution and higher-level affinities of flamingos and grebes. *Comptes Rendus Palevol* **13**, 9–18.

Mayr, G. (2015a) A procellariiform bird from the early Oligocene of North America. *Neues Jahrbuch für Geologie und Paläontologie, Abhandlungen* **275**, 11–17.

Mayr, G. (2015b) The middle Eocene European "ratite" *Palaeotis* (Aves, Palaeognathae) restudied once more. *Paläontologische Zeitschrift* **89**, 503–514.

Mayr, G. (2015c) Cranial and vertebral morphology of the straight-billed Miocene phoenicopteriform bird *Palaelodus* and its evolutionary significance. *Zoologischer Anzeiger* **254**, 18–26.

Mayr, G. (2015d) Eocene fossils and the early evolution of frogmouths (Podargiformes): Further specimens of *Masillapodargus* and a comparison with *Fluvioviridavis*. *Palaeobiodiversity and Palaeoenvironments* **95**, 587–596.

Mayr, G. (2015e) A reassessment of Eocene parrotlike fossils indicates a previously undetected radiation of zygodactyl stem group representatives of passerines (Passeriformes). *Zoologica Scripta* **44**, 587–602.

Mayr, G. (2015f) Skeletal morphology of the middle Eocene swift *Scaniacypselus* and the evolutionary history of true swifts (Apodidae). *Journal of Ornithology* **156**, 441–450.

Mayr, G. (2015g) New remains of the Eocene *Prophaethon* and the early evolution of tropicbirds (Phaethontiformes). *The Ibis* **157**, 54–67.

Mayr, G. (2015h) A new skeleton of the late Oligocene "Enspel cormorant" – from *Oligocorax* to *Borvocarbo*, and back again. *Palaeobiodiversity and Palaeoenvironments* **95**, 87–101.

Mayr, G. (2015i) A new specimen of the Early Eocene *Masillacolius brevidactylus* and its implications for the evolution of feeding specializations in mousebirds (Coliiformes). *Comptes Rendus Palevol* **14**, 363–370.

Mayr, G., & Bertelli, S. (2011) A record of *Rhynchaeites* (Aves, Threskiornithidae) from the early Eocene Fur Formation of Denmark, and the affinities of the alleged parrot *Mopsitta*. *Palaeobiodiversity and Palaeoenvironments* **91**, 229–236.

Mayr, G., & Clarke, J. (2003) The deep divergences of neornithine birds: A phylogenetic analysis of morphological characters. *Cladistics* **19**, 527–553.

Mayr, G., & Daniels, M. (1998) Eocene parrots from Messel (Hessen, Germany) and the London Clay of Walton-on-the-Naze (Essex, England). *Senckenbergiana lethaea* **78**, 157–177.

Mayr, G., & De Pietri, V.L. (2013) A goose-sized anseriform bird from the late Oligocene of France: The youngest record and largest species of Romainvilliinae. *Paläontologische Zeitschrift* **87**, 423–430.

Mayr, G., & De Pietri, V.L. (2014) Earliest and first Northern Hemispheric hoatzin fossils substantiate Old World origin of a "Neotropic endemic." *Naturwissenschaften* **101**, 143–148.

Mayr, G., & Göhlich, U.B. (2004) A new parrot from the Miocene of Germany, with comments on the variation of hypotarsus morphology in some Psittaciformes. *Belgian Journal of Zoology* **134**, 47–54.

Mayr, G., & Gregorová, R. (2012) A tiny stem group representative of Pici (Aves, Piciformes) from the early Oligocene of the Czech Republic. *Paläontologische Zeitschrift* **86**, 333–343.

Mayr, G., & Knopf, C. (2007) A stem lineage representative of buttonquails from the Lower Oligocene of Germany – fossil evidence for a charadriiform origin of the Turnicidae. *The Ibis* **149**, 774–782.

Mayr, G., & Manegold, A. (2006) New specimens of the earliest European passeriform bird. *Acta Palaeontologica Polonica* **51**, 315–323.

Mayr, G., & Manegold, A. (2013) Can ovarian follicles fossilize? *Nature* **499**, E1.

Mayr, G., & Micklich, N. (2010) New specimens of the avian taxa *Eurotrochilus* (Trochilidae) and *Palaeotodus* (Todidae) from the early Oligocene of Germany. *Paläontologische Zeitschrift* **84**, 387–395.

Mayr, G., & Mourer-Chauviré, C. (2000) Rollers (Aves: Coraciiformes s.s.) from the Middle Eocene of Messel (Germany) and the Upper Eocene of the Quercy (France). *Journal of Vertebrate Paleontology* **20**, 533–546.

Mayr, G., & Mourer-Chauviré, C. (2006) Three-dimensionally preserved cranial remains of *Elaphrocnemus* (Aves, Cariamae) from the Paleogene Quercy fissure fillings in France. *Neues Jahrbuch für Geologie und Paläontologie, Monatshefte* **2006**, 15–27.

Mayr, G., & Noriega, J.I. (2015) A well-preserved partial skeleton of the poorly known early Miocene seriema *Noriegavis santacrucensis*. *Acta Palaeontologica Polonica* **60**, 589–598.

Mayr, G., & Pavia, M. (2014) On the true affinities of *Chenornis graculoides* Portis, 1884 and *Anas lignitifila* Portis, 1884 – an albatross and an unusual duck from the Miocene of Italy. *Journal of Vertebrate Paleontology* **34**, 914–923.

Mayr, G., & Peters, D.S. (1998) The mousebirds (Aves: Coliiformes) from the Middle Eocene of Grube Messel (Hessen, Germany). *Senckenbergiana lethaea* **78**, 179–197.

Mayr, G., & Poschmann, M. (2009) A loon leg (Aves, Gaviidae) with crocodilian tooth from the late Oligocene of Germany. *Waterbirds* **32**, 468–471.

Mayr, G., & Richter, G. (2011) Exceptionally preserved plant parenchyma in the digestive tract indicates an herbivorous diet in the Middle Eocene bird *Strigogyps sapea* (Ameghinornithidae). *Paläontologische Zeitschrift* **85**, 303–307.

Mayr, G., & Rubilar-Rogers, D. (2010) Osteology of a new giant bony-toothed bird from the Miocene of Chile, with a revision of the taxonomy of Neogene Pelagornithidae. *Journal of Vertebrate Paleontology* **30**, 1313–1330.

Mayr, G., & Scofield, R.P. (2016) New avian remains from the Paleocene of New Zealand: The first early Cenozoic Phaethontiformes (tropicbirds) from the Southern Hemisphere. *Journal of Vertebrate Paleontology* **36**, e1031343.

Mayr, G., & Smith, R. (2001) Ducks, rails, and limicoline waders (Aves: Anseriformes, Gruiformes, Charadriiformes) from the lowermost Oligocene of Belgium. *Geobios* **34**, 547–561.

Mayr, G., & Smith, T. (2010) Bony-toothed birds (Aves: Pelagornithidae) from the middle Eocene of Belgium. *Palaeontology* **53**, 365–376.

Mayr, G., & Smith, T. (2012a) Phylogenetic affinities and taxonomy of the Oligocene Diomedeoididae, and the basal divergences amongst extant procellariiform birds. *Zoological Journal of the Linnean Society* **166**, 854–875.

Mayr, G., & Smith, T. (2012b) A fossil albatross from the early Oligocene of the North Sea Basin. *The Auk* **129**, 87–95.

Mayr, G., & Smith, T. (2013) Galliformes, Upupiformes, Trogoniformes, and other avian remains (?Phaethontiformes and ?Threskiornithidae) from the Rupelian stratotype in Belgium, with comments on the identity of "*Anas*" *benedeni* Sharpe, 1899. In: U.B. Göhlich & A. Kroh (eds.), *Paleornithological Research 2013 – Proceedings of the 8th International Meeting of the Society of Avian Paleontology and Evolution*. Vienna: Natural History Museum Vienna, pp. 23–35.

Mayr, G., & Weidig, I. (2004) The Early Eocene bird *Gallinuloides wyomingensis* – a stem group representative of Galliformes. *Acta Palaeontologica Polonica* **49**, 211–217.

Mayr, G., & Wilde, V. (2014) Eocene fossil is earliest evidence of flower-visiting by birds. *Biology Letters* **10**, 20140223.

Mayr, G., & Zvonok, E. (2011) Middle Eocene Pelagornithidae and Gaviiformes (Aves) from the Ukrainian Paratethys. *Palaeontology* **54**, 1347–1359.

Mayr, G., & Zvonok, E. (2012) A new genus and species of Pelagornithidae with well-preserved pseudodentition and further avian remains from the middle Eocene of the Ukraine. *Journal of Vertebrate Paleontology* **32**, 914–925.

Mayr, G., Peters, D.S., Plodowski, G., & Vogel, O. (2002a) Bristle-like integumentary structures at the tail of the horned dinosaur *Psittacosaurus*. *Naturwissenschaften* **89**, 361–365.

Mayr, G., Peters, D.S., & Rietschel, S. (2002b) Petrel-like birds with a peculiar foot morphology from the Oligocene of Germany and Belgium (Aves: Procellariiformes). *Journal of Vertebrate Paleontology* **22**, 667–676.

Mayr, G., Mourer-Chauviré, C., & Weidig, I. (2004) Osteology and systematic position of the Eocene Primobucconidae (Aves, Coraciiformes sensu stricto), with first records from Europe. *Journal of Systematic Palaeontology* **2**, 1–12.

Mayr, G., Pohl, B., & Peters, D.S. (2005) A well-preserved *Archaeopteryx* specimen with theropod features. *Science* **310**, 1483–1486.

Mayr, G., Poschmann, M., & Wuttke, M. (2006) A nearly complete skeleton of the fossil galliform bird *Palaeortyx* from the late Oligocene of Germany. *Acta Ornithologica* **41**, 129–135.

Mayr, G., Pohl, B., Hartman, S., & Peters, D.S. (2007) The tenth skeletal specimen of *Archaeopteryx*. *Zoological Journal of the Linnean Society* **149**, 97–116.

Mayr, G., Rana, R.S., Rose, K.D., Sahni, A., Kumar, K., Singh, L., & Smith, T. (2010) *Quercypsitta*-like birds from the early Eocene of India (Aves, ?Psittaciformes). *Journal of Vertebrate Paleontology* **30**, 467–478.

Mayr, G., Alvarenga, H., & Clarke, J. (2011a) An *Elaphrocnemus*-like landbird and other avian remains from the late Paleocene of Brazil. *Acta Palaeontologica Polonica* **56**, 679–684.

Mayr, G., Alvarenga, H.M.F., & Mourer-Chauviré, C. (2011b) Out of Africa: Fossils shed light on the origin of the hoatzin, an iconic Neotropic bird. *Naturwissenschaften* **98**, 961–966.

Mayr, G., Goedert, J.L., & McLeod, S.A. (2013a) Partial skeleton of a bony-toothed bird from the late Oligocene/early Miocene of Oregon (USA) and the systematics of Neogene Pelagornithidae. *Journal of Paleontology* **87**, 922–929.

Mayr, G., Zvonok, E., & Gorobets, L. (2013b) The tarsometatarsus of the middle Eocene loon *Colymbiculus udovichenkoi*. In: U.B. Göhlich & A. Kroh (eds.), *Paleornithological Research 2013 – Proceedings of the 8th International Meeting of the Society of Avian Paleontology and Evolution*. Vienna: Natural History Museum Vienna, pp. 17–22.

Mayr, G., Yang, J., de Bast, E., Li, C.-S., & Smith, T. (2013c) A *Strigogyps*-like bird from the middle Paleocene of China with an unusual grasping foot. *Journal of Vertebrate Paleontology* **33**, 895–901.

Mayr, G., Rana, R.S., Rose, K.D., Sahni, A., Kumar, K., & Smith, T. (2013d) New specimens of the early Eocene bird *Vastanavis* and the interrelationships of stem group Psittaciformes. *Paleontological Journal* **47**, 1308–1314.

Mayr, G., Goedert, J.L., & Vogel, O. (2015) Oligocene plotopterid skulls from western North America and their bearing on the phylogenetic affinities of these penguin-like seabirds. *Journal of Vertebrate Paleontology* **35**, e943764.

McKellar, R.C., Chatterton, B.D., Wolfe, A.P., & Currie, P.J. (2011) A diverse assemblage of Late Cretaceous dinosaur and bird feathers from Canadian amber. *Science* **333**, 1619–1622.

McNamara, M.E., Briggs, D.E.G., Orr, P.J., Field, D.J., & Wang, Z. (2013) Experimental maturation of feathers: Implications for reconstructions of fossil feather colour. *Biology Letters* **9**, 20130184.

Meijer, H.J.M. (2014) A peculiar anseriform (Aves: Anseriformes) from the Miocene of Gargano (Italy). *Comptes Rendus Palevol* **13**, 19–26.

Meijer, H.J.M., & Due, R.A. (2010) A new species of giant marabou stork (Aves: Ciconiiformes) from the Pleistocene of Liang Bua, Flores (Indonesia). *Zoological Journal of the Linnean Society* **160**, 707–724.

Meiri, S., Dayan, T., & Simberloff, D. (2004) Body size of insular carnivores: Little support for the island rule. *The American Naturalist* **163**, 469–479.

Meseguer, J., Chiappe, L.M., Sanz, J.L., Ortega, F., Sanz-Andrés, A., Pérez-Grande, I., & Franchini, S. (2012) Lift devices in the flight of *Archaeopteryx*. *Spanish Journal of Palaeontology* **27**, 125–130.

Miller, A.H. (1953) A fossil hoatzin from the Miocene of Colombia. *The Auk* **70**, 484–489.

Miller, A.H. (1963) The fossil flamingos of Australia. *The Condor* **65**, 289–299.

Miller, A.H., & Compton, L.V. (1939) Two fossil birds from the lower Miocene of South Dakota. *The Condor* **41**, 153–156.

Miller, A.H., & Sibley, C.G. (1941) A Miocene gull from Nebraska. *The Auk* **58**, 563–566.

Miller, E.R., Rasmussen, D.T., & Simons, E.L. (1997) Fossil storks (Ciconiidae) from the Late Eocene and Early Miocene of Egypt. *Ostrich* **68**, 23–26.

Milner, A.C., & Walsh, S.A. (2009) Avian brain evolution: New data from Palaeogene birds (Lower Eocene) from England. *Zoological Journal of the Linnean Society* **155**, 198–219.

Mitchell, J.S., & Makovicky, P.J. (2014) Low ecological disparity in Early Cretaceous birds. *Proceedings of the Royal Society B: Biological Sciences* **281**, 20140608.

Mitchell, K.J., Llamas, B., Soubrier, J., Rawlence, N.J., Worthy, T.H., Wood, J., Lee, M.S.Y., & Cooper, A. (2014) Ancient DNA reveals elephant birds and kiwi are sister taxa and clarifies ratite bird evolution. *Science* **344**, 898–900.

Mlíkovský, J. (1996, ed.) Tertiary Avian Localities of Europe. *Acta Universitatis Carolinae, Geologica* **39**, 1–852.

Mlíkovský, J. (2002) *Cenozoic Birds of the World. Part 1: Europe.* Praha: Ninox Press.

Mlíkovský, J. (2003) Early Miocene birds of Djebel Zelten, Lybia. *Časopis Národního muzea, Řada přírodovědná* **172**, 114–120.

Mlíkovský, J., & Göhlich, U.B. (2000) A new wood-hoopoe (Aves: Phoeniculidae) from the early Miocene of Germany and France. *Acta Societatis Zoologicae Bohemicae* **64**, 419–424.

Montes, C., Cardona, A., Jaramillo, C., Pardo, A., Silva, J.C., Valencia, V., Ayala, C., Pérez-Angel, L.C., Rodriguez-Parra, L.A., Ramirez, V., & Niño, H. (2015) Middle Miocene closure of the Central American Seaway. *Science* **348**, 226–229.

Mortimer, M. (2014). *Gansus zheni is Iteravis.* The Theropod Database Blog. http://theropoddatabase.blogspot.de/2014/12/gansus-zheni-is-iteravis.html. Accessed 7 March 2015.

Mourer-Chauviré, C. (1980) The Archaeotrogonidae from the Eocene and Oligocene deposits of "Phosphorites du Quercy," France. *Natural History Museum of Los Angeles County, Contributions in Science* **330**, 17–31.

Mourer-Chauviré, C. (1981) Première indication de la présence de Phorusracidés, famille d'oiseaux géants d'Amérique du Sud, dans le Tertiaire européen: *Ameghinornis* nov. gen. (Aves, Ralliformes) des Phosphorites du Quercy, France. *Geobios* **14**, 637–647.

Mourer-Chauviré, C. (1983a) Les Gruiformes (Aves) des Phosphorites du Quercy (France). 1. Sous-ordre Cariamae (Cariamidae et Phorusrhacidae). Systématique et biostratigraphie. *Palaeovertebrata* **13**, 83–143.

Mourer-Chauviré, C. (1983b) *Minerva antiqua* (Aves, Strigiformes), an owl mistaken for an edentate mammal. *American Museum Novitates* **2773**, 1–11.

Mourer-Chauviré, C. (1985) Les Todidae (Aves, Coraciiformes) des Phosphorites du Quercy (France). *Proceedings of the Koninklijke Nederlandse Akademie van Wetenschappen, Series B* **88**, 407–414.

Mourer-Chauviré, C. (1987) Les Strigiformes (Aves) des Phosphorites du Quercy (France): Systématique, biostratigraphie et paléobiogéographie. *Documents des Laboratoires de Géologie de Lyon* **99**, 89–135.

Mourer-Chauviré, C. (1988a) Le gisement du Bretou (Phosphorites du Quercy, Tarn-et-Garonne, France) et sa faune de vertébrés de l'Eocène supérieur. II Oiseaux. *Palaeontographica (A)* **205**, 29–50.

Mourer-Chauviré, C. (1988b) Les Aegialornithidae (Aves: Apodiformes) des Phosphorites du Quercy. Comparaison avec la forme de Messel. *Courier Forschungsinstitut Senckenberg* **107**, 369–381.

Mourer-Chauviré, C. (1989) A peafowl from the Pliocene of Perpignan, France. *Palaeontology* **32**, 439–446.

Mourer-Chauviré, C. (1991) Les Horusornithidae nov. fam., Accipitriformes (Aves) à articulation intertarsienne hyperflexible de l'Éocène du Quercy. *Geobios, mémoire spécial* **13**, 183–192.

Mourer-Chauviré, C. (1992a) The Galliformes (Aves) from the Phosphorites du Quercy (France): Systematics and biostratigraphy. *Natural History Museum of Los Angeles County, Science Series* **36**, 67–95.

Mourer-Chauviré, C. (1992b) Une nouvelle famille de perroquets (Aves, Psittaciformes) dans l'Éocène supérieur des Phosphorites du Quercy, France. *Geobios, mémoire spécial* **14**, 169–177.

Mourer-Chauviré, C. (1993) Les gangas (Aves, Columbiformes, Pteroclidae) du Paléogène et du Miocène inférieur de France. *Palaeovertebrata* **22**, 73–98.

Mourer-Chauviré, C. (1994) A large owl from the Palaeocene of France. *Palaeontology* **37**, 339–348.

Mourer-Chauviré, C. (1995) The Messelornithidae (Aves: Gruiformes) from the

Paleogene of France. *Courier Forschungsinstitut Senckenberg* **181**, 95–105.

Mourer-Chauviré, C. (1999) Comments on "Tertiary barn owls of Europe." *Journal für Ornithologie* **140**, 363–364.

Mourer-Chauviré, C. (2000) A new species of *Ameripodius* (Aves: Galliformes: Quercymegapodiidae) from the Lower Miocene of France. *Palaeontology* **43**, 481–593.

Mourer-Chauviré, C. (2002) Revision of the Cathartidae (Aves, Ciconiiformes) from the Middle Eocene to the Upper Oligocene Phosphorites du Quercy, France. In: Z.-H. Zhou & F.-Z. Zhang (eds.), *Proceedings of the 5th Symposium of the Society of Avian Paleontology and Evolution, 1–4 June 2000*. Beijing: Science Press, pp. 97–111.

Mourer-Chauviré, C. (2003) Birds (Aves) from the Middle Miocene of Arrisdrift (Namibia). Preliminary study with description of two new genera: *Amanuensis* (Accipitridae, Sagittariidae) and *Namibiavis* (Gruiformes, Idiornithidae). *Memoir of the Geological Survey of Namibia* **19**, 103–113.

Mourer-Chauviré, C. (2008) Birds (Aves) from the Early Miocene of the Northern Sperrgebiet, Namibia. *Memoir of the Geological Survey of Namibia* **20**, 147–167.

Mourer-Chauviré, C. (2013a) New data concerning the familial position of the genus *Euronyctibius* (Aves, Caprimulgiformes) from the Paleogene of the Phosphorites du Quercy, France. *Paleontological Journal* **47**, 1315–1322.

Mourer-Chauviré, C. (2013b) *Idiornis* Oberholser, 1899 (Aves, Gruiformes, Cariamae, Idiornithidae): A junior synonym of *Dynamopterus* Milne-Edwards, 1892 (Paleogene, Phosphorites du Quercy, France). *Neues Jahrbuch für Geologie und Paläontologie, Abhandlungen* **270**, 13–22.

Mourer-Chauviré, C., & Balouet, J.C. (2005). Description of the skull of the genus *Sylviornis* Poplin, 1980 (Galliformes, Sylviornithidae new family), a giant extinct bird from the Holocene of New Caledonia. *Monografies de la Societat d'Història Natural de les Balears* **12**, 205–118.

Mourer-Chauviré, C., & Cheneval, J. (1983) Les Sagittariidae fossiles (Aves, Accipitriformes) de l'Oligocène des Phosphorites du Quercy et du Miocène inférieur du Saint-Gérand-le-Puy. *Geobios* **16**, 443–459.

Mourer-Chauviré, C., & Geraads, D. (2008) The Struthionidae and Pelagornithidae (Aves: Struthioniformes, Odontopterygiformes) from the Late Pliocene of Ahl al Oughlam, Morocco. *Oryctos* **7**, 169–194.

Mourer-Chauviré, C., & Geraads, D. (2010) The Upper Pliocene avifauna of Ahl al Oughlam, Morocco. Systematics and biogeography. *Records of the Australian Museum* **62**, 157–184.

Mourer-Chauviré, C., Senut, B., Pickford, M., & Mein, P. (1996) Le plus ancien représentant du genre *Struthio* (Aves, Struthionidae), *Struthio coppensi* n. sp., du Miocène inférieur de Namibie. *Comptes rendus de l'Académie des Sciences Paris, série IIa* **322**, 325–332.

Mourer-Chauviré, C., Berthet, D., & Hugueney, M. (2004) The late Oligocene birds of the Créchy quarry (Allier, France), with a description of two new genera (Aves: Pelecaniformes: Phalacrocoracidae, and Anseriformes: Anseranatidae). *Senckenbergiana lethaea* **84**, 303–315.

Mourer-Chauviré, C., Pickford, M., & Senut, B. (2011a) The first Palaeogene galliform from Africa. *Journal of Ornithology* **152**, 617–622.

Mourer-Chauviré, C., Tabuce, R., Mahboubi, M., Adaci, M., & Bensalah, M. (2011b) A phororhacoid bird from the Eocene of Africa. *Naturwissenschaften* **98**, 815–823.

Mourer-Chauviré, C., Tabuce, R., El Mabrouk, E., Marivaux, L., Khayati, H., Vianey-Liaud, M., & Ben Haj Ali, M. (2013a) A new taxon of stem group Galliformes and the earliest record for stem group Cuculidae from the Eocene of Djebel Chambi, Tunisia. In: U.B. Göhlich & A. Kroh (eds.), *Paleornithological Research 2013 – Proceedings of the 8th International Meeting of the Society of Avian Paleontology and Evolution*. Vienna: Natural History Museum Vienna, pp. 1–15.

Mourer-Chauviré, C., Peyrouse, J.-B., & Hugueney, M. (2013b) A new roller (Aves: Coraciiformes s. s.: Coraciidae) from the Early Miocene of the Saint-Gérand-le-Puy area, Allier, France. In: U.B. Göhlich & A. Kroh (eds.), *Paleornithological Research 2013 – Proceedings of the 8th International Meeting of the Society of Avian Paleontology and Evolution*. Vienna: Natural History Museum Vienna, pp. 81–92.

Mourer-Chauviré, C., Pickford, M., & Senut, B. (2015) Stem group galliform and stem group psittaciform birds (Aves, Galliformes, Paraortygidae, and Psittaciformes, family incertae sedis) from the Middle Eocene of Namibia. *Journal of Ornithology* **156**, 275–286.

Moyer, A.E., Zheng, W., Johnson, E.A., Lamanna, M.C., Li, D.Q., Lacovara, K.J., & Schweitzer, M.H. (2014) Melanosomes or microbes: Testing an alternative hypothesis for the origin of microbodies in fossil feathers. *Scientific Reports* **4**, 4233.

Murray, P.F., & Vickers-Rich, P. (2004) *Magnificent Mihirungs: The Colossal Flightless Birds of the Australian Dreamtime.* Bloomington: Indiana University Press.

Mustoe, G.E., Tucker, D.S., & Kemplin, K.L. (2012) Giant Eocene bird footprints from Northwest Washington, USA. *Palaeontology* **55**, 1293–1305.

Myrcha, A., Jadwiszczak, P., Tambussi, C.P., Noriega, J.I., Gaździcki, A., Tatur, A., & del Valle, R.A. (2002) Taxonomic revision of Eocene Antarctic penguins based on tarsometatarsal morphology. *Polish Polar Research* **23**, 5–46.

Nesbitt, S.J., Ksepka, D.T., & Clarke, J.A. (2011) Podargiform affinities of the enigmatic *Fluvioviridavis platyrhamphus* and the early diversification of Strisores ("Caprimulgiformes" + Apodiformes). *PLoS ONE* **6**, e26350.

Nguyen, J.M.T. (2016) Australo-Papuan treecreepers (Passeriformes: Climacteridae) and a new species of sittella (Neosittidae: *Daphoenositta*) from the Miocene of Australia. *Palaeontologia Electronica* **19.1.1A**, 1–13.

Nguyen, J.M.T., Boles, W.E., & Hand, S. (2010) New material of *Barawertornis tedfordi*, a dromornithid bird from the Oligo-Miocene of Australia, and its phylogenetic implications. *Records of the Australian Museum* **62**, 45–60.

Nguyen, J.M.T., Worthy, T.H., Boles, W.E., Hand, S.J., & Archer, M. (2013) A new cracticid (Passeriformes: Cracticidae) from the Early Miocene of Australia. *Emu* **113**, 374–382.

Nguyen, J.M.T., Boles, W.E., Worthy, T.H., Hand, S.J., & Archer, M. (2014) New specimens of the logrunner *Orthonyx kaldowinyeri* (Passeriformes: Orthonychidae) from the Oligo-Miocene of Australia. *Alcheringa* **38**, 245–255.

Norell, M.A., & Makovicky, P.J. (2004) Dromaeosauridae. In: D.B. Weishampel, P. Dodson, & H. Osmólska (eds.), *The Dinosauria*, 2nd edn. Berkeley: University of California Press, pp. 196–209.

Noriega, J.I. (2001) Body mass estimation and locomotion of the Miocene pelecaniform bird *Macranhinga*. *Acta Palaeontologica Polonica* **46**, 247–260.

Noriega, J.I., & Agnolín, F.L. (2008) El registro paleontológico de las Aves del "Mesopotamiense" (Formación Ituzaingó; Mioceno tardío-Plioceno) de la provincia de Entre Ríos, Argentina. *INSUGEO, Miscelánea* **17**, 271–290.

Noriega, J.I., & Alvarenga, H.M.F. (2002) Phylogeny of the Tertiary giant darters (Pelecaniformes: Anhingidae) from South America. In: Z.-H. Zhou & F.-Z. Zhang (eds.), *Proceedings of the 5th Symposium of the Society of Avian Paleontology and Evolution, 1–4 June 2000*. Beijing: Science Press, pp. 41–49.

Noriega, J.I., & Chiappe, L.M. (1993) An early Miocene passeriform from Argentina. *The Auk* **110**, 936–938.

Noriega, J.I., & Cladera, G. (2008) First record of an extinct marabou stork in the Neogene of South America. *Acta Palaeontologica Polonica* **53**, 593–600.

Noriega, J.I., Tambussi, C.P., & Cozzuol, M.A. (2008) New material of *Cayaoa bruneti* Tonni, an early Miocene anseriform (Aves) from Patagonia, Argentina. *Neues Jahrbuch für Geologie und Paläontologie, Abhandlungen* **249**, 271–280.

Noriega, J.I., Areta, J.I., Vizcaíno, S.F., & Bargo, M.S. (2011) Phylogeny and taxonomy of the Patagonian Miocene falcon *Thegornis musculosus* Ameghino, 1895 (Aves: Falconidae). *Journal of Paleontology* **85**, 1089–1104.

Northcote, E. (1982) Size, form and habit of the extinct Maltese Swan *Cygnus falconeri*. *The Ibis* **124**, 148–158.

Novas, F.E., Agnolín, F.L., & Scanferla, C.A. (2010) New enantiornithine bird (Aves, Ornithothoraces) from the Late Cretaceous of NW Argentina. *Comptes Rendus Palevol* **9**, 499–503.

Nudds, R.L. (2014) Reassessment of the wing feathers of *Archaeopteryx lithographica* suggests no robust evidence for the presence of elongated dorsal wing coverts. *PLoS ONE* **9**, e93963.

O'Connor, J.K., & Chang, H. (2015) Hindlimb feathers in paravians: Primarily "wings" or ornaments? *Biology Bulletin* **42**, 616–621.

O'Connor, J.K., & Chiappe, L.M. (2011) A revision of enantiornithine (Aves: Ornithothoraces) skull morphology. *Journal of Systematic Palaeontology* **9**, 135–157.

O'Connor, J.K., & Sullivan, C. (2014) Reinterpretation of the Early Cretaceous maniraptoran (Dinosauria: Theropoda) *Zhongornis haoae* as a scansoriopterygid-like non-avian, and morphological resemblances between scansoriopterygids and basal oviraptorosaurs. *Vertebrata PalAsiatica* **52**, 3–30.

O'Connor, J.K., & Zelenkov, N.V. (2013) The phylogenetic position of *Ambiortus*: Comparison with other Mesozoic birds from Asia. *Paleontological Journal* **47**, 1270–1281.

O'Connor, J.K., & Zhou, Z. (2013) A redescription of *Chaoyangia beishanensis* (Aves) and a comprehensive phylogeny of Mesozoic birds. *Journal of Systematic Palaeontology* **11**, 889–906.

O'Connor, J.K., Wang, X., Chiappe, L.M., Gao, C., Meng, Q., Cheng, X., & Liu, J. (2009) Phylogenetic support for a specialized clade of Cretaceous enantiornithine birds with information from a new species. *Journal of Vertebrate Paleontology* **29**, 188–204.

O'Connor, J.K., Gao, K.-Q., & Chiappe, L.M. (2010) A new ornithuromorph (Aves: Ornithothoraces) bird from the Jehol Group indicative of higher-level diversity. *Journal of Vertebrate Paleontology* **30**, 311–321.

O'Connor, J.K., Chiappe, L.M., & Bell, A. (2011a) Pre-modern birds: Avian divergences in the Mesozoic. In: G. Dyke & G. Kaiser (eds.), *Living Dinosaurs: The Evolutionary History of Modern Birds*. Chichester: John Wiley & Sons, pp. 39–114.

O'Connor, J.K., Zhou, Z., & Zhang, F. (2011b) A reappraisal of *Boluochia zhengi* (Aves: Enantiornithes) and a discussion of intraclade diversity in the Jehol avifauna, China. *Journal of Systematic Palaeontology* **9**, 51–63.

O'Connor, J.K., Chiappe, L.M., Gao, C., & Zhao, B. (2011c) Anatomy of the Early Cretaceous enantiornithine bird *Rapaxavis pani*. *Acta Palaeontologica Polonica* **56**, 463–475.

O'Connor, J.K., Chiappe, L.M., Chuong, C.M., Bottjer, D.J., & You, H. (2012a) Homology and potential cellular and molecular mechanisms for the development of unique feather morphologies in early birds. *Geosciences* **2**, 157–177.

O'Connor, J.K., Sun, C., Xu, X., Wang, X., & Zhou, Z. (2012b) A new species of *Jeholornis* with complete caudal integument. *Historical Biology* **24**, 29–41.

O'Connor, J., Wang, X., Sullivan, C., Zheng, X., Tubaro, P., Zhang, X., & Zhou, Z. (2013a) Unique caudal plumage of *Jeholornis* and complex tail evolution in early birds. *Proceedings of the National Academy of Sciences USA* **110**, 17404–17408.

O'Connor, J.K., Zhang, Y., Chiappe, L.M., Meng, Q., Quanguo, L., & Di, L. (2013b) A new enantiornithine from the Yixian Formation with the first recognized avian enamel specialization. *Journal of Vertebrate Paleontology* **33**, 1–12.

O'Connor, J.K., Averianov, A.O., & Zelenkov, N.V. (2014) A confuciusornithiform (Aves, Pygostylia)-like tarsometatarsus from the Early Cretaceous of Siberia and a discussion of the evolution of avian hind limb musculature. *Journal of Vertebrate Paleontology* **34**, 647–656.

O'Connor, J.K., Zheng X.-T., Wang X.-L., Zhang X.-M., & Zhou Z.-H. (2015a) The gastral basket in basal birds and their close relatives: Size and possible function. *Vertebrata PalAsiatica* **53**, 133–152.

O'Connor, J.K., Wang, M., Zhou, S., & Zhou, Z. (2015b) Osteohistology of the Lower Cretaceous Yixian Formation ornithuromorph (Aves) *Iteravis huchzermeyeri*. *Palaeontologia Electronica* **18.2.35A**, 1–11.

O'Connor, J.K., Li, D.-Q., Lamanna, M.C., Wang, M., Harris, J.D., Atterholt, J., & You, H.-L. (2016) A new Early Cretaceous enantiornithine (Aves, Ornithothoraces) from northwestern China with elaborate tail ornamentation. *Journal of Vertebrate Paleontology* **36**, e1054035.

O'Connor, J.K., Wang, M., & Hu, H. (in press) A new ornithuromorph (Aves) with an elongate rostrum from the Jehol Biota, and the early evolution of rostralization in birds. *Journal of Systematic Palaeontology*.

Olson, S.L. (1976) Oligocene fossils bearing on the origins of the Todidae and the Momotidae (Aves: Coraciiformes). *Smithsonian Contributions to Paleobiology* **27**, 111–119.

Olson, S.L. (1977a) A synopsis of the fossil
Rallidae. In: D.S. Ripley (ed.), *Rails of the
World: A Monograph of the Family Rallidae.*
Boston, MA: Godine, pp. 339–379.

Olson, S.L. (1977b) A Lower Eocene frigatebird
from the Green River Formation of Wyoming
(Pelecaniformes: Fregatidae). *Smithsonian
Contributions to Paleobiology* **35**, 1–33.

Olson, S.L. (1980) A new genus of penguin-like
pelecaniform bird from the Oligocene of
Washington (Pelecaniformes: Plotopteridae).
*Natural History Museum of Los Angeles
County, Contributions in Science* **330**, 51–57.

Olson, S.L. (1984) A hamerkop from the early
Pliocene of south Africa (Aves: Scopidae).
*Proceedings of the Biological Society of
Washington* **97**, 736–740.

Olson, S.L. (1985a) The fossil record of birds. In:
D.S. Farner, J.R. King, & K.C. Parkes (eds.),
Avian Biology, vol. **8**. New York: Academic
Press, pp. 79–238.

Olson, S.L. (1985b) A new genus of tropicbird
(Pelecaniformes: Phaethontidae) from the
Middle Miocene Calvert Formation of
Maryland. *Proceedings of the Biological Society
of Washington* **98**, 851–855.

Olson, S.L. (1987) An early Eocene oilbird from
the Green River Formation of Wyoming
(Caprimulgiformes: Steatornithidae).
*Documents des Laboratoires de Géologie de
Lyon* **99**, 57–69.

Olson, S.L. (1992a) *Neogaeornis wetzeli*
Lambrecht, a Cretaceous loon from Chile (Aves:
Gaviidae). *Journal of Vertebrate Paleontology*
12, 122–124.

Olson, S.L. (1992b) A new family of primitive
landbirds from the Lower Eocene Green River
Formation of Wyoming. *Natural History
Museum of Los Angeles County, Science Series*
36, 137–160.

Olson, S.L. (1995) *Thiornis sociata* Navas, a
nearly complete Miocene grebe (Aves:
Podicipedidae). *Courier Forschungsinstitut
Senckenberg* **181**, 131–140.

Olson, S.L. (1999a) The anseriform relationships
of *Anatalavis* Olson and Parris (Anseranatidae),
with a new species from the Lower Eocene
London Clay. *Smithsonian Contributions to
Paleobiology* **89**, 231–243.

Olson, S.L. (1999b) A new species of pelican
(Aves: Pelecanidae) from the Lower Pliocene of
North Carolina and Florida. *Proceedings of the
Biological Society of Washington* **112**, 503–509.

Olson, S.L. (2001) Why so many kinds of
passerine birds? *BioScience* **51**, 268–269.

Olson, S.L. (2003) First fossil record of a finfoot
(Aves: Heliornithidae) and its biogeographical
significance. *Proceedings of the Biological
Society of Washington* **116**, 732–736.

Olson, S.L. (2008) A new species of large,
terrestrial caracara from Holocene deposits in
southern Jamaica (Aves: Falconidae). *Journal of
Raptor Research* **42**, 265–272.

Olson, S.L. (2009) A new diminutive species of
shearwater of the genus *Calonectris* (Aves:
Procellariidae) from the Middle Miocene Calvert
Formation of Chesapeake Bay. *Proceedings of
the Biological Society of Washington* **122**,
466–470.

Olson, S.L. (2011) A new genus and species of
unusual tern (Aves: Laridae: Anoinae) from the
Middle Miocene Calvert Formation of Virginia.
*Proceedings of the Biological Society of
Washington* **124**, 270–279.

Olson, S.L., & Alvarenga, H.M.F. (2002) A new
genus of small teratorn from the Middle
Tertiary of the Taubaté Basin, Brazil (Aves:
Teratornithidae). *Proceedings of the Biological
Society of Washington* **115**, 701–705.

Olson, S.L., & Eller, K.G. (1989) A new species of
painted snipe (Charadriiformes: Rostratulidae)
from the Early Pliocene at Langebaanweg,
southwestern Cape Province, South Africa.
Ostrich **60**, 118–121.

Olson, S.L., & Feduccia, A. (1979) Flight
capability and the pectoral girdle of
Archaeopteryx. *Nature* **278**, 247–248.

Olson, S.L., & Feduccia A (1980a) *Presbyornis* and
the origin of the Anseriformes (Aves:
Charadriomorphae). *Smithsonian Contributions
to Zoology* **323**, 1–24.

Olson, S.L., & Feduccia A (1980b) Relationships
and evolution of flamingos (Aves:
Phoenicopteridae). *Smithsonian Contributions
to Zoology* **316**, 1–73.

Olson, S.L., & Hasegawa, Y. (1985) A femur of
Plotopterum from the early middle Miocene of
Japan (Pelecaniformes: Plotopteridae). *Bulletin
of the National Science Museum, Series C* **11**,
137–140.

Olson, S.L., & Hasegawa, Y. (1996) A new genus
and two new species of gigantic Plotopteridae

from Japan (Aves: Plotopteridae). *Journal of Vertebrate Paleontology* **16**, 742–751.

Olson, S.L., & Hearty, P.J. (2003) Probable extirpation of a breeding colony of Short-tailed Albatross (*Phoebastria albatrus*) on Bermuda by Pleistocene sea-level rise. *Proceedings of the National Academy of Sciences USA* **100**, 12825–12829.

Olson, S.L., & James, H.F. (1991) Descriptions of thirty-two new species of birds from the Hawaiian Islands: Part I. Non-Passeriformes. *Ornithological Monographs* **45**, 1–88.

Olson, S.L., & Matsuoka, H. (2005) New specimens of the early Eocene frigatebird *Limnofregata* (Pelecaniformes: Fregatidae), with the description of a new species. *Zootaxa* **1046**, 1–15.

Olson, S.L., & Parris, D.C. (1987) The Cretaceous birds of New Jersey. *Smithsonian Contributions to Paleobiology* **63**, 1–22.

Olson, S.L., & Rasmussen, P.C. (2001) Miocene and Pliocene birds from the Lee Creek Mine, North Carolina. *Smithsonian Contributions to Paleobiology* **90**, 233–365.

Olson, S.L., & Walker, C.A. (1997) A trans-Atlantic record of the fossil tropicbird *Heliadornis ashbyi* (Aves: Phaethontidae) from the Miocene of Belgium. *Proceedings of the Biological Society of Washington* **110**, 624–628.

Osmólska, H., Currie, P.J., & Barsbold, R. (2004) Oviraptorosauria. In: D.B. Weishampel, P. Dodson, & H. Osmólska (eds.), *The Dinosauria*, 2nd edn. Berkeley: University of California Press, pp. 165–183.

Ostrom, J.H. (1974) *Archaeopteryx* and the origin of flight. *Quarterly Review of Biology* **49**, 27–47.

Ostrom, J.H. (1976) *Archaeopteryx* and the origin of birds. *Biological Journal of the Linnean Society* **8**, 91–182.

Owen, R. (1873) Description of the skull of a dentigerous bird (*Odontopteryx toliapicus*, Ow.) from the London Clay of Sheppey. *Quarterly Journal of the Geological Society of London* **29**, 511–521.

Pacheco, M.A., Battistuzzi, F.U., Lentino, M., Aguilar, R.F., Kumar, S., & Escalante, A.A. (2011) Evolution of modern birds revealed by mitogenomics: Timing the radiation and origin of major orders. *Molecular Biology and Evolution* **28**, 1927–1942.

Padian, K. (2003) Four-winged dinosaurs, bird precursors, or neither? *BioScience* **53**, 450–452.

Pan, Y., Sha, J., Zhou, Z., & Fürsich, F.T. (2013) The Jehol Biota: Definition and distribution of exceptionally preserved relicts of a continental Early Cretaceous ecosystem. *Cretaceous Research* **44**, 30–38.

Panteleyev, A.V. (2011) [First bird remains from the Paleogene of Crimea]. In: M.S. Batashev, N.P. Makarov, & N.V. Martinovich, (eds.), [*Dedicated to Arkadiy Yakovlevich Tugarinov, a selection of scientific articles*]. Krasnoyarsk: Krasnoyarsk Regional Museum, pp. 83–91 [in Russian].

Panteleyev, A.V., Popov, E.V., & Averianov, A.O. (2004) New record of *Hesperornis rossicus* (Aves, Hesperornithiformes) in the Campanian of Saratov Province, Russia. *Paleontological Research* **8**, 115–122.

Parish, J.C. (2012) *The Dodo and the Solitaire: A Natural History*. Bloomington: Indiana University Press.

Parris, D.C., & Hope, S. (2002) New interpretations of the birds from the Navesink and Hornerstown Formations, New Jersey, USA (Aves: Neornithes). In: Z.-H. Zhou & F.-Z. Zhang (eds.), *Proceedings of the 5th Symposium of the Society of Avian Paleontology and Evolution, 1–4 June 2000*. Beijing: Science Press, pp. 113–124.

Paton, T.A., & Baker, A.J. (2006) Sequences from 14 mitochondrial genes provide a well-supported phylogeny of the charadriiform birds congruent with the nuclear RAG-1 tree. *Molecular Phylogenetics and Evolution* **39**, 657–667.

Paul, G.S. (2002) *Dinosaurs of the Air: The Evolution and Loss of Flight in Dinosaurs and Birds*. Baltimore, MD: Johns Hopkins University Press.

Pavia, M. (2004). A new large barn owl (Aves, Strigiformes, Tytonidae) from the Middle Pleistocene of Sicily, Italy, and its taphonomical significance. *Geobios* **37**, 631–641.

Pavia, M., Göhlich, U.B., & Mourer-Chauviré, C. (2012) Description of the type-series of *Palaeocryptonyx donnezani* Deperet, 1892 (Aves: Phasianidae) with the selection of a lectotype. *Comptes Rendus Palevol* **11**, 257–263.

Pavia, M., Manegold, A., & Haarhoff, P. (2015) New early Pliocene owls from Langebaanweg,

South Africa, with first evidence of the *Athene* south of the Sahara and a new species of *Tyto*. *Acta Palaeontologica Polonica* **60**, 815–828.

Perrichot, V., Marion, L., Néraudeau, D., Vullo, R., & Tafforeau, P. (2008) The early evolution of feathers: Fossil evidence from Cretaceous amber of France. *Proceedings of the Royal Society B: Biological Sciences* **275**, 1197–1202.

Persons, W.S., & Currie, P.J. (2012) Dragon tails: Convergent caudal morphology in winged archosaurs. *Acta Geologica Sinica (English edition)* **86**, 1402–1412.

Peters, D.S. (1983) Die "Schnepfenralle" *Rhynchaeites messelensis* Wittich 1898 ist ein Ibis. *Journal für Ornithologie* **124**, 1–27.

Peters, D.S. (1985) Functional and constructive limitations in the early evolution of birds. In: M.K. Hecht, J.H. Ostrom, G. Viohl, & P. Wellnhofer (eds.), *The Beginnings of Birds*. Eichstätt: Freunde des Jura-Museums Eichstätt, pp. 243–249.

Peters, D.S. (1987) Ein "Phorusrhacide" aus dem Mittel-Eozän von Messel (Aves: Gruiformes: Cariamae). *Documents des Laboratoires de Géologie de Lyon* **99**, 71–87.

Peters, D.S. (1988) Ein vollständiges Exemplar von *Palaeotis weigelti* (Aves, Palaeognathae). *Courier Forschungsinstitut Senckenberg* **107**, 223–233.

Peters, D.S. (1992) A new species of owl (Aves: Strigiformes) from the Middle Eocene Messel oil shale. *Natural History Museum of Los Angeles County, Science Series* **36**, 161–169.

Peters, D.S., & Ji, Q. (1998) The diapsid temporal construction of the Chinese fossil bird *Confuciusornis*. *Senckenbergiana lethaea* **78**, 153–155.

Peters, D.S., & Ji, Q. (1999) Mußte *Confuciusornis* klettern? *Journal für Ornithologie* **140**, 41–50.

Peters, W.S., & Peters, D.S. (2010) Sexual size dimorphism is the most consistent explanation for the body size spectrum of *Confuciusornis sanctus*. *Biology Letters* **6**, 531–532.

Phillips, M.J., Gibb, G.C., Crimp, E.A., & Penny, D. (2010) Tinamous and moa flock together: Mitochondrial genome sequence analysis reveals independent losses of flight among ratites. *Systematic Biology* **59**, 90–107.

Pickford, M., Senut, B., & Mourer-Chauviré, C. (2004) Early Pliocene Tragulidae and peafowls in the Rift Valley, Kenya: Evidence for rainforest in East Africa. *Comptes Rendus Palevol* **3**, 179–189.

Pittman, M., Gatesy, S.M., Upchurch, P., Goswami, A., & Hutchinson, J.R. (2013) Shake a tail feather: The evolution of the theropod tail into a stiff aerodynamic surface. *PLoS ONE* **8**, e63115.

Poplin, F., Mourer-Chauviré, C., & Evin, J. (1983) Position systématique et datation de *Sylviornis neocaledoniae*, Mégapode géant (Aves, Galliformes, Megapodiidae) éteint de la Nouvelle-Calédonie. *Comptes Rendus des Séances de l'Académie des Sciences, série II* **297**, 301–304.

Porras-Múzquiz, H.G., Chatterjee, S., & Lehman, T.M. (2014) The carinate bird *Ichthyornis* from the Upper Cretaceous of Mexico. *Cretaceous Research* **51**, 148–152.

Proctor, N.S., & Lynch, P.J. (1993) *Manual of Ornithology: Avian Structure and Function*. New Haven, CT: Yale University Press.

Prum, R.O. (2002) Why ornithologists should care about the theropod origin of birds. *The Auk* **119**, 1–17.

Prum, R.O. (2010) Moulting tail feathers in a juvenile oviraptorosaur. *Nature* **468**, E1.

Prum, R.O., & Brush, A.H. (2002) The evolutionary origin and diversification of feathers. *Quarterly Review of Biology* **77**, 261–295.

Prum, R.O., Torres, R.H., Williamson, S., & Dyck, J. (1998) Coherent light scattering by blue feather barbs. *Nature* **396**, 28–29.

Prum, R.O., Berv, J.S., Dornburg, A., Field, D.J., Townsend, J.P., Lemmon, E.M., & Lemmon, A.R. (2015) A comprehensive phylogeny of birds (Aves) using targeted next-generation DNA sequencing. *Nature* **526**, 569–573.

Pu, H., Chang, H., Lü, J., Wu, Y., Xu, L., Zhang, J., & Jia, S. (2013) A new juvenile specimen of *Sapeornis* (Pygostylia: Aves) from the Lower Cretaceous of Northeast China and allometric scaling of this basal bird. *Paleontological Research* **17**, 27–38.

Puttick, M.N., Thomas, G.H., & Benton, M.J. (2014) High rates of evolution preceded the origin of birds. *Evolution* **68**, 1497–1510.

Ramirez, J.L., Miyaki, C.Y., & Del Lama, S.N. (2013) Molecular phylogeny of Threskiornithidae (Aves: Pelecaniformes) based on nuclear and mitochondrial DNA. *Genetics and Molecular Research* **12**, 2740–2750.

Rando, J.C., López, M., & Seguí, B. (1999) A new species of extinct flightless passerine (Emberizidae: *Emberiza*) from the Canary Islands. *The Condor* **101**, 1–13.

Rasmussen, D.T., Olson, S.L., & Simons, E.L. (1987) Fossil birds from the Oligocene Jebel Qatrani Formation, Fayum Province, Egypt. *Smithsonian Contributions to Paleobiology* **62**, 1–20.

Rasmussen, D.T., Simons, E.L., Hertel, F., & Judd, A. (2001) Hindlimb of a giant terrestrial bird from the Upper Eocene, Fayum, Egypt. *Palaeontology* **44**, 325–337.

Rasmussen, P.C. (1998) Early Miocene avifauna from the Pollack Farm Site, Delaware. *Delaware Geological Survey Special Publication* **21**, 149–151.

Rayner, J.M.V. (2001) On the origin and evolution of flapping flight aerodynamics in birds. In: J. Gauthier & L.F. Gall (eds.), *New Perspectives on the Origin and Early Evolution of Birds: Proceedings of the International Symposium in Honor of John H. Ostrom*. New Haven, CT: Peabody Museum of Natural History, Yale University, pp. 363–381.

Rees, J., & Lindgren, J. (2005) Aquatic birds from the Upper Cretaceous (lower Campanian) of Sweden and the biology and distribution of hesperornithiforms. *Palaeontology* **48**, 1321–1329.

Regal, P.J. (1975) The evolutionary origin of feathers. *Quarterly Review of Biology* **50**, 35–66.

Rich, P.V. (1980) "New World vultures" with Old World affinities? A review of fossil and recent Gypaetinae of both the Old and the New World. In: M.K. Hecht & F.S. Szalay (eds.), *Contributions to Vertebrate Evolution*, vol. **5**. Basel: Karger, pp. 1–115.

Rich, P.V., & Bohaska, D.J. (1981) The Ogygoptyngidae, a new family of owls from the Paleocene of North America. *Alcheringa* **5**, 95–102.

Rich, P.V., & Haarhoff, P.J. (1985) Early Pliocene Coliidae (Aves, Coliiformes) from Langebaanweg, South Africa. *Ostrich* **56**, 20–41.

Rich, P.V., & McEvey, A. (1977) A new owlet-nightjar from the early to mid-Miocene of eastern New South Wales. *Memoirs of the National Museum of Victoria* **38**, 247–253.

Rich, P.V., & Walker, C.A. (1983) A new genus of Miocene flamingo from East Africa. *Ostrich* **54**, 95–104.

Rietschel, S. (1985) Feathers and wings of *Archaeopteryx*, and the question of her flight ability. In: M.K. Hecht, J.H. Ostrom, G. Viohl, & P. Wellnhofer (eds.), *The Beginnings of Birds*. Eichstätt: Freunde des Jura-Museums Eichstätt, pp. 251–260.

Rinderknecht, A., & Noriega, J.I. (2002) Un nuevo género de Anhingidae (Aves: Pelecaniformes) de la Formación San José (Plioceno-Pleistoceno) del Uruguay. *Ameghiniana* **39**, 183–192.

Sakurai, K., Kimura, M., & Katoh, T. (2008) A new penguin-like bird (Pelecaniformes: Plotopteridae) from the Late Oligocene Tokoro Formation, northeastern Hokkaido, Japan. *Oryctos* **7**, 83–94.

Sallaberry, M., Rubilar-Rogers, D., Suárez, M.E., & Gutstein, C.S. (2007) The skull of a fossil prion (Aves: Procellariiformes) from the Neogene (Late Miocene) of northern Chile. *Revista Geológica de Chile* **34**, 147–154.

Sangster, G. (2005) A name for the flamingo-grebe clade. *The Ibis* **147**, 612–615.

Sanz, J.L., Chiappe, L.M., Pérez-Moreno, B.P., Buscalioni, A.D., Moratalla, J.J., Ortega, F., & Poyato-Ariza, F.J. (1996) An Early Cretaceous bird from Spain and its implications for the evolution of avian flight. *Nature* **382**, 442–444.

Sauer, E.G.F. (1969) Evidence and evolutionary interpretation of *Psammornis*. *Bonner zoologische Beiträge* **20**, 290–310.

Sauer, E.G.F. (1979) A Miocene ostrich from Anatolia. *The Ibis* **121**, 494–501.

Sauer, E.G.F., & Rothe, P. (1972) Ratite eggshells from Lanzarote, Canary Islands. *Science* **176**, 43–45.

Schaller, N.U. (2008) *Structural attributes contributing to locomotor performance in the ostrich*. Unpublished PhD thesis, Ruperto-Carola University, Heidelberg.

Schaller, N.U., D'Août, K., Villa, R., Herkner, B., & Aerts, P. (2011) Toe function and dynamic pressure distribution in ostrich locomotion. *Journal of Experimental Biology* **214**, 1123–1130.

Scher, H.D., & Martin, E.E. (2006) Timing and climatic consequences of the opening of Drake Passage. *Science* **312**, 428–430.

Schweizer, M., Seehausen, O., & Hertwig, S.T. (2011) Macroevolutionary patterns in the diversification of parrots: Effects of climate change, geological events and key innovations. *Journal of Biogeography* **38**, 2176–2194.

Seguí, B. (2002) A new genus of crane (Aves: Gruiformes) from the Late Tertiary of the Balearic Islands, Western Mediterranean. *The Ibis* **144**, 411–422.

Seguí, B., Quintana, J., Fornós, J.J., & Alcover, J.A. (2001) A new fulmarine petrel from the Upper Miocene of the western Mediterranean. *Palaeontology* **44**, 933–948.

Senter, P. (2006) Scapular orientation in theropods and basal birds, and the origin of flapping flight. *Acta Palaeontologica Polonica* **51**, 305–313.

Shapiro, B., Sibthorpe, D., Rambaut, A., Austin, J., Wragg, G.M., Bininda-Emonds, O.R., Lee, P.L.M., & Cooper, A. (2002) Flight of the dodo. *Science* **295**, 1683.

Sheldon, F.H., Jones, C.E., & McCracken, K.G. (2000) Relative patterns and rates of evolution in heron nuclear and mitochondrial DNA. *Molecular Biology and Evolution* **17**, 437–450.

Slack, K.E., Jones, C.M., Ando, T., Harrison, G.L., Fordyce, R.E., Arnason, U., & Penny, D. (2006) Early penguin fossils, plus mitochondrial genomes, calibrate avian evolution. *Molecular Biology and Evolution* **23**, 1144–1155.

Smith, A.G., Smith, D.G., & Funnell, B.M. (1994) *Atlas of Mesozoic and Cenozoic Coastlines.* Cambridge: Cambridge University Press.

Smith, J.V., Braun, E.L., & Kimball, R.T. (2013) Ratite non-monophyly: Independent evidence from 40 novel loci. *Systematic Biology* **62**, 35–49.

Smith, N.A. (2011) Taxonomic revision and phylogenetic analysis of the flightless Mancallinae (Aves, Pan-Alcidae). *Zookeys* **91**, 1–116.

Smith, N.A. (2013) A new species of auk (Charadriiformes, Pan-Alcidae) from the Miocene of Mexico. *The Condor* **115**, 77–83.

Smith, N.A. (2014) The fossil record and phylogeny of the auklets (Pan-Alcidae, Aethiini). *Journal of Systematic Palaeontology* **12**, 217–236.

Smith, N.A., & Clarke, J.A. (2011) An alphataxonomic revision of extinct and extant razorbills (Aves, Alcidae): A combined morphometric and phylogenetic approach. *Ornithological Monographs* **72**, 1–61.

Smith, N.A., & Clarke, J.A. (2015) Systematics and evolution of the Pan-Alcidae (Aves, Charadriiformes). *Journal of Avian Biology* **46**, 125–140.

Smith, N.D. (2010) Phylogenetic analysis of Pelecaniformes (Aves) based on osteological data: Implications for waterbird phylogeny and fossil calibration studies. *PLoS ONE* **5**, e13354.

Smith, N.D., Grande, L., & Clarke, J.A. (2013) A new species of Threskiornithidae-like bird (Aves, Ciconiiformes) from the Green River Formation (Eocene) of Wyoming. *Journal of Vertebrate Paleontology* **33**, 363–381.

Sorenson, M.D., Cooper, A., Paxinos, E.E., Quinn, T.W., James, H.F., Olson, S.L., & Fleischer, R.C. (1999) Relationships of the extinct moa-nalos, flightless Hawaiian waterfowl, based on ancient DNA. *Proceedings of the Royal Society of London, Series B* **266**, 2187–2193.

Steadman, D.W. (2006) *Extinction and Biogeography of Tropical Pacific Birds.* Chicago, IL: University of Chicago Press.

Steadman, D.W. (2008) Doves (Columbidae) and cuckoos (Cuculidae) from the early Miocene of Florida. *Bulletin of the Florida Museum of Natural History* **48**, 1–16.

Stegmann, B. (1964) Die funktionelle Bedeutung des Schlüsselbeines bei den Vögeln. *Journal für Ornithologie* **105**, 450–463.

Stegmann, B. (1965) Funktionell bedingte Eigenheiten am Metacarpus des Vogelflügels. *Journal für Ornithologie* **106**, 179–189.

Stidham, T.A. (2011) The carpometacarpus of the Pliocene turkey (Galliformes: Phasianidae) and the problem of morphological variability in turkeys. *PaleoBios* **30**, 13–17.

Stidham, T.A. (2015) A new species of *Limnofregata* (Pelecaniformes: Fregatidae) from the Early Eocene Wasatch Formation of Wyoming: Implications for palaeoecology and palaeobiology. *Palaeontology* **58**, 239–249.

Stidham, T.A., & Ni, X.-J. (2014) Large anseriform (Aves: Anatidae: Romainvilliinae?) fossils from the Late Eocene of Xinjiang, China. *Vertebrata PalAsiatica* **52**, 98–111.

Stidham, T.A., & Smith, N.A. (2015) An ameghinornithid-like bird (Aves, Cariamae, ?Ameghinornithidae) from the early Oligocene of Egypt. *Palaeontologia Electronica* **18.1.5A**, 1–8.

Stidham, T.A., Lofgren, D., Farke, A.A., Paik, M., & Choi, R. (2014) A lithornithid (Aves: Palaeognathae) from the Paleocene (Tiffanian) of southern California. *PaleoBios* **31**, 1–7.

Stille, P., Steinmann, M., & Riggs, S.R. (1996) Nd isotope evidence for the evolution of the paleocurrents in the Atlantic and Tethys Oceans during the past 180 Ma. *Earth and Planetary Science Letters* **144**, 9–19.

Storer, R.W. (2000) The systematic position of the Miocene grebe *Thiornis sociata* Navás. *Annales de Paléontologie* **86**, 129–139.

Stucchi, M. (2003) Los piqueros (Aves: Sulidae) de la Formación Pisco, Peruì. *Boletín de la Sociedad Geológica del Perú* **95**, 75–91.

Stucchi, M., & Emslie, S.D. (2005) A new condor (Ciconiiformes, Vulturidae) from the late Miocene/early Pliocene Pisco Formation, Peru. *The Condor* **107**, 107–113.

Stucchi, M., & Urbina, M. (2004) *Ramphastosula* (Aves, Sulidae): A new genus from the early Pliocene of the Pisco Formation, Peru. *Journal of Vertebrate Paleontology* **24**, 974–978.

Stucchi, M., Urbina, M., & Giraldo, A. (2003) Una nueva especie de Spheniscidae del Mioceno Tardío de la Formación Pisco, Perú. *Boletín del Instituto Francés de Estudios Andinos* **32**, 361–375.

Stucchi, M., Emslie, S.D., Varas-Malca, R.M., & Urbina-Schmitt, M. (2015) A new late Miocene condor (Aves, Cathartidae) from Peru and the origin of South American condors. *Journal of Vertebrate Paleontology* **35**, e972507.

Suárez, W., & Olson, S.L. (2009) A new genus for the Cuban teratorn (Aves: Teratornithidae). *Proceedings of the Biological Society of Washington* **122**, 103–116.

Subramanian, S., Beans-Picón, G., Swaminathan, S.K., Millar, C.D., & Lambert, D.M. (2013) Evidence for a recent origin of penguins. *Biology Letters* **9**, 20130748.

Suh, A., Paus, M., Kiefmann, M., Churakov, G., Franke, F., Brosius, J., Kriegs, J., & Schmitz, J. (2011) Mesozoic retroposons reveal parrots as the closest living relatives of passerine birds. *Nature Communications* **2**, 443.

Suh, A., Smeds, L., & Ellegren, H. (2015) The dynamics of incomplete lineage sorting across the ancient adaptive radiation of neoavian birds. *PLoS Biology* **13**, e1002224.

Sullivan, C., Wang, Y., Hone, D.W., Wang, Y., Xu, X., & Zhang, F. (2014) The vertebrates of the Jurassic Daohugou Biota of northeastern China. *Journal of Vertebrate Paleontology* **34**, 243–280.

Švec, P. (1982) Two new species of diving birds from the Lower Miocene of Czechoslovakia. *Časopis pro Mineralogii a Geologii* **27**, 243–260.

Tambussi, C.P. (1995) The fossil Rheiformes from Argentina. *Courier Forschungsinstitut Senckenberg* **181**, 121–129.

Tambussi, C.P., & Acosta Hospitaleche, C. (2007) Antarctic birds (Neornithes) during the Cretaceous-Eocene times. *Revista de la Asociación Geológica Argentina* **62**, 604–617.

Tambussi, C.P., & Degrange, F. (2013) *South American and Antarctic Continental Cenozoic Birds: Paleobiogeographic Affinities and Disparities*. Dordrecht: Springer.

Tambussi, C., Ubilla, M., & Perea, D. (1999) The youngest large carnassial bird (Phorusrhacidae, Phorusrhacinae) from South America (Pliocene-Early Pleistocene of Uruguay). *Journal of Vertebrate Paleontology* **19**, 404–406.

Tennyson, A.J.D., Worthy, T.H., Jones, C.M., Scofield, R.P., & Hand, S.J. (2010) Moa's ark: Miocene fossils reveal the great antiquity of moa (Aves: Dinornithiformes) in Zealandia. *Records of the Australian Museum* **62**, 105–114.

Thomas, D.B., & Ksepka, D.T. (2013) A history of shifting fortunes for African penguins. *Zoological Journal of the Linnean Society* **168**, 207–219.

Thomas, D.B., Ksepka, D.T., & Forydce, R.E. (2011) Penguin heat-retention structures evolved in a greenhouse Earth. *Biology Letters* **7**, 461–464.

Tickle, P.G., Ennos, A.R., Lennox, L.E., Perry, S.F., & Codd, J.R. (2007) Functional significance of the uncinate processes in birds. *Journal of Experimental Biology* **210**, 3955–3961.

Tokaryk, T.T., Cumbaa, S.L., & Storer, J.E. (1997) Early Late Cretaceous birds from Saskatchewan, Canada: The oldest diverse avifauna known from North America. *Journal of Vertebrate Paleontology* **17**, 172–176.

Tomek, T., Bochenski, Z.M., Wertz, K., & Swidnicka, E. (2014) A new genus and species of a galliform bird from the Oligocene of Poland. *Palaeontologia Electronica* **17.3.38A**, 1–15.

Tonni, E.P., & Noriega, J.I. (1996) Una nueva especie de *Nandayus* Bonaparte, 1854 (Aves: Psittaciformes) del Plioceno tardío de Argentinia. *Revista Chilena de Historia Natural* **69**, 97–104.

Tonni, E.P., & Noriega, J.I. (1998) Los cóndores (Ciconiiformes, Vulturidae) de la región pampeana de la Argentina durante el Cenozoico tardío: Distribución, interacciones y extinciones. *Ameghiniana* **35**, 141–150.

Torres, C.R., De Pietri, V.L., Louchart, A., & van Tuinen, M. (2015) New cranial material of the earliest filter-feeding flamingo *Harrisonavis croizeti* (Aves, Phoenicopteridae) informs the evolution of the highly specialized filter-feeding apparatus. *Organisms, Diversity & Evolution* **15**, 609–618.

Tsuihiji, T., Barsbold, R., Watabe, M., Tsogtbaatar, K., Chinzorig, T., Fujiyama, Y., & Suzuki, S. (2014) An exquisitely preserved troodontid theropod with new information on the palatal structure from the Upper Cretaceous of Mongolia. *Naturwissenschaften* **101**, 131–142.

Turner, A.H., Pol, D., Clarke, J.A., Erickson, G.M., & Norell, M.A. (2007a) A basal dromaeosaurid and size evolution preceding avian flight. *Science* **317**, 1378–1381.

Turner, A.H., Makovicky, P.J., & Norell, M.A. (2007b) Feather quill knobs in the dinosaur *Velociraptor*. *Science* **317**, 1721–1721.

Turner, A.H., Makovicky, P.J., & Norell, M.A. (2012) A review of dromaeosaurid systematics and paravian phylogeny. *Bulletin of the American Museum of Natural History* **371**, 1–206.

Turvey, S.T., Green, O.R., & Holdaway, R.N. (2005) Cortical growth marks reveal extended juvenile development in New Zealand moa. *Nature* **435**, 940–943.

van Tets, G.F., Rich, P.V., & Marino-Hadiwardoyo, H.R. (1989) A reappraisal of *Protoplotus beauforti* from the early Tertiary of Sumatra and the basis of a new pelecaniform family. *Bulletin of the Geological Research and Development Centre, Paleontology Series* **5**, 57–75.

van Valkenburgh, B. (1999) Major patterns in the history of carnivorous mammals. *Annual Review of Earth and Planetary Sciences* **27**, 463–493.

Varricchio, D.J., & Barta, D.E. (2015) Revisiting Sabath's "larger avian eggs" from the Gobi Cretaceous. *Acta Palaeontologica Polonica* **60**, 11–25.

Vezzosi, R.I. (2012) First record of *Procariama simplex* Rovereto, 1914 (Phorusrhacidae, Psilopterinae) in the Cerro Azul Formation (upper Miocene) of La Pampa Province; remarks on its anatomy, palaeogeography and chronological range. *Alcheringa* **36**, 157–169.

Vickers-Rich, P., Chiappe, L.M., & Kurzanov, S. (2002) The enigmatic birdlike dinosaur *Avimimus portentosus*. In: L.M. Chiappe & L.M. Witmer (eds.), *Mesozoic Birds: Above the Heads of Dinosaurs*. Berkeley: University of California Press, pp. 65–86.

Vinther, J. (2015) A guide to the field of palaeo colour: Melanin and other pigments can fossilise: Reconstructing colour patterns from ancient organisms can give new insights to ecology and behavior. *BioEssays* **37**, 643–656.

Vinther, J., Briggs, D.E.G., Prum, R.O., & Saranathan, V. (2008) The colour of fossil feathers. *Biology Letters* **4**, 522–525

Vinther, J., Briggs, D.E.G., Clarke, J., Mayr, G., & Prum, R.O. (2010) Structural coloration in a fossil feather. *Biology Letters* **6**, 128–131.

Walker, C.A., & Dyke, G.J. (2006) New records of fossil birds of prey from the Miocene of Kenya. *Historical Biology* **18**, 95–98.

Walker, C.A., & Dyke, G.J. (2009) Euenantiornithine birds from the late Cretaceous of El Brete (Argentina). *Irish Journal of Earth Sciences* **27**, 15–62.

Walker, C.A., Buffetaut, E., & Dyke, G.J. (2007) Large euenantiornithine birds from the Cretaceous of southern France, North America and Argentina. *Geological Magazine* **144**, 977–986.

Walsh, S.A., & Hume, J.P. (2001) A new Neogene marine avian assemblage from north-central Chile. *Journal of Vertebrate Paleontology* **21**, 484–491.

Walsh, S.A., & Suárez, M.E. (2006) New penguin remains from the Pliocene of Northern Chile. *Historical Biology* **18**, 115–126.

Wang, M., Mayr, G., Zhang, J., & Zhou, Z. (2012a) Two new skeletons of the enigmatic, rail-like avian taxon *Songzia* Hou, 1990 (Songziidae) from the early Eocene of China. *Alcheringa* **36**, 487–499.

Wang, M., Mayr, G., Zhang, J., & Zhou, Z. (2012b) New bird remains from the Middle Eocene of Guangdong, China. *Acta Palaeontologica Polonica* **57**, 519–526.

Wang, M., Zhou, Z.-H., O'Connor, J.K., & Zelenkov, N.V. (2014a) A new diverse enantiornithine family (Bohaiornithidae fam. nov.) from the Lower Cretaceous of China with information from two new species. *Vertebrata PalAsiatica* **52**, 31–76.

Wang, M., O'Connor, J.K., & Zhou, Z. (2014b) A new robust enantiornithine bird from the Lower Cretaceous of China with scansorial adaptations. *Journal of Vertebrate Paleontology* **34**, 657–671.

Wang, M., Li, D., O'Connor, J.K., & Zhou, Z. (2015a) Second species of enantiornithine bird from the Lower Cretaceous Changma Basin, northwestern China with implications for the taxonomic diversity of the Changma avifauna. *Cretaceous Research* **55**, 56–65.

Wang, M., Zheng, X., O'Connor, J.K., Lloyd, G.T., Wang, X., Wang, Y., Zhang, X., & Zhou, Z. (2015b) The oldest record of ornithuromorpha from the early cretaceous of China. *Nature Communications* **6**, 6987.

Wang, N., Braun, E.L., & Kimball, R.T. (2012) Testing hypotheses about the sister group of the Passeriformes using an independent 30-locus data set. *Molecular Biology and Evolution* **29**, 737–750.

Wang, N., Kimball, R.T., Braun, E.L., Liang, B., & Zhang, Z. (2013) Assessing phylogenetic relationships among Galliformes: A multigene phylogeny with expanded taxon sampling in Phasianidae. *PLoS ONE* **8**, e64312.

Wang, S. (2008) Re-examination of taxonomic assignment of "*Struthio linxiaensis* Hou et al., 2005." *Acta Palaeontologica Sinica* **47**, 362–368.

Wang, S., Hu, Y., & Wang, L. (2011) New ratite eggshell material from the Miocene of Inner Mongolia, China. *Chinese Birds* **2**, 18–26.

Wang, X., & Zhang, Z. (2011) Enantiornithine Birds in China. *Acta Geologica Sinica* **85**, 1211–1223.

Wang, X., O'Connor, J.K., Zhao, B., Chiappe, L.M., Gao, C., & Cheng, X. (2010) New species of Enantiornithes (Aves: Ornithothoraces) from the Qiaotou Formation in Northern Hebei, China. *Acta Geologica Sinica* **84**, 247–256.

Wang, X., Chiappe, L.M., Teng, F., & Ji, Q. (2013) *Xinghaiornis lini* (Aves: Ornithothoraces) from the Early Cretaceous of Liaoning: An example of evolutionary mosaic in early birds. *Acta Geologica Sinica* **87**, 686–689.

Wang, X., O'Connor, J.K., Zheng, X., Wang, M., Hu, H., & Zhou, Z. (2014) Insights into the evolution of rachis dominated tail feathers from a new basal enantiornithine (Aves: Ornithothoraces). *Biological Journal of the Linnean Society* **113**, 805–819.

Wang, Y.-M., O'Connor, J.K., Li, D.-Q., & You, H.-L. (2013) Previously unrecognized ornithuromorph bird diversity in the Early Cretaceous Changma Basin, Gansu Province, Northwestern China. *PLoS ONE* **8**, e77693.

Wang, Y.-M., O'Connor, J.K., Li, D.-Q., & You, H.-L. (2016, in press) New information on postcranial skeleton of the Early Cretaceous *Gansus yumenensis* (Aves: Ornithuromorpha). *Historical Biology*.

Wang, Z., Young, R.L., Xue, H., & Wagner, G.P. (2011) Transcriptomic analysis of avian digits reveals conserved and derived digit identities in birds. *Nature* **477**, 583–586.

Warheit, K.I. (1992) A review of the fossil seabirds from the Tertiary of the North Pacific: Plate tectonics, paleoceanography, and faunal change. *Paleobiology* **18**, 401–424.

Warheit, K.I. (2002) The seabird fossil record and the role of paleontology in understanding seabird community structure. In: E.A. Schreiber & J. Burger (eds.), *Biology of Marine Birds*. Boca Raton, FL: CRC Marine Biology Series, pp. 17–55.

Warheit, K.I., & Lindberg, D.R. (1988) Interactions between seabirds and marine mammals through time: Interference competition at breeding sites. In: J. Burger (ed.), *Seabirds and Other Marine Vertebrates: Competition, Predation, and Other Interactions*. New York: Columbia University Press, pp. 292–328.

Watanabe, J., & Matsuoka, H. (2015) Flightless diving duck (Aves, Anatidae) from the Pleistocene of Shiriya, Northeast Japan. *Journal of Vertebrate Paleontology* **35**, e994745.

Weidig, I. (2010) New birds from the Lower Eocene Green River Formation, North America. *Records of the Australian Museum* **62**, 29–44.

Weir, J.T., & Mursleen, S. (2013) Diversity-dependent cladogenesis and trait evolution in the adaptive radiation of the auks (Aves: Alcidae). *Evolution* **67**, 403–416.

Wellnhofer, P. (2009) *Archaeopteryx: The Icon of Evolution*. München: Friedrich Pfeil.

Wetmore, A. (1944) A new terrestrial vulture from the Upper Eocene deposits of Wyoming. *Annals of Carnegie Museum* **30**, 57–69.

Wijnker, E., & Olson, S.L. (2009) A revision of the fossil genus *Miocepphus* and other Miocene Alcidae (Aves: Charadriiformes) of the Western North Atlantic Ocean. *Journal of Systematic Palaeontology* **7**, 471–487.

Wilson, L.E., & Chin, K. (2014) Comparative osteohistology of *Hesperornis* with reference to pygoscelid penguins: The effects of climate and behaviour on avian bone microstructure. *Royal Society Open Science* **1**, 140245.

Wood, J.R., Rawlence, N.J., Rogers, G.M., Austin, J.J., Worthy, T.H., & Cooper, A. (2008) Coprolite deposits reveal the diet and ecology of the extinct New Zealand megaherbivore moa (Aves, Dinornithiformes). *Quaternary Science Reviews* **27**, 2593–2602.

Worthy, T.H. (2000) The fossil megapodes (Aves: Megapodiidae) of Fiji with descriptions of a new genus and two new species. *Journal of the Royal Society of New Zealand* **30**, 337–364.

Worthy, T.H. (2001) A giant flightless pigeon gen. et sp. nov. and a new species of *Ducula* (Aves: Columbidae), from Quaternary deposits in Fiji. *Journal of the Royal Society of New Zealand* **31**, 763–794.

Worthy, T.H. (2009) Descriptions and phylogenetic relationships of two new genera and four new species of Oligo-Miocene waterfowl (Aves: Anatidae) from Australia. *Zoological Journal of the Linnean Society* **156**, 411–454.

Worthy, T.H. (2011) Descriptions and phylogenetic relationships of a new genus and two new species of Oligo-Miocene cormorants (Aves: Phalacrocoracidae) from Australia. *Zoological Journal of the Linnean Society* **163**, 277–314.

Worthy, T.H. (2012a) A phabine pigeon (Aves: Columbidae) from Oligo-Mioce Australia. *Emu* **112**, 23–31.

Worthy, T.H. (2012b) A new species of Oligo-Miocene darter (Aves: Anhingidae) from Australia. *The Auk* **129**, 96–104.

Worthy, T.H., & Boles, W.E. (2010) *Australlus*, a new genus for *Gallinula disneyi* (Aves: Rallidae) and a description of a new species from Oligo-Miocene deposits at Riversleigh, northwestern Queensland, Australia. *Records of the Australian Museum* **63**, 61–77.

Worthy, T.H., & Holdaway, R.N. (2002) *The Lost World of the Moa: Prehistoric Life of New Zealand*. Bloomington: Indiana University Press.

Worthy, T.H., & Lee, M.S.Y. (2008) Affinities of Miocene waterfowl (Anatidae: *Manuherikia*, *Dunstanetta* and *Miotadorna*) from the St. Bathans fauna, New Zealand. *Palaeontology* **51**, 677–708.

Worthy, T.H., & Scanlon, J.D. (2009) An Oligo-Miocene Magpie Goose (Aves: Anseranatidae) from Riversleigh, northwestern Queensland, Australia. *Journal of Vertebrate Paleontology* **29**, 205–211.

Worthy, T.H., & Scofield, R.P. (2012) Twenty-first century advances in knowledge of the biology of moa (Aves: Dinornithiformes): A new morphological analysis and moa diagnoses revised. *New Zealand Journal of Zoology* **39**, 87–153.

Worthy, T.H., Tennyson, A.J.D., Jones, C., McNamara, J.A., & Douglas, B.J. (2007) Miocene waterfowl and other birds from central Otago, New Zealand. *Journal of Systematic Palaeontology* **5**, 1–39.

Worthy, T.H., Tennyson, A.J.D., Hand, S.J., & Scofield, R.P. (2008) A new species of the diving duck *Manuherikia* and evidence for geese (Aves: Anatidae: Anserinae) in the St Bathans Fauna (Early Miocene), New Zealand. *Journal of the Royal Society of New Zealand* **38**, 97–114.

Worthy, T.H., Hand, S.J., Worthy, J.P., Tennyson, A.J.D., & Scofield, R.P. (2009) A large fruit pigeon (Columbidae) from the Early Miocene of New Zealand. *The Auk* **126**, 649–656.

Worthy, T.H., Tennyson, A.J.D., Archer, M., & Scofield, R.P. (2010a) First record of *Palaelodus* (Aves: Phoenicopteriformes) from New Zealand. *Records of the Australian Museum* **62**, 77–88.

Worthy, T.H., Hand, S.J., Nguyen, J.M.T., Tennyson, A.J.D., Worthy, J.P., Scofield, R.P., Boles, W.E., & Archer, M. (2010b) Biogeographical and phylogenetic implications of an early Miocene wren (Aves: Passeriformes:

Acanthisittidae) from New Zealand. *Journal of Vertebrate Paleontology* **30**, 479–498.

Worthy, T.H., Tennyson, A.J.D., & Scofield, R.P. (2011a) An early Miocene diversity of parrots (Aves, Strigopidae, Nestorinae) from New Zealand. *Journal of Vertebrate Paleontology* **31**, 1102–1116.

Worthy, T.H., Tennyson, A.J.D., & Scofield, R.P. (2011b) Fossils reveal an early Miocene presence of the aberrant gruiform Aves: Aptornithidae in New Zealand. *Journal of Ornithology* **152**, 669–680.

Worthy, T.H., Worthy, J.P., Tennyson, A.J.D., Salisbury, S.W., Hand S.J., & Scofield, R.P. (2013a) Miocene fossils show that kiwi (*Apteryx*, Apterygidae) are probably not phyletic dwarves. In: U.B. Göhlich & A. Kroh (eds.), *Paleornithological Research 2013 – Proceedings of the 8th International Meeting of the Society of Avian Paleontology and Evolution*. Vienna: Natural History Museum Vienna, pp. 63–80.

Worthy, T.H., Worthy, J.P., Tennyson, A.J.D., & Scofield, R.P. (2013b) A bittern (Aves: Ardeidae) from the Early Miocene of New Zealand. *Paleontological Journal* **47**, 1331–1343.

Worthy, T.H., Hand, S.J., & Archer, M. (2014) Phylogenetic relationships of the Australian Oligo-Miocene ratite *Emuarius gidju* Casuariidae. *Integrative Zoology* **9**, 148–166.

Wright, T.F., Schirtzinger, E.E., Matsumoto, T., Eberhard, J.R., Graves, G.R., Sanchez, J.J., Capelli, S., Müller, H., Scharpegge, J., Chambers, G.K., & Fleischer, R.C. (2008) A multilocus molecular phylogeny of the parrots (Psittaciformes): Support for a Gondwanan origin during the Cretaceous. *Molecular Biology and Evolution* **25**, 2141–2156.

Xu, X., & Guo, Y. (2009) The origin and early evolution of feathers: Insights from recent paleontological and neontological data. *Vertebrata PalAsiatica* **47**, 311–329.

Xu, X., & Zhang, F. (2005) A new maniraptoran dinosaur from China with long feathers on the metatarsus. *Naturwissenschaften* **92**, 173–177.

Xu, X., Wang, X.-L., & Wu, X.-C. (1999) A dromaeosaurid dinosaur with a filamentous integument from the Yixian Formation of China. *Nature* **401**, 262–266.

Xu, X., Zhou, Z.-H., & Prum, R.O. (2001) Branched integumental structures in *Sinornithosaurus* and the origin of feathers. *Nature* **410**, 200–204.

Xu, X., Zhou, Z., Wang, X., Kuang, X., Zhang, F., & Du, X. (2003) Four-winged dinosaurs from China. *Nature* **421**, 335–340.

Xu, X., Clark, J.M., Mo, J., Choiniere, J., Forster, C.A., Erickson, G.M., Hone, D.W.E., Sullivan, C., Eberth, D.A., Nesbitt, S., Zhao, Q., Hernandez, R., Jia, C.-K., Han, F.-L., & Guo, Y. (2009a) A Jurassic ceratosaur from China helps clarify avian digital homologies. *Nature* **459**, 940–944.

Xu, X., Zhao, Q., Norell, M., Sullivan, C., Hone, D., Erickson, G., Wang, X.-L., Han, F.-L., & Guo, Y. (2009b) A new feathered maniraptoran dinosaur fossil that fills a morphological gap in avian origin. *Chinese Science Bulletin* **54**, 430–435.

Xu, X., Zheng, X., & You, H. (2009c) A new feather type in a nonavian theropod and the early evolution of feathers. *Proceedings of the National Academy of Sciences USA* **106**, 832–834.

Xu, X., Zheng, X., & You, H. (2010) Exceptional dinosaur fossils show ontogenetic development of early feathers. *Nature* **464**, 1338–1341.

Xu, X., You, H., Du, K., & Han, F. (2011) An *Archaeopteryx*-like theropod from China and the origin of Avialae. *Nature* **475**, 465–470.

Xu, X., Zhou, Z., Dudley, R., Mackem, S., Chuong, C.-M., Erickson, G.M., & Varricchio, D.J. (2014) An integrative approach to understanding bird origins. *Science* **346**, 1253293.

Xu, X., Zheng, X., Sullivan, C., Wang, X., Xing, L., Wang, Y., Zhang, X., O'Connor, J., Zhang, F., & Pan, Y. (2015) A bizarre Jurassic maniraptoran theropod with preserved evidence of membranous wings. *Nature* **521**, 70–73.

You, H.-L., Lamanna, M.C., Harris, J.D., Chiappe, L.M., O'Connor, J., Ji, S.-A., Lü, J.-C., Yuan, C.-X., Li, D.Q., Zhang, X., Lacovara, K.J., Dodson, P., & Ji, Q. (2006) A nearly modern amphibious bird from the Early Cretaceous of northwestern China. *Science* **312**, 1640–1643.

Yuri, T., Kimball, R.T., Harshman, J., Bowie, R.C., Braun, M.J., Chojnowski, J.L., Han, K.-L., Hackett, S.J., Huddleston, C.J., Moore, W.S., Reddy, S., Sheldon, F.H., Steadman, D.W., Witt, C.C., & Braun, E.L. (2013) Parsimony and model-based analyses of indels in avian nuclear genes reveal congruent and incongruent phylogenetic signals. *Biology* **2**, 419–444.

Zachos, J., Pagani, M., Sloan, L., Thomas, E., & Billups, K. (2001) Trends, rhythms, and aberrations in global climate 65 Ma to present. *Science* **292**, 686–693.

Zanno, L.E., & Makovicky, P.J. (2011) Herbivorous ecomorphology and specialization patterns in theropod dinosaur evolution. *Proceedings of the National Academy of Sciences USA* **108**, 232–237.

Zelenitsky, D.K., Therrien, F., Ridgely, R.C., McGee, A.R., & Witmer, L.M. (2011) Evolution of olfaction in non-avian theropod dinosaurs and birds. *Proceedings of the Royal Society B: Biological Sciences* **278**, 3625–3634.

Zelenitsky, D.K., Therrien, F., Erickson, G.M., DeBuhr, C.L., Kobayashi, Y., Eberth, D.A., & Hadfield, F. (2012) Feathered non-avian dinosaurs from North America provide insight into wing origins. *Science* **338**, 510–514.

Zelenkov, N.V. (2011a) Diving ducks from the Middle Miocene of western Mongolia. *Paleontological Journal* **45**, 191–199.

Zelenkov, N.V. (2011b) *Ardea sytchevskayae* sp. nov., a new heron species (Aves: Ardeidae) from the Middle Miocene of Mongolia. *Paleontological Journal* **45**, 572–579.

Zelenkov, N.V. (2012a) Neogene geese and ducks (Aves: Anatidae) from localities of the Great Lakes Depression, western Mongolia. *Paleontological Journal* **46**, 607–619.

Zelenkov, N.V. (2012b) A new duck from the Miocene of Mongolia, with comments on the Miocene evolution of ducks. *Paleontological Journal* **46**, 520–530.

Zelenkov, N.V. (2013) Cenozoic phoenicopteriform birds from Central Asia. *Paleontological Journal* **47**, 1323–1330.

Zelenkov, N.V., & Averianov, A.O. (2016) A historical specimen of enantiornithine bird from the Early Cretaceous of Mongolia representing a new taxon with a specialized neck morphology. *Journal of Systematic Palaeontology* **14**, 319–338.

Zelenkov, N.V., & Dyke, G.J. (2008) The fossil record and evolution of mousebirds (Aves: Coliiformes). *Palaeontology* **51**, 1403–1418.

Zelenkov, N.V., & Kurochkin, E.N. (2009a). Neogene phasianids (Aves: Phasianidae) of Central Asia: 1. Genus *Tologuica* gen. nov. *Paleontological Journal* **43**, 208–215.

Zelenkov, N.V., & Kurochkin, E.N. (2009b) Neogene Phasianids (Aves: Phasianidae) of Central Asia: 2. Genera *Perdix*, *Plioperdix*, and *Bantamyx*. *Paleontological Journal* **43**, 318–325.

Zelenkov, N.V., & Kurochkin, E.N. (2010) Neogene phasianids (Aves: Phasianidae) of Central Asia: 3. Genera *Lophogallus* gen. nov. and *Syrmaticus*. *Paleontological Journal* **44**, 328–336.

Zelenkov, N.V., & Kurochkin, E.N. (2012) Dabbling ducks (Aves: Anatidae) from the middle Miocene of Mongolia. *Paleontological Journal* **46**, 88–95.

Zelenkov, N.V., Volkova, N.V., & Gorobets, L.V. (2016) Late Miocene buttonquails (Charadriiformes, Turnicidae) from the temperate zone of Eurasia. *Journal of Ornithology* **157**, 85–92.

Zhang, F., & Zhou, Z. (2000) A primitive enantiornithine bird and the origin of feathers. *Science* **290**, 1955–1959.

Zhang, F., & Zhou, Z. (2004) Leg feathers in an Early Cretaceous bird. *Nature* **431**, 925.

Zhang, F., Zhou, Z., Xu, X., & Wang, X. (2002) A juvenile coelurosaurian theropod from China indicates arboreal habits. *Naturwissenschaften* **89**, 394–398.

Zhang, F., Zhou, Z., Xu, X., Wang, X., & Sullivan, C. (2008a) A bizarre Jurassic maniraptoran from China with elongate ribbon-like feathers. *Nature* **455**, 1105–1108.

Zhang, F., Zhou, Z., & Benton, M.J. (2008b) A primitive confuciusornithid bird from China and its implications for early avian flight. *Science in China, Series D: Earth Sciences* **51**, 625–639.

Zhang, F., Kearns, S.L., Orr, P.J., Benton, M.J., Zhou, Z., Johnson, D., Xu, X., & Wang, X. (2010) Fossilized melanosomes and the colour of Cretaceous dinosaurs and birds. *Nature* **463**, 1075–1078.

Zhang, Y., O'Connor, J., Di, L., Qingjin, M., Sigurdsen, T., & Chiappe, L.M. (2014) New information on the anatomy of the Chinese Early Cretaceous Bohaiornithidae (Aves: Enantiornithes) from a subadult specimen of *Zhouornis hani*. *PeerJ* **2**, e407.

Zhang, Z., Hou, L., Hasegawa, Y., O'Connor, J., Martin, L.D., & Chiappe, L.M. (2006) The first

Mesozoic heterodactyl bird from China. *Acta Geologica Sinica* **80**, 631–635.

Zhang, Z., Zheng, X., Zheng, G., & Hou, L. (2010) A new Old World vulture (Falconiformes: Accipitridae) from the Miocene of Gansu Province, northwest China. *Journal of Ornithology* **151**, 401–408.

Zhang, Z., Huang, Y., James, H.F., & Hou, L. (2012a) A marabou (Ciconiidae: *Leptoptilos*) from the Middle Pleistocene of northeastern China. *The Auk* **129**, 699–706.

Zhang, Z., Feduccia, A., & James, H.F. (2012b) A Late Miocene accipitrid (Aves: Accipitriformes) from Nebraska and its implications for the divergence of Old World vultures. *PLoS ONE* **7**, e48842.

Zhang, Z., Chen, D., Zhang, H., & Hou, L. (2014) A large enantiornithine bird from the Lower Cretaceous of China and its implication for lung ventilation. *Biological Journal of the Linnean Society* **113**, 820–827.

Zhao, T., Mayr, G., Wang, M., & Wang, W. (2015) A trogon-like arboreal bird from the early Eocene of China. *Alcheringa* **39**, 287–294.

Zheng, X., Zhang, Z., & Hou, L. (2007) A new enantiornitine bird with four long rectrices from the Early Cretaceous of northern Hebei, China. *Acta Geologica Sinica* **81**, 703–708.

Zheng, X., Xu, X., Zhou, Z., Mioao, D. & Zhang, F. (2010) Comment on "Narrow primary feather rachises in *Confuciusornis* and *Archaeopteryx* suggest poor flight ability." *Science* **330**, 320-c.

Zheng, X., Martin, L.D., Zhou, Z., Burnham, D.A., Zhang, F., & Miao, D. (2011) Fossil evidence of avian crops from the Early Cretaceous of China. *Proceedings of the National Academy of Sciences USA* **108**, 15904–15907.

Zheng, X., Wang, X., O'Connor, J., & Zhou, Z. (2012) Insight into the early evolution of the avian sternum from juvenile enantiornithines. *Nature Communications* **3**, 1116.

Zheng, X., Zhou, Z., Wang, X., Zhang, F., Zhang, X., Wang, Y., Wie, G., Wang, S., & Xu, X. (2013a) Hind wings in basal birds and the evolution of leg feathers. *Science* **339**, 1309–1312.

Zheng, X., O'Connor, J., Huchzermeyer, F., Wang, X., Wang, Y., Wang, M., & Zhou, Z. (2013b) Preservation of ovarian follicles reveals early evolution of avian reproductive behaviour. *Nature* **495**, 507–511.

Zheng, X., O'Connor, J., Wang, X., Wang, M., Zhang, X., & Zhou, Z. (2014a) On the absence of sternal elements in *Anchiornis* (Paraves) and *Sapeornis* (Aves) and the complex early evolution of the avian sternum. *Proceedings of the National Academy of Sciences USA* **111**, 13900–13905.

Zheng, X., O'Connor, J.K., Huchzermeyer, F., Wang, X., Wang, Y., Zhang, X., and Zhou, Z. (2014b) New specimens of *Yanornis* indicate a piscivorous diet and modern alimentary canal. *PLoS ONE* **9**, e95036.

Zheng, X.-T., O'Connor, J.K., Wang, X.-L., Zhang X.-M., & Wang, Y. (2014c) New information on Hongshanornithidae (Aves: Ornithuromorpha) from a new subadult specimen. *Vertebrata PalAsiatica* **52**, 217–232.

Zhou, S., Zhou, Z.-H., & O'Connor, J.K. (2012) A new toothless ornithurine bird (*Schizooura lii* gen. et sp. nov.) from the Lower Cretaceous of China. *Vertebrata PalAsiatica* **50**, 9–24.

Zhou, S., Zhou, Z., & O'Connor, J.K. (2013) Anatomy of the basal ornithuromorph bird *Archaeorhynchus spathula* from the Early Cretaceous of Liaoning, China. *Journal of Vertebrate Paleontology* **33**, 141–152.

Zhou, S., Zhou, Z., & O'Connor, J. (2014a) A new piscivorous ornithuromorph from the Jehol Biota. *Historical Biology* **26**, 608–618.

Zhou, S., O'Connor, J.K., & Wang, M. (2014b) A new species from an ornithuromorph (Aves: Ornithothoraces) dominated locality of the Jehol Biota. *Chinese Science Bulletin* **59**, 5366–5378.

Zhou, Z. (2004) The origin and early evolution of birds: Discoveries, disputes, and perspectives from fossil evidence. *Naturwissenschaften* **91**, 455–471.

Zhou, Z. (2014) The Jehol Biota, an Early Cretaceous terrestrial Lagerstätte: New discoveries and implications. *National Science Review* **1**, 543–559.

Zhou, Z., & Martin, L.D. (2011) Distribution of the predentary bone in Mesozoic ornithurine birds. *Journal of Systematic Palaeontology* **9**, 25–31.

Zhou, Z., & Zhang, F. (2001) Two new ornithurine birds from the Early Cretaceous of western Liaoning, China. *Chinese Science Bulletin* **46**, 1258–1264.

Zhou, Z., & Zhang, F. (2002a) A long-tailed, seed-eating bird from the Early Cretaceous of China. *Nature* **418**, 405–409.

Zhou, Z., & Zhang, F. (2002b) Largest bird from the Early Cretaceous and its implications for the earliest avian ecological diversification. *Naturwissenschaften* **89**, 34–38.

Zhou, Z., & Zhang, F. (2003a) *Jeholornis* compared to *Archaeopteryx*, with a new understanding of the earliest avian evolution. *Naturwissenschaften* **90**, 220–225.

Zhou, Z., & Zhang, F. (2003b) Anatomy of the primitive bird *Sapeornis chaoyangensis* from the Early Cretaceous of Liaoning, China. *Canadian Journal of Earth Sciences* **40**, 731–747.

Zhou, Z., & Zhang, F. (2004) A precocial avian embryo from the Lower Cretaceous of China. *Science* **306**, 653–653.

Zhou, Z., & Zhang, F. (2005) Discovery of an ornithurine bird and its implication for Early Cretaceous avian radiation. *Proceedings of the National Academy of Sciences USA* **102**, 18998–19002.

Zhou, Z., & Zhang, F. (2006) A beaked basal ornithurine bird (Aves, Ornithurae) from the Lower Cretaceous of China. *Zoologica Scripta* **35**, 363–373.

Zhou, Z., & Zhang, F. (2007) Mesozoic birds of China – a synoptic review. *Frontiers of Biology in China* **2**, 1–14.

Zhou, Z.-H., Wang, X.-L., Zhang, F.-C., & Xu, X. (2000) Important features of *Caudipteryx* – evidence from two nearly complete new specimens. *Vertebrata PalAsiatica* **38**, 243–265.

Zhou, Z.-H., Clarke, J.A., Zhang, F.-C., & Wings, O. (2004) Gastroliths in *Yanornis*: An indication of the earliest radical diet-switching and gizzard plasticity in the lineage leading to living birds? *Naturwissenschaften* **91**, 571–574.

Zhou, Z., Chiappe, L.M., & Zhang, F. (2005) Anatomy of the Early Cretaceous bird *Eoenantiornis buhleri* (Aves: Enantiornithes) from China. *Canadian Journal of Earth Sciences* **42**, 1331–1338.

Zhou, Z., Clarke, J., & Zhang, F. (2008) Insight into diversity, body size and morphological evolution from the largest Early Cretaceous enantiornithine bird. *Journal of Anatomy* **212**, 565–577.

Zhou, Z.-H., Zhang, F.-C., & Li, Z.-H. (2009) A new basal ornithurine bird (*Jianchangornis microdonta* gen. et sp. nov.) from the Lower Cretaceous of China. *Vertebrata PalAsiatica* **47**, 299–310.

Zhou, Z., Zhang, F., & Li, Z. (2010) A new Lower Cretaceous bird from China and tooth reduction in early avian evolution. *Proceedings of the Royal Society B: Biological Sciences* **277**, 219–227.

Zusi, R.L., & Warheit, K.I. (1992) On the evolution of intraramal mandibular joints in pseudodontorns (Aves: Odontopterygia). *Natural History Museum of Los Angeles County, Science Series* **36**, 351–360.

Index

Note: Numbers in italics indicate figures.